Dennis, the book I mentioned. Very complex! But really quite good.

Don't expect to read it in one sitting (!); but let me know what you think —

Judy

COGNITIVE SYSTEMS
ENGINEERING

WILEY SERIES IN SYSTEMS ENGINEERING

Andrew P. Sage

ANDREW P. SAGE and JAMES D. PALMER
Software Systems Engineering

WILLIAM B. ROUSE
Design for Success: A Human-Centered Approach to Designing Successful Products and Systems

LEONARD ADELMAN
Evaluating Decision Support and Expert System Technology

ANDREW P. SAGE
Decision Support Systems Engineering

YEFIM FASSER and DONALD BRETTNER
Process Improvement in the Electronics Industry

WILLIAM B. ROUSE
Strategies for Innovation: Creating Successful Products, Systems, and Organizations

ANDREW P. SAGE
Systems Engineering

LIPING FANG, KEITH W. HIPEL, and D. MARC KILGOUR
Interactive Decision Making: The Graph Model for Conflict Resolution

WILLIAM B. ROUSE
Catalysts for Change: Concepts and Principles for Enabling Innovation

JENS RASMUSSEN, ANNELISE MARK PEJTERSEN, and L. P. GOODSTEIN
Cognitive Systems Engineering

JAMES E. ARMSTRONG and ANDREW P. SAGE
An Introduction to Systems Engineering
[forthcoming 1995]

COGNITIVE SYSTEMS ENGINEERING

Jens Rasmussen
Annelise Mark Pejtersen
L. P. Goodstein
Risø National Laboratory
Roskilde, Denmark

A Wiley-Interscience Publication
JOHN WILEY & SONS, INC.
New York / Chichester / Brisbane / Toronto / Singapore

Library of Congress Cataloging in Publication Data:

Rasmussen, Jens, 1926–
    Cognitive systems engineering / Jens Rasmussen, Annelise Mark
Pejtersen, L. P. Goodstein.
        p.   cm.
    "A Wiley-Interscience Publication."
    ISBN 0-471-01198-3
    1. Systems engineering.   2. Expert systems (Computer science)
I. Pejtersen, Annelise Mark.   II. Goodstein, L. P.   III. Title.
TA168.R346   1994
658.4'038—dc20   93-46162
                                                            CIP

Printed in the United States of America

10 9 8 7 6 5 4 3 2

Contents

Preface

SCOPE AND USE OF THE BOOK

The theme of this book is *cognitive systems engineering*—a term with three components, which in combination reflect the goal and scope of the book in relation to the multifarious nature of people and their interaction with the modern world. The first component is cognitive, because thinking, problem solving, and decision making have come to play a greater role in our daily lives than physical strength and dexterity. The second component is engineering, because we are devoted to developing concepts, methods, and tools for analyzing and designing (i.e., engineering) useful and acceptable systems (the third component) to help humans as they carry out their daily endeavors. As such, we seek an optimal combination of a solid conceptual basis with a practical and realistic sense for "what it is really like out there."

Integrated information networks and computerized work stations are becoming the interface between people and their daily activities. Not only is the work place of the individual changing but, in addition, the cooperative patterns supporting work, as well as the underlying organizational structure, are being affected. This situation coupled with the fast change of pace in available technology makes it difficult to develop appropriate information systems through an incremental updating of existing concepts. Instead, design (and redesign) has to be based on a conceptual framework capable of supporting the analysis of work systems, the prediction of the impact of new information systems, as well as the evaluation of the ultimate user–system interaction. By necessity, such a framework has to integrate modeling concepts from many different disciplines, such as engineering, psychology, and the cognitive, management, information, and computer sciences.

This book continues and expands the conceptual foundations regarding human information processing and human–system interaction described in one of Rasmussen's previous books (Rasmussen, 1986) in an attempt to combine theory and application. It is based on experiences gained over several decades in a multidisciplinary research setting from field studies of work, analyses of accidents and incidents, and experiments with new system concepts. The material, we feel, has relevance for students and researchers in cognitive psychology, design engineering, and the organizational and decision sciences—not to mention human factors and information and computer science. In addition, practitioners in the field who are concerned with human–work problems—including areas such as risk, safety, and security, as well as effectiveness, efficiency, and well being—will discover useful things—if, as we hope, they come with a reasonably open mind. This book is profusely illustrated to enhance our understanding and, thereby, further our belief in the importance of the actor (in this case the reader) being able to resonate directly with features of their reading environment.

This book can be read from start to finish although, hopefully, not at one sitting. With some effort, it should also be usable as a "jump-in/jump-out" text for particular courses or to support special projects.

THE CONTENT OF THE BOOK

After a brief mention of the background for the book, the introductory chapter (Chapter 1) deals with two basic issues that underlie the total approach—the all-important modeling problems associated with modern dynamic work situations and, in conjunction with this, matters related to the representations utilized in these models. The introduction concludes with an overview of the proposed framework as a two-stringed analytic activity within a given work domain comprising activities, agents, and their interaction. Then, three chapters (Chapters 2–4) describe the underlying dimensions of this framework in detail. These consist of work domain, activity analysis, information processing strategies, functional work organization, social organization and, lastly, actors' cognitive resources. For each topic, important methodological issues connected with data collection and performance analysis are also discussed.

Chapter 5 and the following chapters deal essentially with the design and evaluation of information systems and, in particular, ecological information systems based on the described work analysis as the underlying approach for configuring human–work interface systems. While the word "ecological" can directly trigger associations to organizations like Greenpeace and currently warm topics such as rain forests and carbon dioxide (CO_2) pollution, our use of the term has distinctly different roots and connotations. Our interpretation of ecology comes from the American psychologist J. J. Gibson's views on direct perception and the human ability to read the functional properties and possibilities for actions directly from the natural environment. It is this particular feature of his work that we attempt to transfer to the manner in which information is presented to users–operators in their various human contrived work situations. Thus our approach is not a literal application of all aspects of Gibsonean psychology, but rather reflects an inspiration that has been fruitful for our own work with system design.

While Chapter 5 deals with the various aspects of human–work coupling and introduces the ecological interface concept, Chapter 6 treats in some detail problems with mismatches in this coupling that can result in so-called human errors. Implications with regard to relations between adaptation and error, the design of error-tolerant systems, and so on, are discussed in detail.

The next two chapters (Chapters 7 and 8) are concerned with information system design and evaluation. As intimated in the main text, a dilemma exists concerning the interrelation between the activities of analysis, design and evaluation, and their support. Designers are to be considered as experts, that is, experienced decision makers who function in a natural professional setting and have a deep understanding of their particular context. For them, design is not strictly an orderly and normative procedural process but comprises highly creative elements, which are both constrained and inspired from many different directions. In addition, design is more often than not a *re*design of an existing system that could be interpreted as lessening the need for a new top-down synthesis and instead embarking on more of a side-ways "migration" within a familiar context. Therefore, designers are primed into a kind of "recognition-based" decision making because they have been there before. On the other hand, systems are often created that receive the verdict: "does not meet original goals; does not satisfy user needs." This verdict can occur because desired changes frequently have very unexpected and widespread consequences if their potential effects are not adequately studied. The upshot of all of this, as reflected in this book, is

1. The need to recognize actual *design* processes for what they are, and attempt to impart to designers the need for attaining a broader, yet personal, intuition for the implications of the total "activities ↔ agents" interaction by furnishing them with appropriate support.

2. The complementary realization that any realistic (and often necessary) *evaluation* of these systems must build more directly on the systematic and rational basis provided by the framework. Thus in Chapter 7, the relations between (re)design and design guidance are stressed. Then the concept of design maps for supporting designers is introduced as a natural consequence and extension of the analytic framework; that is, the use of maps of the "design territory" as an alternative to conventional context-free procedural design guidelines.

In Chapter 8, system evaluation is dealt with—at first in general terms. The framework is then again employed—this time to furnish the basis for structuring evaluation efforts and for dealing with issues regarding the generalization of results.

The following chapters, 9–12, serve as a demonstration of the use of the analytic framework in the design and evaluation of a library system. Finally, Chapter 13 represents an attempt to establish a first version of an annotated catalog of display types for various *W*ork domain/*T*ask situation/*U*ser profile combinations (WTUs), which reflects our interest in supporting ecological information presentation and, hopefully, will incite designers to collect their own domain maps and display catalogs.

Use of Examples

The examples used to demonstrate concepts and to argue for the framework are selected so as to span its dimensions and scope. Consequently, they should be considered as borderline cases chosen because they forcefully illustrate an important concept or relationship. However, several colleagues found that we placed too much emphasis on the adaptive properties of a modern work organization. They pointed out that cultural differences and attitudinal variations (e.g., between Europe, Japan, and the United States) can greatly color readers' tacit understanding of the points being made. In addition, differences between domains such as high hazard process plants, libraries, and hospitals are of course nontrivial. All of this is true. However, it should be remembered that the examples have been chosen to clearly illuminate the conceptual capacity of the presented framework and not to present some empirical data to the reader. (This having been said, however, we have found from accident analyses that work organization, even in high hazard plants, is more adaptive than is usually realized.)

In conclusion, the book is intended to (1) deal with human–work design in a systematic way, (2) demonstrate the cognitive systems engineering approach so that readers hopefully can begin to apply the ideas in their own areas, and (3) indicate the multidisciplinary nature of the subject and point out the needs for further work, cooperation, and consensus.

COGNITIVE SYSTEMS ENGINEERING AS A DISCIPLINE

Several aspects of modern work systems make it necessary to supplement established Human Factors and Human–Computer Interaction activities with a cross-disciplinary cooperation among researchers from several fields. Important factors that emphasize this need are the rapid pace of changes in products and services as the result of the technology applied, as well as the large scale and high degree of integration of modern work systems. As a consequence, the work content, the tools, the cooperative patterns, and the social organization are becoming strongly interdependent and subject to changes in response to challenges from a dynamic environment. Consequently, the study of work and the design of information systems have to be concerned with work performance in its sociotechnical context, and thus from a multidisciplinary point of view.

Fortunately, current interest in cognitive phenomena within several of the traditional, academic disciplines—psychology, sociology, information and computer science, and so on —is leading to an increasing convergence in research paradigms that can prepare the ground for the interdisciplinary cooperation necessary for the development of cognitive systems engineering as a viable approach.

A Cross-Disciplinary Market Place

When reading this book, it should be kept in mind that the presentation is based on the view that cognitive systems engineering should be considered as a cross-disciplinary market place and not as a distinct research discipline. This view of cognitive systems engineering has some important implications for the way that the concepts are defined and presented in the book.

A cross-disciplinary research effort depends on the development of a common frame of reference that can serve to transfer research findings among the involved academic disciplines and to generalize and communicate results to designers of advanced information systems.

Problem-driven research in a complex context (e.g., cognitive systems engineering) is, by nature, cross-disciplinary. The focus will be on a selection of those academic approaches that are most useful for the problems at hand. Therefore, the cross-disciplinary problem is to identify or develop compatible paradigms in the involved disciplines and to become familiar with their different terminologies. Of course, this creates very difficult, but also very stimulating problems—first in the efforts to establish a common framework and terminology within cognitive engineering and, second, in maintaining concepts and terms that can serve the communication within the communities of the individual academic disciplines.

In problem-driven research, the criterion is whether conceptual models are useful and have predicitive power. For example, in the present context of understanding and controlling the implications of modern information technology, the ultimate test of a concept is the validation of its predictive power when used for the design of a system prototype.

However, it is of course true that different requirements regarding research emerge from teaching within the continuity of an established discipline as opposed to dealing with problems created by a turbulent technological environment. This difference is neither a question of basic or applied research nor of the degree of rigor and conceptual clarity. It is solely a question of a difference in aim and scope.

Consistency in Terminology

As a consequence of the cross-disciplinary context of cognitive systems engineering, the odds are that a reader familiar with one of the disciplines involved in the topics discussed in the book will find that the terminology is ambiguous. Indeed we have often been involved in discussions of the need to define certain concepts and to develop a consistent terminology.

However, we found it counterproductive to aim for a particular terminology. In a complex field of research covering human activities in a work context, we cannot completely define concepts and models, as is the ideal in the natural sciences. As discussed in detail in the introduction to the book (Chapter 1), the complexity of work systems often forces us to use a representation in terms of objects and causal connections among events, not well-defined relationships among quantitative variables. The concepts of objects and events are "prototypical" in the sense that they cannot be defined by an exhaustive list of attributes. Completeness removes regularity. Regularity in terms of causal relations is found between kinds of events, not between particular, individually defined events. When concepts and terms cannot be defined by an exhaustive set of attributes, they will be defined by the context in which they are used. Therefore, if a reader disagrees with the use of a term or a concept, he or she should attempt to identify a set of preconditions in the context that would make sense rather than focus on eventual differences in conception.

Rather than demand a particular terminology for cognitive systems engineering that might separate it from other disciplines, it is important to maintain a flexibility of mind that serves communication with all the involved, academic disciplines.

The Message Is in the Structure Not the Data

In discussions based on causal representations of complex systems, it is always possible to find counterexamples to an argument. The messsage of a statement is more likely to lie in the preconditions that make a statement acceptable than in the isolated statement itself. Consequently, the important message of the conceptual framework presented in the book is its structure, not its data.

Jens Rasmussen
Annelise Mark Pejtersen
L. P. Goodstein

Acknowledgments

The material presented in this book has been evolving through several decades in a cross-disciplinary research environment at the Risø National Laboratory. We would like to gratefully acknowledge contributions from several quarters. First of all, we wish to thank those who let us watch while they worked and who spent time and effort—on top of their own work—to discuss with us and share the secrets of their expertise. Their openness and interest was of great value. Likewise, the professional staff in various work domains who participated—often with few resources and hard work—in the development and evaluation of a number of prototype systems has contributed greatly to the validation of the design concepts put forward in this book. We should also add to the list the contributions of our colleagues in the group at Risø throughout the years; unfortunately they are too numerous to list. Finally, the many international contacts from which we have benefited—often in the form of a new impulse or adjustment of course after we, in our eagerness to learn the secrets of professions that were not our own, involved colleagues in lengthy and, for us, inspiring discussions at Risø, at conferences, or during our visits to their institutions.

In addition to those at Risø, several international colleagues kindly devoted considerable effort during the writing of this book to discuss and review draft versions and thereby, hopefully, helped us to avoid the most obvious mistakes when we tried to make sense of their professional domains within our context. During the book's conception phase, Kim Vicente, now at Toronto University, had a significant influence during his several stays at Risø, particulary with respect to the development of ecological system concepts. The influential discussions during many years with Berndt Brehmer, Uppsala, and James T. Reason, Manchester, are gratefully acknowledged. We also acknowledge the artistic contribution to the design of the multimedia interface of the library system by Steen Agger and Henrik Jensen, both from the Royal Academy of Fine Arts, Copenhagen. Also, the cross-disciplinary discussion forum emerging during the EEC Basic Research Project MOHAWC had an important impact. Recently, during the writing phase of this book, the interest, criticism, and suggestions of John Flach—Wright State, Ohio; Jacques Leplat—Paris; and Neville Moray—Illinois have been extremely valuable and are gratefully acknowledged.

The underlying research has been supported through the years by the Risø National Laboratory and by the Royal School of Librarianship in Denmark during Annelise Mark Pejtersen's former affiliation with the school. In recent years, international cooperation in cognitive engineering has been promoted through the EEC Esprit Program's Basic Research Action project MOHAWC, while the conception and preparation of individiual chapters of the book has been supported by special grants from the Danish Technical Research Council, the Danish Council for the Humanities, the Nordic Council for Information Science, and the Danish Fund for Public Libraries.

Finally, we acknowledge the helpful editorial advice of the series editor, Andrew Sage, and the professional efforts of John Hanley and the staff at Wiley in the production of this volume.

Chapter 1

Introduction

BACKGROUND

Information technology is changing human conditions of work in several respects. Levels of mechanization and automation are steadily increasing and computers are used in planning and control within most work domains. Consequently, computer-based interfaces are being inserted between humans and their work, and advanced communication networks serve to integrate the operation of large-scale distributed systems. The thesis of this book is that a new approach to the analysis of work is needed together with the development of a frame of reference that can serve the strategical planning, design, and evaluation of sociotechnical systems based on advanced technology.

When computer-based work stations are becoming the interface between people and their work, and information networks connect people in cooperative efforts, one pronounced effect is a diversification of work. In addition, when elementary work routines are mechanized and automated, the tasks of the individual move to a higher conceptual level. Based on analyses of users' needs, designers of information systems will have to select and integrate data into information representing the particular work domain at higher conceptual levels, while display formats will have to be developed from hypotheses about the nature of the mental models and strategies brought to work by the staff. Attempts to optimize system design from misconceived hypotheses about the users or about the actual work content will be detrimental to system performance. Consequently, for the design of information systems and user interfaces, expertise is needed with respect to the particular work domain for which the system is going to be used, as well as with respect to human cognition. As a result, the often heard requirement that human factors should be an integral part of the basic design process, and not merely an add-on polish to the interface, becomes increasingly important.

With respect to the methods and frameworks used for the analysis and description of modern work systems, the consequence is a need for a tight cross-disciplinary cooperation. We need a shift towards compatible paradigms in the involved provinces of the human sciences. Fortunately, this necessary shift is being facilitated by concurrent shifts of emphasis within several relevant disciplines toward a cognitive point of view, and by an increasing interest in descriptive models based on field studies instead of normative models based on theoretical operations research. Another cardinal issue for a cross-disciplinary integration is a shift from a structural perspective on work systems toward a functional perspective.

Thus, a new perspective is required for analysis, modeling, and design of integrated work systems. This is the substance matter of cognitive systems engineering and it is the message of this book. The present development within advanced manufacturing systems will be used as an example to illustrate the extent of change of work systems caused by information technology and the direction in which the analytical methodology has to change. It should, however, be considered that similar changes are found in many other work domains.

INFORMATION TECHNOLOGY AND ADVANCED MANUFACTURING

To illustrate the conceptual issues caused by introduction of information technology in a complex, modern work domain, the present change toward flexible manufacturing systems operating under dynamic market conditions will be discussed.

Manufacturing is presently changing from long-term planning of production for stable markets toward more flexible, responsive system operating in highly turbulent and competitive markets. In this situation, computer integration of complete manufacturing systems are considered to provide the necessary flexibility and ability to respond quickly and effectively to specific customer requirements. As Drucker (1988) observed, design, production planning, and production can no longer take place as a rather slow sequence of separate processes in different departments but need to be organized as simultaneous activities within a task force oriented, high-tempo organization. Such organizations require effective information systems, decentralization of decision making, and advanced communication systems. These can only be designed on the basis of a reliable model of the function of the staff members in the *new organization* and, notably, in the *actual* work organization, *not* the one represented in the formal organization charts showing the allocation of legal and financial responsibility.

In modern work systems, the effects of one drastic technological change have not stabilized before another change appears, and the multiple influences of the technological development can no longer be studied separately in an otherwise stable system. We are no longer faced with the application of new technological means for solving problems that have been empirically identified in the current systems. Instead, a continuous influx of simultaneous changes will result from the introduction of new tools that place established functions in completely new relationships, together with other changes arising from dynamic markets and governmental practices.

The increasing reliance on information technology has caused a drastic and widely recognized need for an explicit consideration of the cognitive and organizational aspects of manufacturing systems. A round table discussion on the use of "second generation management for fifth generation technology" (Savage and Appleton, 1988) concluded that the introduction in the United States of computer integrated manufacturing (CIM) is in a state of confusion. The reason for this has been the focus on computer technology while the "tough management problems have, for the most part, been ignored." It is found that computers are used to *interface* existing functions in the organization, not to *integrate* them. Therefore, the label CIM presently represents *C*omputer *I*nterfaced rather than *I*ntegrated *M*anufacturing. Also, from management research (Drucker, 1988), the warning is given that significant changes are needed in organizations. It is one of Drucker's main theses that the center of gravity in employment is moving rapidly from manual and clerical workers to knowledge workers who resist the command-and-control model that business took from the military 100 years ago. He also finds that both the number of management levels and the number of managers can be sharply cut because whole layers of management neither make decisions nor lead. Instead, they serve to propagate "the faint, unfocused signals that pass for communication in the traditional preinformation organization." These arguments point to the need to change the organizational structure of companies operating in a dynamic and turbulent environment in order to keep-up with the required fast responses.

In addition, the need to reconsider the *management and planning strategies* required by modern work systems has recently been stressed by several authors. In discussing "a new paradigm of work organization" Aoki (1988) distinguishes between *strategic* and *operational planning*. Operational planning is concerned with the need of a firm to adapt its operating tasks to evolving technical and human emergencies (the malfunction of machines, defective products, absenteeism of workers, etc.) and changing market circumstances, while strategic planning is concerned with long-term planning of resources and methods. Aoki contrasts the requirements of operational planning with the traditional western, hierarchical organization and stresses the need for horizontal communication, for creating adequate staff competence through opportunities for learning on the job and a diversification of job content, and of the evolution of proper criteria functions within the staff including profit-sharing schemes. In this way, Aoki argues for an explicit distinction between two management strategies, that is, the control by open-loop preplanning and by closed-loop feedback.

The two planning strategies discussed by Aoki have very different requirements to the speed of response of an organization. Centralized, long-term planning of resources and financial strategies match the traditional hierarchical and formal organization, while the fast response required for operational planning call for more horizontal interaction. The resulting interaction between a formal hierarchical organizations for strategic planning and a complex, often informal organization for operational control, have also been observed from studies of the activity on the flight deck of aircraft carriers (Rochlin, La Porte, and Roberts, 1987). These studies show clear differences between the formal, rank organization, and the high-tempo organization evolving during the take-down of aircraft on the flight deck and illustrate their complex interaction.

The basic point is that the control of the *productive function* (to supply goods or services) is a critical ingredient in a turbulent environment. Such control is *not* primarily based on a concern for plans and detailed schedules but on a competence, distributed across the organization, to make decisions depending on local conditions. Local decision making is only possible under certain conditions, such as (1) propagation of adequate criteria functions, the decision makers must be able to generate a local target from overall criteria; (2) availability of up-to-date information about the state of affairs in the system; and (3) adequate resource margins facilitating fast changes in work processes. Given these conditions, capable people at all levels (including those on the "floor") will generate the proper schedules. (Below we will see that the preconditions stated here for adequate distributed decision making are in fact defining a closed-loop, adaptive control strategy.) However, to ensure adequate resources for flexible local planning, long-term, feedforward planning of resources can be necessary and, frequently, an important problem can be to formulate the proper dynamic interaction between long-term resource planning and short-term production control. (see also Chapter 3).

From the economie's point of view, Kaplan (1989) argues that the difficulties many companies have in adapting to a dynamic environment is caused by their inadequate accounting systems, which have not followed the requirements of modern technology. Traditionally, accounting has been made in terms of deviations from prognoses, plans, and budgets. What is needed is operational control based on an accurate and timely performance measuring system. In this way, Kaplan also stresses the need to explicitly consider the information need of a closed-loop production planning strategy. In Kaplan (1990), an approach to the design of cost systems is proposed that introduces the measurement system vital for a transition from centralized long-term planning, to short-term, feedback planning.

In conclusion, the design of modern work stations and suitable support systems cannot be based solely on preplanned, effective work procedures issued to actors who then function within a formal, stable organization. Instead, design has to be concerned with supplying objectives and resources to individuals who then solve their problems in a dynamic work space and flexibly change their preferred cooperative patterns. In turn, this requires that the designers possess an intimate understanding of the constraints and mechanisms of the work place involved and of its behavior-shaping features.

The Pace and Depth of Change: An Example

An example will demonstrate the fast pace and deep penetration of the change of organizational structure and the required control strategy that can be necessary in a company during the transient following the decision to switch from preplanned production to a stable market to customer-driven production (see Fig. 1.1). In this particular company, production planning before 1974 was built around the efficient production of one model of an advanced piece of equipment that was batch produced on the basis of sales prognoses. During this period, centralized, operations analytic resource planning tools were developed. Production management was focused on control with reference to budgets and the company was organized in the usual hierarchical structure. Around 1974, a decision was made to concentrate on customer and order-controlled production. Several different models and products were to be produced in smaller series in response to specific orders. Consequently, a high number of possible combinations of product specifications is found with a related need to keep a low volume of "parts in process" (i.e., a very dynamic situation with many order changes and requirements for faster responses to spare parts orders). As a result, a fundamental change in planning conditions from a plan-driven long-term control to an high tempo, order-driven control was called for.

In this new situation, preplanned production control is less effective. However, when problems with the operations analytic tools that are used are actually experienced, they are frequently perceived as indicating a lack of detail in planning and inadequate information input to the system. Central planning offices, therefore, may respond by striving to improve their tools and the centralized control. This is counterproductive in a dynamic environment. Consequently, to an increasing degree, the production scheduling is taken over by informal networks on the floor operating in a feedback mode on the basis of direct observations of the actual state of affairs with reference to the perceived priorities of the individual orders. These networks control the production up-stream from the output end—often in spite of the general planning edicts. In other words, an informal, distributed decision control function at the floor

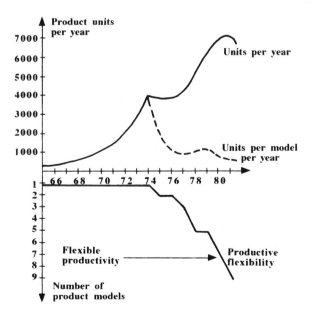

FIGURE 1.1. The change of pace in a manufacturing company: following one single strategical decision (adopted from Barfod, 1983, with permission).

compensates for the weaknesses of the central, long term planning in the dynamic environment.

Figure 1.1 illustrates the fast pace of change caused by one strategical decision and the discussion above illustrates the deep penetration of the change within the organizational structure and management patterns. These two aspects, when taken together, make it difficult to adjust an organization by evolutionary, piecemeal adjustments. Instead, a thorough analysis of the mechanisms governing the organizational adaptation is required, followed by a careful design of the organizational changes.

BASIC MODELING PROBLEMS

From a methodological point of view, such an analysis of the adaptive mechanisms of a work system requires an increased attention for abstract functional relations in addition to the usual structural perspective focused on causal interaction among parts. In the following sections we will discuss some simple examples to illustrate the implications of these two perspectives to set the stage for the presentation of the modeling and design framework presented in the following chapters.

Functional Abstraction Versus Structural Decomposition

When we conceive of a new work system, we normally use the structural perspective. To serve a particular purpose, we choose among the available parts, tools, and productive processes, and we select staff members having an appropriate background and education. We then aggregate these elements into a productive structure and instruct the actors how to apply the tools and productive processes for the purpose of work. In other words, we arrange the elements in cause-and-effect chains according to their individual input–output characteristics so as to have the intended overall effect. The problem we face in modeling systems incorporating human actors is, however, that humans do not have stable input–output characteristics that can be studied in isolation, and we cannot develop models of human machine systems by aggregating input–output models developed in isolation. When the system is put to work, the human elements change their charateristics; they adapt to the functional charateristics of the working system, and they modify system charateristics to serve their particular needs and preferences. In other words, to understand system behavior when adaptation has taken place, we have to look at the entire system, and instead of decomposing functions according to the structural elements, we have to abstract from these elements and, at a purely functional level, to identify and to separate the relevant functional relations (see Fig. 1.2).

In addition, we have to identify the adaptive mechanisms that generate the observed behavioral trajectories and work practices. In other words, design by aggregation of input–output relations identified for the individual structural elements in isolation makes it *possible* for the resulting work system to function. However, to make the functioning *effective*, the design must not result in a system that constrains the behavior of human actors to only one possible work process. Instead, the design

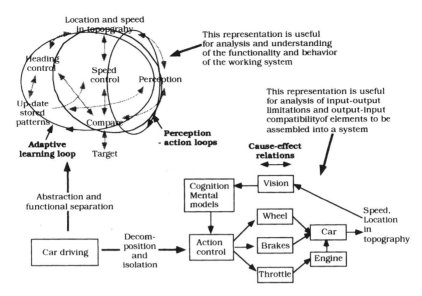

FIGURE 1.2. The behavior of a system can be analyzed in two ways. One is by decomposition into parts, the behavior of which is known in isolation. The behavior of the system is described in terms of cause-and-effect relations. Another is by abstraction to a functional level, and separation of the loops of interaction of interest.

should define for the actors a space bounded by the goal and resource constraints. Within this space they should be allowed to adapt freely according to subjective criteria, such as effort, time spent, or joy of discovery (see Fig. 1.3). Analysis of existing work systems to understand their behavior and to get a basis for design of new systems, therefore, should not be focused on decomposition into their structural elements, but on a functional abstraction and separation of functional patterns and the criteria that generate behavior.

A familiar example illustrates this point. When a novice is driving a car, it is based on an instruction manual identifying the controls of the car and explaining the use of instrument readings, that is, when to shift gears, what distance to maintain to the car ahead (depending on the speed), and how to use the steering wheel. In this way, the function of the car is controlled by discrete rules related to separate observations, and navigation depends on continuous observation of the heading error and correction by steering wheel movements. This aggregation of car characteristics and instructed input–output behavior makes it *possible* to drive; it initiates the novice by synchronizing them to the car functions. However, when driving skill evolves, the picture changes radically. Behavior changes from a sequence of separate acts to a complex, continuous behavioral pattern. Variables are no longer observed individually. Complex patterns of movements are synchronized with situational patterns and navigation depends on the perception of a field of safe driving (see Fig. 1.10). The drivers are perceiving the environment in terms of their driving goals. At this stage, the behavior

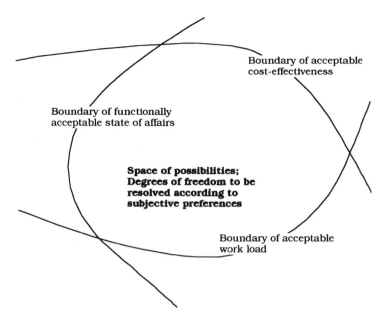

FIGURE 1.3. The actors' bounded work space in an adaptive system.

of the system cannot be decomposed according to the structural elements. A description must be based on abstraction into functional relationships. This is, in fact, the control theoretic or "general systems" perspective focused on global, functional relationships. However, the point is not to introduce the mathematical modeling of control theory. The central issue is to consider the functional abstraction underlying control theory and to understand the implications of different control strategies on system behavior and design requirements. We will discuss the functional representation in detail below, but first we will have a closer look at the structural, causal representation that is important for the initial conception of new systems.

THE STRUCTURAL CAUSE-AND-EFFECT PERSPECTIVE

The functional, relational perspective is necessary when analyzing an operating adaptive system and to predict responses to changes of such systems in which closed-loop interaction prevent reliable cause-and-effect reasoning. For design of new systems it is, as already mentioned, necessary to identify the elements available, their individual input–output characteristics and mutual output–input compatibility, and to define the limits of their capability. For this purpose, decomposition and representation by causal relations are useful and, therefore, it is necessary to analyze the properties of this perspective in some detail.

The distinction between a structural analysis in terms of decomposition and causality and a functional analysis in terms of abstraction and separation is a classical theme in the philosophy of science. From Aristotle and onward scientific theories were formulated in terms of causal connections between events. The breakthrough of modern natural science (due to Galileo and Newton) resulted in the replacement of observations of *events* with measurements of *variables*, and of *causal* laws connecting events with mathematical *relations* among variables.

The quantitative, mathematical representations of physical sciences and engineering have been so successful that often the qualitative concept of causality has been discredited by scientists. In his classical essay on the notion of cause, Russell (1913) finds the concept of causality to be so diffuse that it should be banished from science. However, his conclusion that causal explanations should be replaced by relational, mathematical representations appears to be too radical. The latter depend on the possibility of establishing–defining isolated relationships and amount to a modeling under well-controlled circumstances. Therefore, this mode of representation is not applicable for the analysis of complex courses of events (e.g., when the structure of a complex of technical systems breaks down during an accident). The complexity is too high, as is the number of unknown situation attributes, and thus each situation is unique. Likewise, a relational representation is not suited for describing the local interaction of human decision makers and technical systems. Consequently, a representation in terms of a causal flow of events is still an important tool for modeling human interaction with complex work systems.

Relational and Causal Representations

An understanding of the difference between the two modes of representation is important. The two modes, causal and relational representations, are supplementary. These modes are based on fundamentally different methods of generalization and serve different purposes in scientific analysis.

A relational representation of physical phenomena is based on mathematical equations relating physical, measurable variables. The generalization depends on a selection of relationships that are separated or "practically isolated" (Russell, op. cit.). This is possible when they are isolated by nature (e.g., in the planetary system) or because a system is designed so as to isolate the relationship of interest (e.g., in scientific experiments or in a machine supporting a physical process in a controlled way). In this representation, material objects are only implicitly represented by sets of coefficients in the mathematical equations. The representation is particularly well suited for the analysis of optimal conditions and theoretical limits of physical processes in a technical system which, by its very design, is carefully separated from the complexity of the outside world.

A causal representation is expressed in terms of regular connections of events. In his essay, Russell discusses the ambiguity of the terms used to define causality—the *necessary connection of events in time sequences*. The concept of an "event," for instance, is elusive; The more one strives to make the definition of an event accurate, the less is the probability that it is ever repeated. In this way, the regularity of causal connections disappears when attempts are made to define the concepts objectively. The weakness in Russell's appeal for objective definitions of causal concepts is its root in the quantitative, mathematical representation. He argues that in order to qualify the argument that a stone thrown against a pane of glass breaks it, one has to specify the weight and velocity of the stone. This argument is basically wrong. Events and causal connections cannot be defined by lists of objective attributes. An attempt to qualify a causal statement objectively by events in conjunction with the conditions that are jointly sufficient and individually necessary for a given effect to occur is, as Russell observed, without end. Completeness removes regularity. However, the solution is neither to give up causal explanations nor to seek objective definitions. Regularity in terms of causal relations is found between kinds of events, not between particular, individually defined events.

The behavior of the complex, real world, is a continuous, dynamic flow that can only be explained in causal terms after a decomposition into discrete events. Therefore, the concept of a causal interaction of events and objects depends on a categorization of human observations and experiences. Perception of occurrences as events in a causal connection does not depend on categories that are defined by lists of objective attributes, but on categories that are identified by examples or *prototypes* (Rosch, 1975). This is the case for objects as well as for events. Everybody knows what "a cup" is. To define it objectively by a list of attributes that separates cups from jars, vases, and bowls is no trivial problem, as demonstrated in numerous attempts to design computer programs for picture analysis. The problem is that the property of being "a cup" is not a feature of an isolated object, but depends on the context of human needs and experience. In the same way, the identification of events depends on the relationship in which they appear in a causal statement. An objective definition, therefore, will be circular. For a discussion of the role of decomposition and categorization in human reasoning, see Bruner et al. (1956).

A classical example of a causal statement is "the short circuit caused the fire in the house" (Mackie, 1965). In fact, this statement only interrelates two prototypes: the kind of a short circuit that can cause a fire in that particular kind of a house. The explanation that the short circuit caused a fire would be immediately accepted by an audience from a region where open wiring and wooden houses are commonplace, but not in an area where brick houses are prevalent. If this explanation is not accepted, a search for more information becomes necessary. Short circuits normally blow fuses; therefore, a further analysis of the conditions present in the electric circuitry is necessary together with more information on the path of the fire from the wiring to the house. A path through unusually inflammable material could have been present. In addition, a clarification of the short circuit's cause may be needed.

The explanation depends on a decomposition and search for *unusual* conditions and events. The normal and usual conditions will be taken for granted (i.e., implicit in the intuitive frame of reference). Therefore, the level of decomposition needed to make a causal explanation understood and accepted depends entirely on the intuitive background of the intended audience. If a causal statement is not accepted, formal logical analysis and deduction will not help. It will be easy to give counterexamples that cannot easily be falsified. Instead, a further search and a more detailed decomposition are necessary until a level is found where the prototypes and relations match intuition.

In the same way that it is impossible to define the meaning of words by a linguistic analysis of one separate sentence, it is impossible by analysis to define the elements in a causal statement separated from its context. In effect, causal explanations are only suited for communication among individuals who share prototypical definitions; that is, know examples that represent the determining features of objects and events because they have similar experience and, therefore, common "tacit knowledge" (Polanyi, 1967). The great effort spent to formalize causality and to cope logically with counterfactual statements (e.g., see Sosa, 1975) is a particular line of argument that is not immediately relevant in the present context. This formalization is based on "causality" defined by class membership from *empirical facts*, while our causal arguments are based on causality defined by *functional analysis* (see the implications of this distinction for medical and technical diagnosis, discussed in Chapter 3, and for risk management, discussed in Chapter 6).

Causal Analysis and Design

A causal description is an analog representation including physical objects as separate elements by a one-to-one mapping. Generalization implies categorization and the identification of *prototypical objects and events*. The great value of causal reasoning is its immediate relationship to the material world—to physical objects and their configuration. The representation is, therefore, very easy to update in correspondence with changes in the real world. This is not the case for relational representations in which a complex set of parameters and coefficients must be changed in order to incorporate physical changes.

The prototypical nature of the representational elements, objects, and events make them very effective for representing new configurations during design because the particular example of a class entering a thought process will be shaped by the context into which it is implanted. Design involves matching an object that is not yet in existence to a context that cannot be completely described (Alexander, 1964). Therefore, qualitative models and causal representations are necessary for creative activity, whether it is design in the classic sense or just problem solving in a dynamic work environment. When expressed in terms of types or classes, the elements of a particular model and the patterns of the context are adaptive and able to shape themselves according to the changing context of the mental experiments of the designer. This feature is very likely the reason why early industrial inventors and designers were actually from professions well versed in visual representation and manipulation of the physical world (map makers, painters, etc.), because they were experts in shaping new visual complexes from modified samples of a visual alphabet (see Chapter 7).

The role of causal representation in the intuition of inventors is illustrated by Hindle (1981), who discusses the imagination of the steam boat designers in terms of "steam boat images":

> The mental images of the steam boat builders have to be the historian's most important quest. He can never reach them directly, but illustrations known to have been used are

basic sources, original drawings come very close to the mental conceptions, and machinery components embody the concepts.

The illustrations presented by Hindle are all drawings of the physical anatomy and the mechanical linkages of components of the machinery discussed; that is, the figures are visualizations of perceptual images. Inventions apparently emerge by imaging familiar elements in new roles and combinations. A steam engine, for instance, in the early period has been described as being a backward running pump, and the first steam boats were based on engine-driven paddles. Hindle mentions that many early inventors of machinery had spatial representational abilities from other activities—as map makers, architects, artists, and so on. On a voyage to the United States Morse visualized the use of electromagnets and needles for telegraphy during a shipboard discussion of Ampere's experiments and the question of whether the speed of response of the needle depends on the length of the wire in Ampere's experiment. Hindle notes that Morse's sketches

> reveal no input of analytical science or projected circuit parameters and quantitative performance. Indeed, they resemble remarkably the spatial images recorded in his travel notebooks of the scenes, costumes, and technology he encountered.

Even when inventions are based on a creative imagination concerning the interaction among objects in a new constellation, the derivation of optimal operation of the involved processes and the limits of acceptable operation will be linked to functional abstraction and mathematical analyses of the relationships among quantitative variables that can be measured.

In this way, qualitative causal representations are useful for guiding reasoning during design or operational decision making. On the other hand, mathematical reasoning related to formal analyses of relations among variables is particularly useful for optimizing a design and finding its theoretical limits. The complementary nature is similar to that found between the use of intuitive judgment and formal proof by mathematicians (see Hadamard, 1945).

Implications of the Causal Modeling Perspective on Terminology

It follows from this discussion that causal analysis and representation are important tools for the design of complex systems. It is important in the discussions of the presented approach to realize that categorization is the basic precondition for causal modeling and that models and relationships are only valid when taken as expressions for prototypical members of the classes. This, in turn, creates problems when predicted, prototypical behavior is compared with a particular actual behavior. This problem is discussed in more detail with reference to empirical validation of models and design prototypes in Chapter 8.

In addition, in causal explanations describing objects that interact in chains of events, neither the objects nor the events can be defined objectively. Their identification depends on a frame of reference that is taken for granted. Thus, causal explanations are ambiguous except for individuals who share the underlying context. In a period of rapidly changing technology and the involvement of researchers of different professional backgrounds in cross-disciplinary interaction, the ambiguity caused by the very nature of causal explanations is an important problem.

In a cross-disciplinary cooperation, one often gets involved in discussions about the need to clearly define certain concepts. Such discussions are important to clarify conceptual differences, but it can be counterproductive to aim for a rigorous set of concepts in a cross-disciplinary cooperation. Because causal representations are "prototypical" and cannot be defined by an exhaustive list of attributes, the *concepts of cognitive systems engineering often will be defined by the context* in which they are used. Furthermore, since cognitive systems engineering is a "melting pot" kind of activity, the terminology will be influenced by the, often conflicting, use of terms from several established disciplines. Indeed, when viewed as a conceptual market place, it will be counterproductive for cognitive systems engineering if attempts were made to establish a separate discipline and develop a consistent professional terminology. Given that the fast pace of change of technology requires a close cooperative research among several technical and human sciences, we should instead develop a flexibility of mind that supports communications among professions. The result will be a hybrid, often ambiguous, terminology. To understand an unusual use of a concept in a discussion with a person with another background, it is important to remember that counterexamples to a causal statement can always be found. Thus in a cross-disciplinary interaction, it is necessary to try to accept the context of an argument and to identify the preconditions that would make sense out of an unfamiliar use of a concept rather than to focus on differences in conception. Therefore, the important message of the framework discussed in the present text is its *structure*, not the terminology employed.

THE FUNCTIONAL, RELATIONAL PERSPECTIVE

Even if human decision making is based on discrete representations in terms of objects and causal chains of events, and if qualitative models such as those found in representations developed within decision theory and cognitive psychology have to be applied for analysis of separate system processes, the need to apply an abstract, functional perspective for the analysis of systems, which includes adaptive human actors, was discussed and illustrated above by the car driving example. One basic modeling problem is that models in terms of causal chains of events and decision chains cannot reliably be used to analyze the behavior of closed-loop systems. For instance, the occasional instability of Watt's steam engine governor was not understood until Maxwell (1868) made his analysis of it in terms of differential state equations. This introduces the control theoretic or "general systems" perspective focused on global, functional relationships for analysis of the behavior of closed loop systems for which causal reasoning breaks down. The point is not, however, to introduce the mathematical modeling of control theory for our purpose. In that respect, the general theory of systems including humans failed because it was focused on simple, linear servomechanisms (see, e.g., Rosenbluth, Wiener and Biegelow, 1943). The central issue is to consider the functional abstraction underlying control theory and to understand the implications of different control strategies on system behavior and design requirements.

A note on terminology may be needed here. In the present context, *control* is taken to be the function of bringing the operational state of a work system into some desired state. In a sense, it also includes the concept "decision making," a term preferred by decision theorists for the situation when somebody, isolated from the dynamics of the particular environment, formulates a "problem," collects some data, and makes a "decision" about a sensible act. In the control perspective, "decision making" during work is more of a continuous activity in a dynamic interaction with the environment. Recently, decision theorists have also been concerned with the dynamic aspects and are increasingly studying "dynamic decision making" (Brehmer, 1992) or "natural decision making" (Klein, et al. 1993). Consequently, the approaches of the different disciplines tend to converge. Also, the "management" concept of organizational theory represents a control function, and several approaches have been taken toward a control theoretic formulation (cf. the general system theory advocated by Beer, 1966, and the "fifth discipline" discussed by Senge, 1990a,b).

Control Strategies in Work

Control of the behavior of a work system can be based on several different control strategies, which have very different characteristics with respect to the information on which they operate. The relevance of the strategies for control of a particular work system depends on several conditions, such as the stability of system parameters compared to the dynamics of the environment in which it operates, the availability of predictive models of system performance, and the access to explicit information about objectives and the current state of affairs. As mentioned earlier, when designing sociotechnical systems, it is very important not to constrain the actors to follow only one predetermined strategy but to give them the opportunity to chose and to shift strategy according to their situation. Before we discuss the analysis and design of work systems, we therefore find it important to review the characteristics of the relevant control strategies. For that purpose, and for the benefit of the readers not familiar with the control jargon, a couple of very simple examples will be considered to support an intuition of their properties.

Open-Loop Control

The open-loop strategy is illustrated schematically in Figure 1.4. A simple example of the use of this control strategy is shooting at a target with conventional artillery. Aiming involves a precalculation of the parabolic trajectory of the shell from an assessment of the distance to the target and the initial velocity of the shell. If the target is moving, its velocity must be estimated and the aim corrected so as to predict its location at the arrival time of the shell. In addition, compensation for the effects of

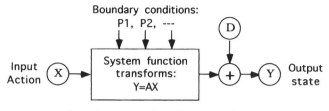

Open-loop control: $Y = A_{P1,P2,--Pn} \cdot X + D$

FIGURE 1.4. Schematic representation of the open-loop control strategy. The output of an open-loop system is a direct transformation of the input, depending on the boundary conditions. The use depends on (1) reliable predictive model of the processes, A, (2) a stable system and operating conditions, and (3) no significant disturbances, D.

disturbances such as wind and rain, is necessary. In short, the shooting task is preplanned by a detailed design of the functional trajectory of the shot. Any change after the trigger has been released will cause the shell to miss its target.

Another simpler example is connected with baking bread. By means of a recipe (written or memorized), the necessary ingredients are mixed, the dough is allowed to rise, and then it is placed in the oven at a given temperature for a given period of time. This is a strictly open-loop operation. No corrections are possible after the bread is placed in the oven. This mode of control is only relevant under some very restrictive conditions. The productive transformation (A in Fig. 1.4) must be stable and well known and the disturbances (D) insignificant, that is, the productive functions must be well separated from the influence from a dynamic environment.

This strategy has been extensively applied in the past for manufacturing activities organized according to the "scientific management" paradigm. It has been successful as long as the pace of change of technology and market conditions has been slow and the manufacturing process could be isolated from disturbances from the environment. In that case, the behavior of the involved people could be constrained to predictable patterns by work place design based on time and motion studies. Production could be planned for a very stable consumer need (Henry Ford even decoupled the process from consumer influence: "The customer can have the Ford T model in the color she wants, as long as this is black"). In this mode, management is based on monitoring the correspondence of performance with reference to plans, budgets, and schedules. This need is served by the traditional formal accounting and reporting schemes.

The introduction of computer-based resource management schemes and flexible manufacturing systems together with effective market analysis methods have extended the strategy into the recent period with less stability and more dynamic markets. As we saw in a preceding section, however, the application of this "second generation" management strategy is now causing considerable problem in industrial manufacturing due to the increasing pace of technological change and the operation in an increasingly turbulent and competitive environment.

Closed-Loop Control

The switch toward a more flexible production control strategy, advocated by Aoki (1988) in a preceding section, involves a switch toward a closed-loop, feedback strategy.

To appreciate the difference between an open- and a closed-loop strategy, compare aiming of the artillery cannon to the use of an active, target seeking missile that can itself observe the location of the target (by radar, television, or heat sensing). To plan a shot, it is only necessary to identify the target to the missile, which then "locks on" to it. The location of the target observed by the missile is compared with the projection of its own current goal and the missile continuously adjusts its direction of travel to intersect the target. In addition to specifying the target, the planner–designer only needs information on "limits," such as the maximum range of the missile, its top velocity, and its maneuvering capability, which all have to be adequate for the desired category of chases.

A corresponding example from the culinary department concerns the preparation of a casserole dish comprising a combination of whatever edible ingredients the chef thinks are suitable. No recipe exists; "add and taste" is the control strategy that is limited only by the size of the pot, the availability of possible ingredients, the arrival of the guests, and the patience of the chef. That is, the activity is controlled *during the process* from an observation of the effects to ensure the intended output.

Closed-loop systems have some very important features and the design of open- and closed-loop systems depends on very different strategies, which are illustrated by Figure 1.5. For a closed-loop feedback system, the input actions X are not planned in advance but depend on a continuous comparison of the actual state with the target state. The actions are defined by the observed discrepancy, and the effect of the actions will propagate through the system so as to compensate for the disturbance D, and for variations of the boundary conditions. If the factor A—called the gain of the productive system—is high (i.e., a small discrepancy will have a large influence on the output), then the output state is independent of A, while the measuring or observation function β is critical. Therefore, the focus during design of a closed-loop feedback system is *not* on a detailed planning of the internal behavior of the transformation A, but on (1) formulation of the *required* target states Y_R, on (2) the function β representing the observation and the interpretation of the *actual* output state in terms comparable to the target states, and on (3) the design of an internal system configuration that has the resources necessary to cope with all the relevant target states, boundary conditions, and disturbances. Independent of variations of the characteristics of the transformation itself, the system will adjust its internal function *during the process* so as to remove the difference between target and actual states as long as the resources are adequate. Note that the states of the productive, forward path are *not* a simple transformation of the target states, but only reflect the propagation of the discrepancy between actual and intended states and are heavily influenced by the disturbances and varying boundary conditions that are subject to compensation. That is, the behavior of the system elements within the loop cannot be described by simple linear cause-and-effect relationships, such an argument would be circular.

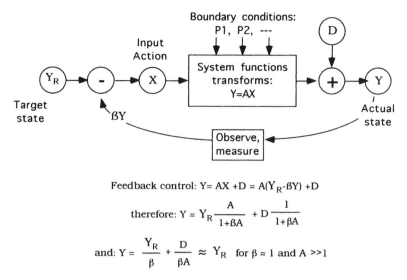

Feedback control: $Y = AX + D = A(Y_R - \beta Y) + D$

therefore: $Y = Y_R \dfrac{A}{1+\beta A} + D \dfrac{1}{1+\beta A}$

and: $Y = \dfrac{Y_R}{\beta} + \dfrac{D}{\beta A} \approx Y_R$ for $\beta \approx 1$ and $A \gg 1$

FIGURE 1.5. In a closed-loop system, a subtraction generates an error signal (X) that brings the output to the target irrespective of the function. The use depends on (1) explicitly stated objectives and targets, (2) reliable measurements of output without excessive delays, and (3) adequate capability to control.

It is clear that management of closed-loop, feedback systems cannot be based only on monitoring the correspondence with plans and budgets. Instead, continuous monitoring of performance with reference to goals is required (cf. Kaplan's request for new accounting schemes mentioned in the preceding sections).

Control by Trial and Error

The two open- and closed-loop strategies are simple boundary cases. When the gunners in the open-loop case observe that they missed the target, they will of course realign their aim to have a better result next time. That is, they will very likely be able to hit the target after some shots. This trial-and-error strategy is effective when past experience is relevant for future operations and the *cost of trials is less than the value of the lessons learned*. Thus after observing a missed target, the gunner, from above, reconsiders the parameters of the aiming calculation and gives it another try. Similarly, only after tasting the bread can modifications be planned for the next try—in the form of changes to the menu, through analytic considerations of the bread making process, or just common sense.

By this strategy, a satisfactory control evolves by survival of the fittest procedure and several common system types fall into a variation of this category—case handling, as exemplified by municipal welfare offices with their myriad of services, appeal channels, and so on. Similar situations exist in other work domains. Within the welfare system, a kind of parallel processing occurs comprising the contact meetings between clients and their respective case handlers. These are more or less procedu-ralized sessions dictated by paragraphs in the law as tempered by municipal practices and the style of the case handler. Control by trial-and-error is exerted by national politicians who adjust the laws every X years to correct for experienced problems and to reflect their updated political intentions. On a shorter time scale, control adjust-ments can come from municipal politicians who, in their capacity as budget watchers, can suddenly adjust one or more of the "parameters" in the case handlers' decision base, and thereby alter the outcome of subsequent client–case handler negotiations. In some cases, whether the cost of these alterations is less than the value of the lessons learned is a debatable issue.

Whether the trial-and-error strategy is considered a closed- or open-loop strategy, depends on the time scale and the goal considered. Over a longer time scale and for performance on average, it is a closed-loop strategy. When we consider the individual outcome and disturbances during one trial, it is an open-loop strategy. Again, it is not a characteristic of the *structure of system*, but of the *functional relationship* con-sidered.

Adaptive and Self-Organizing Control

In general, simple feedback loops serve to stabilize an *output state* such as the magnitude of a critical variable according to some set-point value (e.g., maintaining the aiming of a missile at a moving target or the temperature of a room independently of outside temperature). Devices such as high-fidelity amplifiers maintain good performance by incorporating a high quality feedback loop to compensate for any unstable characteristics of the components (e.g., transistors that vary with tempe-rature, age, and working conditions).

In more complex feedback arrangements, system parameters can be adjusted automatically on the basis of observations of complex performance measures, such as stability or efficiency. For example, aircraft control parameters can be adjusted to compensate for changes in aerodynamic parameters with speed. Figure 1.6 illustrates the control of production parameters to optimize cost and safety. In this case, the control strategies involve *adaptation*, as exemplified by systems that maintain opera-tion at an optimum via a continuous search strategy. Finally, adaptive control systems have been used that change *system structure*. These systems are said to have *self-organizing* features since they can examine the available resources and reconfigure themselves when substandard performance is observed due to unforeseen environ-mental conditions or internal system failures.

Control Strategies in Work

The examples discussed in the previous sections are chosen to illustrate some dif-ferent functional relations in control of system behavior. These examples are abstrac-tions and actual work systems cannot be classified according to this simple distinction

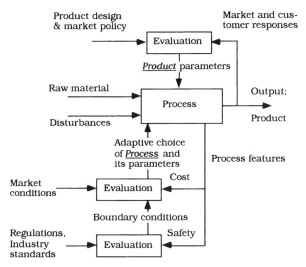

FIGURE 1.6. The actual work systems involve several nested closed-loop loops with adaptive and self-organizing features.

because, in most real systems, several control strategies are interacting in a way, depending on circumstances. When baking bread, for instance, repeated trials will lead the baker to correlate the feel of the dough and its smell to successful results, and thus introduce closed-loop correction during the process. Similarly, preparing the casserole dish under closed-loop control will lead the cook to adopt a stable practice and the performance will attain considerable open-loop features. Furthermore, in an actual work system, many control loops will be active simultaneously in a complex, nested structure and involve many human actors functioning as local controllers. A simple illustration of such nested closed-loop, adaptive control functions is found in Figure 1.6. In conclusion, the control strategies are abstractions and the work system cannot be classified accordingly.

It should also be mentioned that Figure 1.5 and the missile example, for illustrative purposes represent *continuous* feedback control involving propagation and transformation of continuous, quantitative variables and signals within the loop. There can be strong differences between this case and work systems in general. Indeed the simple examples above can perhaps indicate this. While some of them are of a continuous nature, most actual systems place actors in various kinds of discrete, sequential, even time-shared transactional work situations. Thus the "continuous" propagation of quantitative variables will be replaced by propagation of the effects of "occasional" or "periodic" discrete decisions "when required." That is, the critical feedback operation including action–evaluation–decision making–action can take on many (subtle) forms and involve different time scales. It should be realized, however, that the basic features of the closed-loop control are independent of the continuous–discrete nature of the underlying implementation.

MODELING AND DESIGN OF ADAPTIVE WORK SYSTEMS

In order to specify the framework for analyzing and designing adaptive work systems we will review some of the representation issues discussed in the previous sections. One central conclusion is that we need an analysis from the functional perspective to understand the mechanisms generating behavior of the work system through adaptation, while we will use the causal perspective to identify and select the parts and components from which we assemble systems. This latter perspective is necessary to judge input–output capabilities of the elements and their mutual compatibility.

Analysis of Adaptive Systems

Analysis of the behavior of adaptive systems from a functional perspective and prediction of the responses to major changes, such as introduction of new support systems, pose some special problems.

1. *Work Systems Are Goal Directed and Adaptive.* The systems we study are goal directed; they serve various purposes within a given environment in order to survive. A work system exists because the system and the environment share certain goals. By recursion, this is the case both for the entire system and for any subset in terms of teams and individuals.

In well-adapted, stable systems, goals and constraints to a large extent are implicit; they are embedded in practice and customs. By contrast, a changing environment and the resulting discretionary tasks require explicit considerations of goals and constraints. Goals vary, requirements and opportunities change, and the means and tools to pursue goals and to adapt to changes depend on the actual situation. Thus, in a dynamic environment, an effective organization of work depends on *self-organizing* and *adaptive* "mechanisms," which enable the system to change its properties in order to maintain a match with current needs when its internal conditions and/or the environment change.

2. *Systems Evolve Over Time.* Adaptation is a kind of goal following behavior. Performance is frequently modified to keep some measure related to the relevant performance criteria for the system near an optimum. Control of adaptation is distributed across all individuals, teams, and organizations. In other words, a distributed, self-organizing mechanism will shape the functional structure of the system, the role allocation among people, and the performance of the individuals. Thus, a framework useful for the analysis of work and for the prediction of responses to changes in work conditions, must reflect the mechanisms underlying the evolution of work practice. Adaptation is evolutionary; it is not planned by anybody through rational analyses. Its properties (i.e., the structure and performance of the system) emerge from a *"survival of the fittest"* of the possible structures and patterns of behavior. This happens to a large degree as a result of trial-and-error experiments, planned and unplanned, conscious and unconscious.

3. *Modeling Depends on Identification of Behavior-Shaping Constraints.* Human actors are basically goal directed adaptive mechanisms. Great diversity in behavioral patterns is found among the members of an organization. No two individuals are occupied by the same activity; no two patterns of movements are the same. The variety of options with respect to "what to do when and how" is immense. In order to predict why a particular fragment of behavior is chosen instead of another possible pattern, we have to understand how the action alternatives in a particular situation are eliminated (i.e., the degrees of freedom reduced), such that one unique sequence of behavior can manifest itself. As long as action alternatives remain, behavior is indeterminate until a choice is made.

To identify the kind of behavior to expect among all the possible options for action, we have to identify (1) the constraints that shape behavior by guiding the choices taken by the individual together with (2) the subjective performance criteria that are applied by the individual actors to reduce any remaining degrees of freedom. A problem in identifying behavior-shaping constraints is that they will not all be active at the time of the behavior they control. Behavior has a prehistory. Patterns of behavior evolve; they are shaped by prior decisions and choices. Once a segment of behavior is planned by situation and goal analysis and by a consideration of the alternative options for action, the behavior-shaping constraints are compiled into cue–action patterns (see also Chapter 3) and will not be active in subsequent situations when the particular pattern of behavior is reused. However, it is necessary to identify these "hidden" constraints in order to predict and understand the behavioral pattern even if the constraints are no longer "needed" for shaping the behavior. This identification can be difficult because often the constraints are no longer known to the actors and, therefore, have to be inferred from the work requirements, the resource profiles of the actors, and their subjective performance criteria.

4. *User-Involvement: Doing and Knowing.* It follows from this discussion that knowledge about goals, constraints, and the internal functional properties of the work environment is only necessary for the initial planning of an activity and for the exploration of the boundaries of acceptable performance in a *new* "territory," not for the control of behavior during repeated encounters with the same situation.

In stable work environments, know-how and established work practice are normally learned by novices from the older staff members and optimized empirically on a trial-and-error basis. In this situation, the basic understanding of goal structures and internal functionality will tend to deteriorate. Some kind of functional knowledge, however, is still useful for rationalizations and explanations of the need for the various work activities. Therefore, a kind of *actor logic* (i.e., a collection of myths about goals and reasons) can evolve and replace knowledge about the rationale actually underlying the system design. Such informal, mythical knowledge will not be reliable when disturbances or changes require analytical, knowledge-based planning. Therefore, it would be a mistake to base work support systems on this kind of "actor logic." Furthermore, this kind of "knowledge" will not be a reliable source of information about the work domain. Any such representation must be based on analyses of the actual functionality of the domain itself and, for this purpose, a formulation of the intentions and reasons behind the work domain structure is necessary. Therefore, inferences from field studies are often required. For decision support during changing work conditions ways need to be found to bring such knowledge about the work domain to the actors' disposal.

Design of Adaptive Systems

To conclude this discussion, we will have a closer look at the models required for the design of adaptive closed-loop sociotechnical systems, which rely more heavily on a dynamic adaptation to changing conditions than an open-loop, preplanning of activities. The following specific requirements can be formulated:

1. *Representation of the Propagation of System Objectives.* A critical modeling problem is obviously to have faithful representations of the *objective functions* of the system, that is, the goals or targets that the various controllers (decision makers, actors) seek to attain–maintain. Decisions and control actions are not taken by systems or organizations but by a number of individual decision makers carrying out their functions at various levels and locations within the work system. Consequently, an important feature of feedback design of sociotechnical systems is the identification of the roles of the decision makers and the propagation of values and performance criteria that shape their local objectives.

2. *Representation of the "Measuring Function" Available to Actors.* Performance of adaptive, feedback systems depends entirely on the quality of the performance measuring functions available to the controllers (decision makers). Actors are guided by the perceived discrepancy between the actual state of affairs and their objectives (the desired state of affairs). The car driving example discussed in the introductory section illustrates the importance of this issue. The perception of the state of affairs in the work environment, of the goals to attend, and of the means available for interaction changes dramatically with respect to the level of functional abstraction, to the degree of integration and holistic "gestalt," and to temporal characteristics. Also Flach (1990) has stressed the role of perception as being the 'measuring' function of feedback control. Computer based interfaces are used for work stations in many work domains and offer system designers a high degree of freedom for design of interfaces in support of the actor–work coupling. This is, consequently, a central issue in the following chapters.

3. *Representation of System Capabilities and Resources.* Detailed representations of the operational relations of the work systems (i.e., the processes necessary to cope with the individual tasks) are not required for design of adaptive systems. Instead, an identification and categorization of all available productive resources are necessary together with a definition of their capabilities and limiting properties. In other words, a framework for representing categories of work resources is an important modeling issue.

4. *Performance Under Boundary Conditions.* Models representing the first three aspects of work systems are adequate for the design and analysis of the normal work performance lying within the capability limits of the system and its actors. For some

systems, in particular those posing hazards to the staff, the environment, or the investment, it will be necessary to analyze performance when "saturation" occurs within the feedback loop (i.e., when the resource limits of the system are violated). In this situation, the feedback loop is broken and difficult modeling problems are often found because suitable representations of the actual processes under nonnormal system conditions are required.

The cognitive systems engineering approach advocated in the following chapters is not meant to replace the rational, operations analytic approaches that are needed for many purposes (such as resource preplanning and financial accounting) and that can be based entirely on an objective, transactional analysis. The emphasis here is on those aspects of system operation that require a designed-in match between system requirements and human cognitive characteristics. Thus the effort is intended to serve as a supplementary, cross-disciplinary framework that integrates the results of the various relevant disciplines into activities connected with system analysis, planning, and design.

Design for a Fast Change of Pace

A final important issue to consider is that of designing in a multidimensional specification space during periods of rapid technological change.

During relatively stable periods, new work systems can evolve through incremental changes and an optimal match of work system to staff characteristics can move *step-wise* to new locations in the multidimensional design space. However, during a fast change, a *radical leap* can be required simultaneously along several dimensions and a new design may turn out to be a failure if just one dimension of the change has not been adequately considered (see Fig. 1.7 for a greatly simplified two-dimensional example). Thus when several, radical changes are considered in a new design, a good predictive model is necessary to guide the jump to a new optimum. A cautious, piecemeal introduction of changes may very well facilitate a plunge into the valleys of space.

The necessary conclusion is that we need a conceptual framework for designing advanced information systems that can serve to identify and describe the individual dimensions of the match between work systems and staff characteristics and their mutual interactions.

Conclusion

To conclude, we need a framework to represent all the aspects of work systems including the functions of technical equipment, the resources and preferences of the staff members, together with the functional structure of the work organization and management—all in compatible terms. Thus, we need a cross-disciplinary framework serving to coordinate models and findings from several disciplines. This framework should support the analysis and design of adaptive, evolutionary systems from an identification of the system constraints and human performance criteria that shape

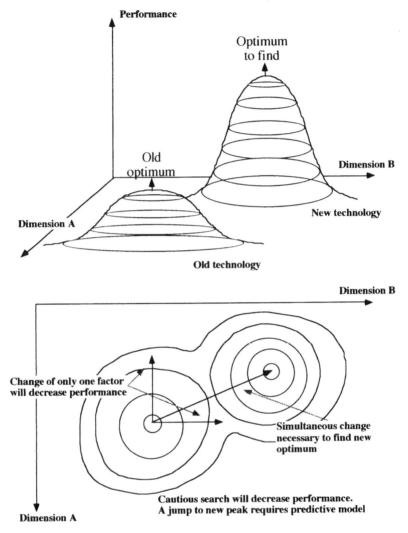

FIGURE 1.7. Design in a multidimensional specification space (in this case two dimensional). A missing consideration of only one dimension may cause an otherwise optimal design to fail.

system behavior during the evolution of practice. This is necessary to predict responses to changes in a dynamic work environment.

OVERVIEW OF A NEW MODEL FRAMEWORK

It is clear from the discussion in the previous sections, that analysis and design of modern, dynamic work systems cannot be based on analysis and design of work systems in terms of stable task procedures. Instead, analysis of work systems must be in terms of the behavior shaping goals and constraints that defines the boundaries of a space within which actors are free to improvise guided by their local and subjective performance criteria.

To create a systematic framework that makes it possible to relate the material and conceptual characteristics of a work environment to the cognitive characteristics and the subjective preferences of the staff is no simple matter considering that, as described previously, it is essential that all the relevant dimensions are adequately included. Furthermore, in order to be well focused and to give a rapid convergence to the number of facets that should be considered during the analysis, a well-defined point of view has to be established.

In order to bridge from a description of the behavior-shaping constraints in work domain terms to a description of human resource profiles and subjective preferences, several different perspectives of analysis and languages of representation are necessary (see Fig. 1.8). These perspectives change from that of domain expertise, through descriptions of activities from the point of view of work psychologists and decision theorists, to a representation of cognitive activities as applied in cognitive science and, finally, to a representation of the resource profile and preferences of the actor in

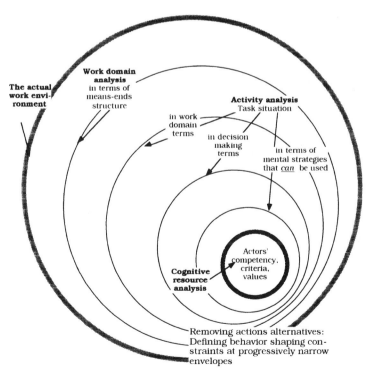

FIGURE 1.8. The shifts in language necessary to relate properties of a work environment to the cognitive resource profiles of the actors.

terms of cognitive psychology. In this way, the approach is intended to combine several sciences spanning from engineering to the basic human sciences.

It will be necessary to adopt an economic strategy of analysis, that is, one that converges rapidly by eliminating the degrees of freedom in the set of behavior-shaping constraints. Such a strategy should be based on a stepwise zooming-in on the repertoire of action alternatives from which an actor can freely choose. The conceptual structure used is like the top-down partition strategy used in the game of "twenty questions." The first level of such a convergent strategy will be to prepare the stage of human action by means of a topographic delimitation of the work space and an explicit identification of the objectives and operational resources (i.e., the goals, constraints, and means for action that are available to the actors). Next, a delimitation in time to determine the task situation will be made, followed by a new delimitation and shift in representation language to describe the decision task. The following step involves a focus on the mental activities and a related shift in language, in order to have a description compatible with a representation of the actor's cognitive resource profile and performance criteria. This framework supports a stepwise narrowing down of the degrees of freedom faced by an actor when considering the alternative possible ways to meet work requirements and the options among which a choice must be made. Shifts in language of description will be necessary depending on the basic source of the constraints, going from the context of the work domain, the situation calling for human intervention, the structure of the related control domain onto human cognitive, and emotional factors.

Cognitive Systems Analysis

Figure 1.9 illustrates the two concurrent analyses of work and work performance encompassed by the proposed framework. One analysis shown in the upper path serves to identify the activities with which an actor is faced. Another, shown below, serves to identify the role and characteristics of the individual actors and their mutual relationships. These will be described in more detail in subsequent sections. In addition, the interrelationships between these two concurrent aspects of analysis should be kept in mind.

1. *Identification of Activities.* The upper path is concerned with the work requirements that have to be compared to the actors' resources and preferences in order to determine the individual actor's likely choice of performance. It is important to stress that this analysis is not aimed at a normative prescription, but at an identification of the repertoire of options from which the individual actors will make their choice. Therefore, it is based on an ethnologic approach to the analysis of the actual basis for performance. The selection of the performance variants to include and the formulation chosen of task and strategies include implicitly the actors' subjective formulation of their actual goal, the way they view their task, and their possible "cognitive styles." This selection is done by including the repertoire of "possible" formulations of tasks and strategies that, judged from field studies, are relevant and can be used by an actor dependent on his/her subjective interpretations. This analysis is based on interdisci-

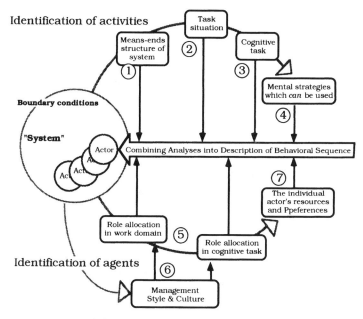

FIGURE 1.9. Overview of the framework. The behavioral trajectory unfolds from the interaction of the task requirements, as identified by the upper analytical sequence, and the individual agent's role and resource profile, as identified by the lower sequence.

plinary studies, such as classical operations research together with ethnologic and anthropologic approaches to the analysis of work (Rasmussen and Jensen, 1974; Pejtersen, 1979a; Rochlin, LaPorte, and Roberts, 1987; Bucciarelli, 1988). Such analyses pose some tricky methodological problems. Study of the normal performance of a feedback system will not reveal internal mechanisms and their properties. Special methods of analysis are required. As Simon (1969) notes:

> The behavior of an adaptive organism largely reveals characteristics of the environment, given the goals of the organism. Only when adaptation breaks down, will the internal mechanisms be visible.

Therefore, for analysis of adaptive work systems, studies of limiting performance can be necessary including errors and system failures. Often special interview techniques are required.

2. *Identification of the Actors.* The lower line of analysis aims at a description of the role, the resource profile, the subjective preferences of the individual actors, and an identification of the cooperative work structure. The work domain is considered as a more or less loosely coupled system controlled by the distributed decision making of cooperating actors. The analysis is focused on a determination of the criteria that control the adaptive, dynamic division of work among the individual actors, that is, on the role allocation and the content of the communication necessary

to control the concerted action. In addition, the preferred forms of communication as influenced by the adopted management style are analyzed. The point of view of the analysis is the evolutionary nature of the actual (informal) cooperative structures and work organizations.

The interaction between the two lines of analysis is complex and iterative. Role allocation interacts with the description of the structure of the work domain and the nature of the task situation. The description of the mental strategies that can be used must be compatible with the description of the individual's resources and preferences. Finally, when a match between possible strategies and preferences has identified the chosen strategy, it has to be "folded back" onto the higher levels of analysis and the work domain in order to determine the actual behavioral sequence. If we wish to put meat on the conceptual bones, it will be necessary to add details of the actual situation, which have been removed during the analysis, back onto the conceptual framework. When the cognitive strategy that is likely to be used is identified, the implications for the decision task and the related information requirements will be inferred. Next, the relevant set of cognitive activities involved in an actual situation will be determined to establish the likely work procedure in actual work domain terms and, finally, the involved means–ends networks in the work domain can be identified, as well as the coupling to other activities and actors.

The Perspectives of Representation

The following section presents an overview of the dimensions of analysis that are described in subsequent chapters.

Work Domain, Task Space. This dimension of the framework serves to delimit the system to be analyzed, and thus it represents the landscape within which the work takes place. It serves to make explicit the goals, constraints, and productive resources found in a particular work system. The representation within this perspective is a kind of general inventory of system elements and, therefore, in the short perspective, independent of particular situations and tasks. It is based on the identification of the productive resources in terms of categories of functional elements and their means–ends relations, which are available for the actors to "design" their local activity. The analysis gives further structure to this identification through its decomposition into modular elements along a part–whole dimension and through its identification of these potential means and ends at several levels of functional abstraction. These levels include representations of physical configuration and anatomy, physical work processes, general functions, abstract value functions and, finally, goals and constraints with reference to the environment.

An analysis within this dimension of the framework will identify the structure and general content of the global knowledge base of the work system. This analysis of the basic means–ends structure is particularly useful in a redesign situation for identifying those primary goals and constraints that may have become latent (forgotten) during earlier stable periods and also for finding alternative possible, but customarily neglected, means–ends relations. In a modern work organization, a complex network is relating multiple objectives and constraints to a wide variety of functional resources and it is very easy during design to miss an important dimension (see Fig. 1.7). It is increasingly realized that multiple perspectives of analysis are necessary for design of modern organizations and work systems (e.g., see Mitroff and Linstone, 1993) and the present analysis and representation of the means–ends structure is well suited to ensure adequate consideration of the various perspectives.

Activity Analysis in Domain Terms. This dimension serves a further delimitation of the analysis through a focus on the degrees of freedom left for meaningful activities by the constraints posed in time and functional space by a particular task situation. It instantiates that subset of the basic means–ends network (from step 1, Fig. 1.9) which is relevant for a particular task. However, it is important to recall that "a typical task sequence" often does not exist in a modern, advanced work setting. As a result, generalizations cannot be made in terms of work procedures revealed by means of a classical task analysis but instead should be made in terms of the objectives, functions, and resources active in prototypical work situations, and the related information requirements. A set of such prototypical work situations can be used in various combinations to characterize the activities to be considered for information system design.

Activity Analysis in Decision Terms. For this dimension of the analysis, a shift in representational language is made. For each of the activities defined under step 2 (Fig. 1.9), the relevant tasks are identified in terms of decision making functions, such as situation analysis, goal evaluation, planning, or actual execution. This representation breaks down work activities into subroutines that can be related to the cognitive activities of the involved people. The analysis also serves to identify prototypical "states of knowledge" that connect the different decision functions and serve as the nodes used for communication in cooperative activities. The information gained in this analysis will identify the knowledge items from the work domain representation that are relevant in a particular situation. In addition, it assists in identifying the queries that are likely to be made by decision makers for retrieving information.

Mental Strategies. A further analysis of the decision functions requires another shift in language in order to be able to compare task requirements with the cognitive resource profiles of the individual actors and their subjective performance criteria. For this purpose, the mental strategies, which *can* be used for each of the decision functions, are identified by detailed analyses of the actual work performance (e.g., by protocol analysis). In the current context, strategies are defined to be categories of mental information processes. Each strategy is based on a particular kind of mental model, a set of tactical rules, and a related mode of interpretation of observations. The characteristics of the various strategies are identified with reference to subjective performance criteria, such as time needed, cognitive strain, amount of information required, and cost of failure. Knowledge about the available effective strategies is important for the user interface design because it supplies the designer with several coherent sets of mental models, data formats, and tactical rule sets, which can be used by actors with varying expertise and competence.

Division and Coordination of Work. In order to identify the actors actually involved in the prototypical decision situations, it is necessary to find the principles and
criteria governing the allocation of roles among the groups and individuals involved.
The work domain is considered to be a loosely coupled system under the control of
a set of cooperating actors, and the allocation of roles can refer, for example, to
subspaces in the work domain or subroles in decision functions. This allocation of
roles to actors is dynamically dependent on the circumstances and is governed by
criteria such as actor competency, access to information, minimizing the communication needed for coordination, sharing of work load, and complying with regulations
(e.g., union agreements). The constraints posed by the work domain and the criteria
for role allocation specify the content of the communication necessary for concerted
activity.

Social Organization. While the role allocation determines the *content* of communication necessary for coordination, the "management culture" of the organization
determines the *form* of communication—whether the coordination depends on orders
from an individual actor, on consensus in a group of decision makers, or on negotiations among the involved actors. The management structure will heavily influence the
subjective performance criteria of the actors as a whole and therefore, indirectly, the
formulation of goals and constraints as well as the performance criteria of the
individual actors. The identification of the communication conventions underlying
the social organization is necessary to determine the communication formats for an
integrated information system. In particular, the communication of social values,
subjective criteria, and intentions, all of which are necessary in various work settings
for the coordination of activities, the resolution of ambiguities, and for recovery from
misinterpretations of messages is crucial. An identification of this communication is
important for allocating functions to an information system, to face-to-face communications, and for the design of the information communication formats.

Cognitive Resources and Subjective Preferences. At this stage, the degrees of
freedom in work performance of the individual have been delimited through an
identification of the work-dependent behavior-shaping constraints down to the level
of mental strategies that can be employed for the decision functions allocated to each
individual actor. In order to judge which strategy is likely to be chosen by an actor
in a given situation, the resource requirements of the various strategies have to be
compared to the cognitive resource profiles of the actors. Therefore, this perspective
of analysis is focused on the level of expertise and the performance criteria of the
individual actors.

USE OF THE FRAMEWORK FOR SYSTEM DESIGN

From the discussion in the introductory sections of Chapter 1, it is clear that the
overriding perspective of the work analysis is one of analysis and design. A work
analysis is made to ultimately change a given system of work for the better. Thus, the
analysis cannot take the structure and patterns of the current behavior of the system

of work for granted. On the contrary, the analyst has to "take it apart"; he/she must uncover the hidden rationale behind current practices as well as the accidental choices of the past, the procedures turned into rituals, the formalized mistakes. What is necessary in order to meet current and future requirements of the work environment? What could be done differently and better? What should be discarded as mere relics? In a sense, then, work analysis can be likened to "reverse engineering" in that the analyst approaches the given system as the result of a design process, then takes it apart so as to put it together again, perhaps differently. The analyst investigates the system to learn what it does and how and to decide what could be done differently. In a similar sense, work analysis can be compared to psychoanalysis in that it seeks to uncover the "unconscious" mechanisms of a system of work so as to enable the system to overcome fixed patterns of behavior. Thus in a discussion of the methodology of organizational studies, Selznick (1957) argues that rather than following the lead of experimental psychologists who study routine psychological processes, the analyst should imitate the clinical psychologist who examines the dynamic adaptation of the organism over time. Instead of focusing on the day-to-day decisions made in organizations, Selznick recommends the analyst to concentrate on those critical decisions which, once made, result in a change in the structure itself. That is, the fundamental approach of work analysis is to question the rationality of current patterns of behavior. The crucial question is not "what" the actors are doing at the time of analysis, but "why," together with the alternatives for "how."

For this purpose, the framework aims to be able to compare behavior-shaping characteristics of different work domains and, in particular, to be able to predict behavior in response to changes such as the introduction of new information systems from an analysis in an existing work place. It will not be possible to predict particular trajectories of work of an individual. Based on a description following the framework, however, it will be possible to define the boundaries given by the work requirements, the resources for work, and the individual's physical and mental resources—all of which will affect the individual trajectories depending on the situation and on subjective performance criteria and preferences (cf. Bateson, 1979: the category can be known, the particular escapes prediction). In this sense, the framework implies a theory of adaptive behavior. In response to changes in work conditions, adaptation is active and the behavior of the actors cannot be described in terms of work procedures or decision scenarios, because these will only be records of observed cases, not predictive models of performance.

Activity under varying conditions will depend on continuous adaptation and improvisation, on the ability to reconfigure patterns of behavior, to modify effective routines, to combine elementary routines into new patterns, and to generate new work procedures on demand. Therefore, only an analysis and representation based on a framework serving an explicit representation of the behavior-shaping constraints can be used for predicting the kinds of behavior to expect in response to changes. The set of constraints will define the boundaries of the space in which the actors can navigate freely according to their individual resources and subjective performance criteria and still meet the work requirements. In this way, the framework serves to delimit the field within which behavior will unfold. Hypotheses about the likely kinds of performance of an actor can be stated if conditioned with proper assumptions about the resources and subjective preferences of the individual actor. Predictions of the actual trajectories of performance are less important for the design of information support systems. Of importance is an evaluation of the capability of the actor to perform given certain individual resources and subjective preferences (i.e., whether an actor can meet the requirements and will accept the conditions). Basically, this is a feedback design approach: One specifies the criteria functions and makes sure the capability is present; then acceptable performance will emerge given the actors are motivated. The capability includes competence in using the means available and ability to explore the options present.

In order to predict a likely work scenario, a hypothesis is formed about the actor's selection among the strategies that can be used in the work situation and the work function considered. This strategy, which is an abstract representation of a type of behavior, should then be folded back onto the work domain. This means that the abstract construct in cognitive terms is instantiated by dressing it up with the details of the material work content and situational features. The prediction will be prototypical in the sense that it will miss all the details of the situation that cannot be known, such as disturbances, distractions from the general social interaction, the frequent shifts in strategy in response to resource demand conflicts, and to situation dependent shifts and subjective preferences. Nevertheless this does not affect the reliability of the prediction of the effects of introducing new information systems on work performance. It does, however, influence the possibility of empirical verification of a prediction. Being prototypical, prediction involves categories of behavior, while actual performance trajectories (as well as simulation of predicted scenarios) represent particular instances. In this case, comparison is no simple matter (see Rasmussen, Brehmer, and Leplat, 1991). A typical feature of prototypical classification is the difficulty in defining members; the most effective way is generally to point to a typical example. In the same way, the best way to evaluate predicted work scenarios may be to show the scenario to a group of "substance matter experts" (i.e., workers in the field) and ask whether they find it "reasonable" and natural, given the context (cf. von Neuman's test of intelligence).

The approach to model the likely behavior in a work context by a representation of the behavior shaping constraints and the actor's criteria functions has some similarity to a model of car driving proposed by Gibson and Crooks (1938). To explain skilled car driving, they identify the "field of safe travel and minimum stopping zone of a driver in traffic" (see Fig. 1.10). According to this approach,

> steering is a perceptually governed series of reactions by the driver of such sort as to keep the car headed into the middle of the field of travel . . .

Such fields

> are fields within which a certain behavior is possible. When they are perceived as such by the driver, the seeing and doing are merged into the same experience.

Gibson's account deals only with locomotion in a physical environment but similar constraint-based descriptions can be used at higher levels of abstraction with respect

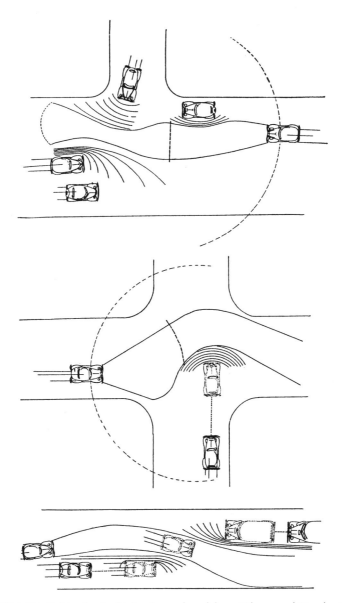

FIGURE 1.10. Fields of safe driving (reproduced from Gibson and Crooks, 1938).

to descriptions of the environment and to higher levels of cognitive control. The present framework can be understood as an attempt toward structuring such a transfer. In addition, the ecological interfaces discussed in Chapters 4 and 5 can be considered as further developments of the resulting ideas in connection with the information system interface design.

Chapter **2**

Work Domain Analysis

INTRODUCTION

The first facet of the model framework to be considered is a representation of a work domain that is useful for system design and analysis. As argued in Chapter 1, we consider a work system as a functionally coupled entity that adapts to the opportunities and requirements posed by its environment under the control of its human actors. When we follow the cognitive systems engineering approach, which emphasizes a feedback design strategy, the aim is *not* to represent the actual coupling and functioning of the work system in particular situations, but instead to produce a generalized representation of the "work domain" in terms of its inventory of objectives, functions, activities, and resources—all of which constitute the elements of the landscape in which the staff operates. That is, we are concerned with categories of goals, functions, resources and their general properties and, in particular, their capabilities for action as well as their limiting characteristics, which can constrain these actions.

DELIMITATION OF THE WORK SYSTEM

The first delimitation necessary for a work analysis is a definition of the work system of interest within the total environment. This identification depends on a pragmatic choice of a boundary around the object of analysis that is relevant for the actual design problem. This choice depends on the circumstances. In some cases, considerations of rather local conditions may be adequate, for instance, in the design of individual universal tools (a word processor). However, it is important to remember that the effects of adapting to the changes introduced as the result of a redesign very likely will propagate to all the work activities that are closely coupled to the activity subject to redesign. Consider the introduction of a particularly user-friendly word processor that may very well lead to drastic changes in the role patterns of professionals and their secretaries.

In many cases, a representation of the work domain must include an entire company due to the potentially tight coupling across activities originating in common strategic planning and resources. An overview of the entire organization in terms of the relevant means–ends relations can be required. This requirement would be necessary, for example, to uncover executive management goals, policies, and constraints that may very well have been embedded in established company practices. In a work analysis for redesign, they must be identified and formulated explicitly in order to judge their influence when they are reactivated during the adaptation to eventual changes.

LEVELS OF ABSTRACTION: THE MEANS–ENDS RELATIONS

The importance of an explicit representation of the work domain is illustrated nicely by Simon's fable about the ant maneuvering along a sand beach: the complexity of the observed track is caused by the complexity of the beach, not by the complexity of the ant. Similarly, the path taken by a human actor through a work space can only be explained on the basis of the complexity of the work space together with the goals and resources of the actor. An example of the path taken through a computer system during trouble shooting by a maintenance engineer is shown in Figure 2.1. The

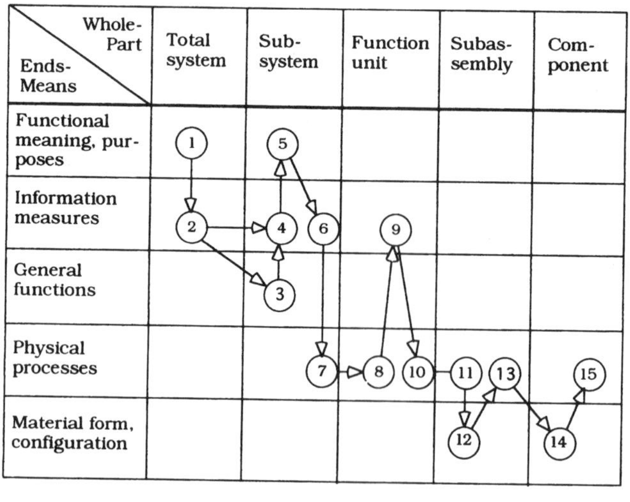

Ends-Means \ Whole-Part	Total system	Sub-system	Function unit	Subas-sembly	Com-ponent
Functional meaning, purposes	①	⑤			
Information measures	②	④ ⑥	⑨		
General functions		③			
Physical processes			⑦ ⑧ ⑩	⑪ ⑬	⑮
Material form, configuration			⑫	⑭	

FIGURE 2.1. Domain of computer repair. The figure illustrates the trajectory of an engineer through the work space during computer repair. Compare to Simon's ant: "Viewed as a geometric figure, the ant's path is irregular, complex, hard to describe. But the complexity is really a complexity in the surface of the beach, not a complexity in the ant" (Simon, 1969).

engineer's trajectory reflects the computer functions and elements mentioned in his/her verbal protocol. It illustrates the changes in concepts he/she employed during the task in terms of the *level of abstraction* and the *decomposition* chosen. Operation in a work space involves an exploration of the available *means* for achieving the immediate *ends*. This exploration includes a *span of attention* dimension connected to *part–whole* considerations, which can range from local components and tools to global features. Thus exploration of a map framed by the dimensions of means–ends and part–whole is a general feature of navigation in a work domain when actors are involved in discretionary tasks.

This example illustrates another feature of a work domain representation. This has to do with the need to include several levels of abstraction and decomposition. The previous chapter argues that several levels of decomposition and conceptual language are necessary to relate characteristics of a work system to the cognitive characteristics of human actors. This is also the case where it is important to have a consistent representation of the potential relationships between the global goals and constraints of a given domain and the roles of the individuals, tools, and other material work resources. This is clearly reflected in the domain representation of Figure 2.1, which shows the shifting focus (vertical axis) and span (horizontal axis) of the actor's attention as reflected in his/her verbal protocol.

Chapter 1 argued that one important way of modeling complex systems including human–system interaction relies on the use of causal representations of the relationships among prototypical objects and events. This is, of course, also the case for the representation of a work domain. No two work domains are identical. A work domain of today will be different tomorrow. Therefore, only descriptions in terms of general categories, represented by prototypical examples, will make sense.

The overall structure of the means–ends space is illustrated in Figure 2.2. It shows

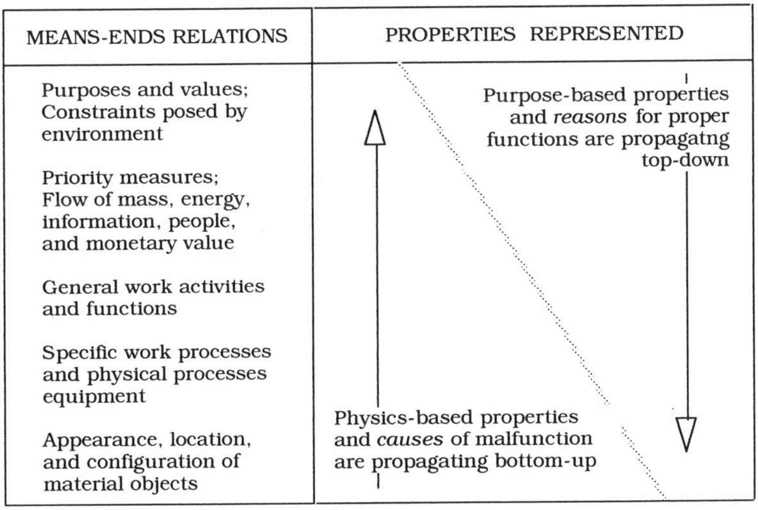

MEANS-ENDS RELATIONS	PROPERTIES REPRESENTED
Purposes and values; Constraints posed by environment	Purpose-based properties and *reasons* for proper functions are propagatng top-down
Priority measures; Flow of mass, energy, information, people, and monetary value	
General work activities and functions	
Specific work processes and physical processes equipment	
Appearance, location, and configuration of material objects	Physics-based properties and *causes* of malfunction are propagating bottom-up

FIGURE 2.2. Levels of abstraction in system representation.

how the functional and material features of the work system dominate the representation at the lower levels, while the intentional features, that is, the objectives that govern the control of the system functions, dominate at the higher levels. The down arrow indicates that changes in these objectives (e.g., company goals, management policy, or general market conditions) will propagate top-down through the levels, as reasons or rules of rightness (Polanyi, 1958). On the other hand, changes in the material, physical basis of the system, such as the introduction of new tools and equipment or breakdowns in major machinery, will propagate bottom-up through the levels as a kind of upwards work activation.

In the following section, the categories of functional resources and objectives represented at the various levels of abstraction are discussed in more detail, based on the overview of Figure 2.2. We will start at the middle level and work our way up and down from this center. A review of these categories is given in Figure 2.3.

General Work Activities and Functions

It is convenient to begin at this level, which comprises representations of work functions and activities having general and familiar terms for the particular domain. In an office environment, these would include marketing, personnel administration, accounting, case handling, document handling, and so on. In a manufacturing environment, general functions would comprise design, production planning, production, maintenance while, in an industrial process plant, frequently used functional labels include steam generation, power conversion, heat exchange, and so on. The important point to be made here is that the denotation of activity or function at this level is *independent* of the underlying processes involved as well as their physical

MEANS-ENDS RELATIONS	PROPERTIES REPRESENTED
Purposes and Constraints	Properties necessary and sufficient to establish relations between the performance of the system and the reasons for its design, that is, the purposes and constraints of its coupling to the environment. *Categories are in terms referring to properties of environment.*
Abstract Functions and Priority measures	Properties necessary and sufficient to establish priorities according to the intention behind design and operation: Topology of flow and accumulation of mass, energy, information, people, monetary value. *Categories in abstract terms, referring neither to system nor environment.*
General Functions	Properties necessary and sufficient to identify the 'functions' which are to be coordinated irrespective of their underlying physical processes. *Categories according to recurrent, familiar input-output relationships.*
Physical Processes and Activities	Properties necessary and sufficient for control of physical work activities and use of equipment: To adjust operation to match specifications or limits; to predict response to control actions; to maintain and repair equipment. *Categories according to underlying physical processes and equipment.*
Physical Form and Configuration	Properties necessary and sufficient for classification, identification and recognition of particular material objects and their configuration: for navigation in the system. *Categories in terms of objects, their appearance and location.*

FIGURE 2.3. The classes within the means–ends hierarchy.

implementation. Normally, several subfunctions are involved in a work activity and their performance must be coordinated to meet a particular purpose. Indeed, as stated in Figure 2.3, a description at this level comprises the properties necessary and sufficient to identify the functions and activities to be coordinated. To accomplish this, a compatible representation is necessary that utilizes a common language and is independent of the underlying implementations. For this purpose, input–output relations—"blackbox" models of performance—are used.

Physical Processes

At the next lower level, the representation is focused on the *physical processes* involved in equipment functioning and in human activities to establish and maintain the general functions and activities. As stated in Figure 2.3, this requires a specification of the properties necessary and sufficient for controlling the human work activities and the use of equipment. Appropriate representations of the various technical elements are necessary to describe the physical processes and the behavior of equipment and tools, to determine their limiting characteristics under different conditions of use, and to select the means for controlling their performance. Descriptions of the physical processes and actions involved in human activities are necessary to identify the information necessary and to evaluate physical work load, ergonomic problems of equipment design, and so on.

References to their potential use at the general functional level should be more or less implicit in the process descriptions. At this level, representation can take on many forms (e.g., verbal descriptions, mathematical equations, and plots of relations between variables and diagrams).

Abstract Functions and Priorities

The next higher level above the general functions covers *abstract functions and priorities*; that is, the representation of concepts that are necessary for setting priorities and allocating resources to the various general functions and activities at the level below. In order to grade the importance of allocating resources to the different work functions, it is necessary to compare the influences of the various functions on the higher level objectives by means of *value* measures that can be applied independently of their functional role. Suitable measures at the abstract level can generally be assumed to follow some kind of conservation law because such values are not supposed to vanish from a system. Thus priorities in most work systems are related to monetary values, material, energy or people, and none of these is expected to disappear in an uncontrolled fashion, either by virtue of the first law of thermodynamics or as the consequence of social conventions and laws. In this way, compliance with conservation laws will be guaranteed. Efficiency and other measures of value, such as reliability, probability, and information theoretic measures, are quite naturally connected to distribution, flow, and the accumulation of values at this level.

For public service organizations, the quality of service of the various functions is an important priority measure. In a commercial organization, the accounting system is geared to control the flow and accumulation of products and monetary values. In a technical system, such as a power plant, the overall control of operation is concerned with the flow and distribution of energy from the source (an oil burning furnace or a nuclear reactor) to the consumers (of electricity or district heating). In general, the management of energy flows is important not only for meeting economical but also safety objectives since, by nature, large accidents can only result from a loss of control of large accumulations of energy. Similarly, in manufacturing companies, attention to financial aspects must be coupled to a direct management of material flows in order to be able to account for environmental impacts from waste disposal and final product "decommissioning" (product life-cycle management).

Since conservation principles related to abstract variables depend on measures that are applicable across a wide variety of general functions, the categories represented at this level will be rather independent of the particular kind of work system.

Interrelationships Between Levels. We have now briefly described the middle three levels of abstraction. Let us pause a minute and discuss their interrelationships. If we consider the level of general work activities and functions as the object of our current attention (i.e., the WHAT), then it would be natural in the course of an analysis (or during system operation for that matter) to want to ask two questions— relating to WHY and to HOW. Let us take the latter first.

Each of the *general* functions is implemented by means of one or more coordinated *physical* functions. Each of these can consist of a set of alternate combinations of physical equipment and people. For example, the general function of document preparation in an office can be based on the physical processes of a person writing on paper with a pencil, typing on a typewriter, or using a word processor. The utility grid, a diesel generator, or a stand-by battery can supply electric power to ensure cooling in a process plant. Thus the description at the physical functional level gives answers to the question HOW when it is asked at the general functional level.

On the other hand, if, at the general functional level, we ask WHY, then we must go up one level to find the answers in the description at the abstract functional level that describes the goals and constraints that propagate downwards to the lower levels.

Let us now continue on to the remaining two levels of abstraction.

Goals and Constraints

The highest level of functional abstraction represents the goals, purposes, and constraints governing the interaction between the work system under consideration and its environment—in other words, the system's "functional meaning" in terms of referring to properties and functions of the environment. Policies and strategies are formulated at this level as the basis for the objectives that have to be communicated to the lower levels of the work system, its departments, and groups, as well as the individual staff members. For a commercial organization, financial objectives are in focus. In relation to this focus are policies with respect to market conditions, competitors, and so on. For a public institution, the quality of service rendered, productivity, and economy are major concerns. In addition, constraints imposed by laws and regulations concerning financial operations, employee career patterns, and work conditions, union relations, environmental protection, and so on, have to be included.

Many public institutions, such as hospitals and universities, have a very complex set of objectives. The primary goal of the public hospitals in Europe, not to mention the profit motive in the privately run institutions, is to treat patients within the constraints of the financial resources made available by the government. However, objectives related to research and teaching are also important. Furthermore, given the current financial situation for medical care, public opinion with respect to priorities (whether treating the many people with bad hips should come before a few expensive heart transplants) is very influential. In general, policy making that can influence company image and public opinion is becoming increasingly important, for instance, with respect to environmental impacts. A trend can be seen towards including "ethical" accounting results (Bogetoft and Pruzan, 1991) in annual reports to enable both positive effects (influence on regional employment and economic activity, support of cultural events), as well as negative effects (pollution, hazards) to be evaluated. Thus, when looking for decision criteria during a work analysis, the analyst should also monitor eventual influences from the public debate as reflected in the daily newspapers. Presently, this is the case at hospitals when they set priorities for patients on waiting lists and for manufacturing companies that contemplate the use of potentially dangerous organic solvents.

Physical Configuration and Anatomy

The lowest level of abstraction represents an inventory of the material resources. These include the tools, equipment, and the staff members, together with a description of their material characteristics, form, location, and so on. In work, this level of representation is necessary to identify objects from their appearance, shape, color, size, and to find them in the material landscape. In any form of work system, this level is necessary to find one's way and to direct attention to objects that are potential sources of information or that can be manipulated, such as tools or control keys and knobs. However, the template used for searching in the material world will normally not be defined at this level, unless one is looking for a particular, individual object. In general, the template is defined from a higher functional level; for example, when thirsty, one usually looks for something useful to drink from, not an object with a particular shape or color (unless it has to be one's own coffee mug).

Thus the verbal labels identifying the tools and equipment at this level refer normally to the processes or functions being served at higher levels of the means–ends network. For instance, consider the terms "typewriter," "pump," and "screw driver." In addition to verbal descriptions, this level also includes representations in the form of pictures, topographic maps, drawings of buildings and equipment, and so on.

Figure 2.4 illustrates that the WHAT, WHY, and HOW interrelationships between levels that were illustrated earlier for the inner three levels are of course relevant for all five levels. Thus any work function (WHAT should be done) can also be considered as a goal (WHY) for another function at a lower level, as well as a means

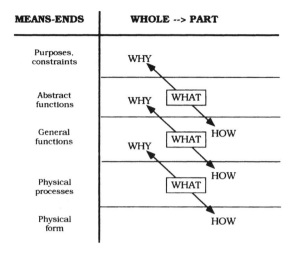

FIGURE 2.4. WHY, WHAT, and HOW in the means–ends space.

(HOW) for still another function at a higher level. This introductory discussion of means–ends mappings will be extended below.

THE STRUCTURE OF THE MEANS–ENDS NETWORK

As described earlier, several levels of representation forming a means–ends structure are necessary to relate the two extremes comprising system's global objectives and the proper utilization of the system's material resources with each other.

From the above description of the means–ends levels, it is seen that a shift from one level of abstraction to the next involves a change in both the concepts used to represent the functional elements and in the format used to reflect the functional structure. It is important to remember that the different levels contain information about the same physical world. However, the information used for representation at the various levels is chosen to serve as a set of conceptual transformational links between the representation of the material world of resources on the "bottom" and the representation of the ultimate goals and constraints at the "top." For example, when moving from one level of abstraction to the next higher level, the change in the represented system properties is not merely a removal of details of information about the original level. More fundamentally, information is added on higher level principles governing the cofunctioning of the various elements at the lower level. If traced to the top level, it is clear these higher level principles derive originally from the overall purpose of the system (i.e., from the reasons and intentions behind the design).

The means–ends relations of a work domain made explicit by the representation described here is implicitly found in many systematic analyses of work systems. It is most clearly seen in analysis of well structured technical domains; see for a clear example the represention of production processes used in Alting's (1978, 1944) systematic theory of manufacturing.

An Example: Life of a City

A city can serve as a familiar example of a "work domain" to illustrate the concepts discussed in the previous sections. Of special interest is the role of the functional and the intentional properties at all levels represented by the means–ends hierarchy (see Fig. 2.5). The material objects present in a city are represented at the bottom level of the means–ends hierarchy. The intentional element at this level lies in the selection of objects necessary to supply the requisite variety for the higher level functions. For example, in a city, people and houses, furniture and tools, cars and street lamps are plentiful, whereas large boulders, wild animals, and village ponds are avoided or, at least, kept at controlled locations. The next level represents the processes of a city; these include housing and feeding people, moving goods and people, and shaping and assembling products by chemical and physical production processes. The intentional aspect at this level concerns the selection and control of these work processes so as to serve the functions at the next higher level. Here, more general functions of a city are found including transport, trade, health care, administration, and public education. The intentional aspects here have to do with the coordination of the lower level processes so as to integrate them into productive and effective entities. At the level of abstract functions, the implications of the lower functions are represented in terms of the values gained and the resources absorbed. The intentional aspects here represent the efforts to set priorities and to direct the flow of money, people, and goods so as to serve the top level goals (such as the prosperity of the city, the well-being of the people, and the mayor's reputation).

This example illustrates how the functional aspects of the material world are

MEANS-ENDS RELATIONS	FUNCTIONALITY	INTENTIONALITY
Purposes and Constraints	Goals and objectives.	Explore opportunities and constraints.
Abstract Functions	Implications of the functions in terms of values and resources absorbed.	Setting priorities and directing flow of money, people, goods to serve the higher level goals.
General Functions	Functions of a city: transport, trade, health care, administration, public education.	Coordinate lower level processes to serve the various functions.
Physical Processes and Activities	Processes of a city: moving goods and people, sleeping, feeding, shaping and assembling products, chemical and physical production processes.	Control of the configuration and boundary conditions.
Physical Form and Configuration	Material objects: people and houses, furniture and tools, cars and street lamps.	Selection of objects.

FIGURE 2.5. Functionality and intentionality of a city.

represented at all levels, but in increasingly abstract and global terms. The intentional component becomes increasingly influential at the higher levels and also more complex as global functions come into play. It can also be seen that the intentional part of the domain representation constitutes a hierarchical control function that serves to coordinate the behavior of the material world at all levels. Again, the functional effects of the physical processes propagate upward through the levels, whereas an explication of intentional control criteria propagate downward. Decision making at each level will deal with discrepancies between the functional state of affairs and the intentions derived from the ultimate goals. Therefore, it is necessary that the representations of the functional implications match the presentation of the intentional explications at each level.

For the individual decision maker, the intentional element is a very real part of the domain. It is embedded as behavior-shaping constraints in the institutional practice and the accepted rules of conduct of the municipal system. In this way, the intentional system of a city is reasonably well established. However, it is still flexible; public debates, political parties, and the efforts of the local government serve to modify the structure in response to the results of "experiments" made by social scientists, political activists, and people violating the rules of conduct.

Levels of Decomposition: Part–Whole Relations

Another dimension of the representation of the work space is the level of decomposition. As stated earlier, the resolution of the representation will change with the span of attention of the actor. In the computer repair example of Figure 2.1, it is evident that the search for a fault starts with an observation of the global functioning of the entire system and ends with a search for a failed physical component. Thus a "zooming in" takes place through a stepwise decomposition of the system.

In general, the relationship between abstraction and decomposition is very complex. In Figure 2.1, the decomposition shown cuts across all abstraction levels. This is a special case because the figure represents the means–ends configuration relevant for one particular task situation: repair. Normally, the domain representation is intended to be independent of task and situation and, in such a case, a decomposition must be considered separately for each level. The abstraction–decomposition relationship is basically nonlinear. For a particular task situation, a change in the level of abstraction followed by decomposition will not give the same result as a decomposition followed by a change in level of abstraction.

Another feature is the need to shift the level of abstraction of the language used for description at a certain level of decomposition. As an example, the general task of writing a text cannot be meaningfully decomposed into functional elements unless the level of description is changed to the physical writing process itself in order to be able to deal with whether pen and paper, a typewriter, or a word processor is used.

Choosing the proper levels of decomposition as well as representation of the productive, functional resources is rather straightforward since this is related to the usual methods of functional analysis. In addition, as discussed in the introduction, categorical representations are adequate for feedback system designs.

This is not the case for decomposing the objective functions—an important task since it serves to define the targets of the feedback loops operating within the work system. This process is very complex because decomposition has to be followed by a transformation to lower abstraction levels to define goals and targets for increasingly detailed work functions and processes. In addition, moving down the levels of abstraction increases the degrees of freedom available for choosing among the available resources and, consequently, local and subjective objectives and criteria have to be considered (see Fig. 2.6). The identification of the criteria that will be likely to govern local objectives is a major problem in work analysis because, as discussed elsewhere, such criteria are likely to be implicit in existing work practices. In general, objectives and goals must be defined for each level in the means–ends space and the objective of a function at one level is found at the next higher level.

As mentioned, the actual, individual goals of the smaller units are not found by a decomposition of the overall goal but are developed independently from subjective

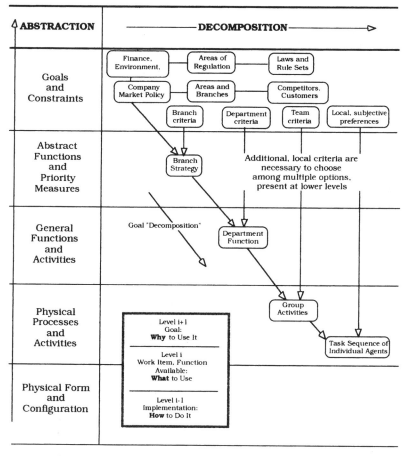

FIGURE 2.6. The complex relationship between levels of abstraction and decomposition.

preferences. Basically, this is a consequence of some of the intentional structure of a work system being embedded in the rules of conduct of the social system and some of it being brought to bear by the individual actors in order to resolve the remaining degrees of freedom. This aspect also stresses that decomposition has to be considered separately at each of the levels in the means–ends space.

This discussion implies that the propagation of goals and constraints top-down through the levels of the means–ends hierarchy is not a transformation by deduction but rather more of an evolutionary selection. High level goals, such as "acquire a high salary for union members" and, at the same time, "make sure that the working life qualities are satisfactory," require reformation and operationalizing down through the levels. When going downwards through the levels, deductive transformations are normally not possible due to the large increase in options as more details are added. Instead, empirical trials (and errors) and/or selections from rules of the trade are made. In stable environments, high level goals and constraints become suppressed and give way to operational constraints in the form of stored rules of action at the lower levels. However, it is incorrect to suppose that these higher level goals and constraints do not shape behavior. In fact, this shaping has been of an evolutionary nature through selection during periods of adaptation. Thus, one can speak of the "survival of the fittest rules" in the same way as "survival value" shapes biological behavior without affecting any particular observed behavior directly.

COUPLING TO THE ENVIRONMENT

Coupling to the environment is found to a varying degree at all levels of abstraction in the means–ends hierarchy. For example, some enterprises are able to "decouple" their activities from the environment at the lower levels by means of buffer inventories of raw materials and products. This makes a delimitation of the object of analysis rather straightforward. With current trends towards more tightly coupled systems—as for just-in-time production—analyses including the entire domain of a particular company may be inadequate because modern integrated planning strategies have to include product development and manufacturing by subcontractors and parts suppliers. Likewise, timely updating of product lines and designs for a "customer driven" production philosophy requires intimate knowledge about the functional characteristics of the customers' work domains in which the products have to function. This implies that it can be necessary to include several traditionally separate domains in an analysis. Such examples are shown in Figures 7.1 and 7.2 for a product design and Figure 2.7 for a health care system.

COUPLING TO THE ACTORS: CATEGORIES OF WORK SYSTEMS

A representation of the work domain should identify the entire network of functional resources and objectives, that is, the means–ends relations relevant for the activities

	PATIENT		HOSPITAL		
	PRIVATE LIFE	HEALTH	CURE	CARE	ADMINISTRATION
GOALS AND CONSTRAINTS	Working relations and conditions; Family relations; Goals and constraints of plans and committments;	Effects of illness and treatment on person's ability to meet subjective goals and criteria	Cure patient; Research, Training MDs; Public opinion ; Legal, economic, and ethical constraints	Patient well being, physical and psychic care; Public opinion, economic and legal constraints	Laws and regulations of society, associations and unions; Workers protection regulations etc.
PRIORITY MEASURES, FLOW OF VALUES AND MATERIAL	Personal economy. Probability of unemployment, cure, etc.	Probability of cure, priority measures, pace versus side-effects, etc.	Categories of diseases: Cost of treatments, patient suffering, research relevance	Flow of patients according to category; Treatment, and load on staff and facilities	Distribution of funds on activities; Flow of material and personnel to diseases, departments,
GENERAL FUNCTIONS AND ACTIVITIES	Work functions; Family relations; Living conditions;	State of health; Deseases and possible treatments;	Cure, diagnostics, surgery, medication, etc. Research, clininical, experiments,	Board and lodging; Hygiene; Social Care, Physical support, transportation, etc.	Personnel and material administration, Accounting, sales and purchase;
PHYSICAL ACTIVITIES IN WORK, PHYSICAL PROCESSES OF EQUIPMENT	Physical work activities, spare Time and sports activities; Home-work; Transportation, etc.	Specific organic disorders and possible treatment. Previous illness and cures;	Specific research and treatment procedures; Use of tools and equipment	Monitoring, treating, moving, cleaning and serving patients; Psychic Care.	Processes in the administrative functions. Office and planning procedures;
APPEARANCE, LOCATION & CONFIGURATION OF MATERIAL OBJECTS	Patient identification, age, address,profession education, family members, etc.	Physical state of patient, weight, height, precvious treatments, etc.	Material resources, patients, personnel, equipment; Medicine, tools, etc.	Facilities and equipment in patient quarters, kitchens, etc. Inventory of linnen, food, etc.	Inventory of employees, patients, buildings, equipment, etc.

FIGURE 2.7. An example of a work system including several professional domains of activity: The work domain of integrated health care systems.

to be considered. In Ashby's (1960) terms, this is the world of "possibilities" or "the requisite variety" necessary to cope with all the requirements and situations that can appear during work. The representation defines the functional inventory of the work system or, alternately, the functional territory within which the actors will navigate. The domain representation is a map that identifies all the functional resources of work in terms of categories for which the functional capabilities and limitations can be identified with reference to the various objectives they can serve. In this way, the work system representation defines the relevant kinds of information sources to be considered for information system design, and thereby gives structure to the "resident knowledge base" for the entire work system. From this, the necessary working knowledge can then be selected and activated in a particular work situation.

In the work domain description, the substance matter of a work system is represented at several levels of abstraction that represent goals and requirements, general functions, physical processes, and activities, as well as material resources. During work, various actors observe the state of affairs and make decisions about actions that will bring the state into correspondence with the current objective at all levels of the domain representation. As described previously, at each of these levels, any work function (WHAT should be done) can be seen both as a *goal* (WHY it is relevant) for a function at a lower level, as well as a *means* for a function at a higher level (HOW this higher level requirement can be met).

The need for human decision making is present only because of the *many-to-many* mappings among objectives and resources at the various levels. In any work domain, there are many degrees of freedom and options for choice. In an actual situation, the choice is guided by functional (product) criteria as well as subjective performance (process) criteria. In addition, choosing a resource to serve a function (WHAT) can influence or affect more than one objective and constraint (WHY), since actions usually have side effects in addition to meeting their primary goal. Thus all possible paths must be considered when judging the upward propagation of the effects of actions at the given work level. Likewise, a function can be effected by alternative means at the level below (HOW) and, moreover, several functions can require the same means and thus compete for its use. This results in the multiple mappings among means and ends. However, in any particular work situation, only a particular sector of the means–ends hierarchy will be relevant. Thus the work domain representation is a useful map and memory aid in the effort to identify the relevant relationships controlling the propagation of inherent–natural changes, as well as those caused by system design modifications.

To ensure effective information retrieval support consistent with the above, it is necessary to make information about work domain properties accessible to decision makers from several perspectives. Thus answers will be required to queries such as: "what" is available, "how" can this be accomplished, or "why" is this required? This means that the search terms are not objectively defined by the *collection* of stored items but by the means–ends *relations* among them. As discussed in the previous section, the representation of the productive, functional features at each level should be comparable to that of the intentional aspects in order to facilitate decision making and choosing among options.

The Means–Ends Network as Behavior Shaping Constraints

It is important that the work domain representation captures the basic features of a work system that shape the intentional activity of its staff. Basically, work depends on purposeful transformations of items found in the work environment. In modern applications, these transformations are not normally made directly by the actors. Instead, their actions serve to set boundary conditions for active forces within the environment that then bring about the intended changes. In general, activities in work seek to control the utilization of energy, material, human, information sources within designed boundary conditions. For example: An oil furnace heats the water in an industrial boiler and steam is produced that rotates a turbine, but only after the operators have created the proper boundary conditions for the physical processes by connecting the plant equipment properly. Or: Instructions to a truck driver will move your goods to the chosen destination, but only when he/she intends to do business with you.

The success of actions to control the state of affairs depends on the presence of a high degree of regularity of the responses of the environment to these actions, (see

Fig. 2.8). This regularity, in turn, is the result of stable internal constraints that link system variables in a predictable way. The two simple examples given above illustrate two different sources of regularity. The first depends on laws of nature, while the second depends on human intentions to follow a socially accepted pattern of behavior.

To sum up, regular patterns of response by a work system to work activities are a prerequisite for human actors acting purposefully. Thus actors' opportunities to plan depend on their possessing knowledge about the sources of such regularity or, in other words, the internal constraints shaping the system's behavior. Control of the state of affairs within a work system involves operations on and through its internal constraints and can take place via the causal constraints of the physical part of a system or the intentional structure of the people involved.

Source of Regularity of Behavior of the Work Domain

The weight of the intentional constraints compared with the functional, causal constraints can be used to characterize the regularity of different work domains. In this respect, the properties of different work domains represent a continuum. At one end are located tightly coupled, technical systems, the regularity of which have their origins in stable laws of nature. At the other extreme are the systems in which the entire intentional structure depends on an actual user's own subjective preferences and values. In the middle are a wide variety of systems that owe their regularity to influences from formal, legal constraints, as well as institutional and social practises. In general, the functionality will be built around expected work situations (see Fig. 2.9), which attempts to depict this. Thus the relationship between the causal and intentional structuring of a work system and the degree to which the intentionality of the system is embedded in the system or brought to play by the individual actor is an important characteristic.

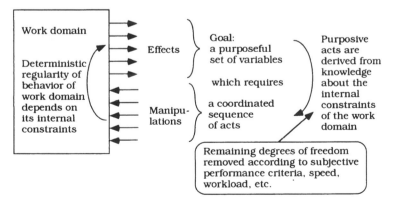

FIGURE 2.8. The predictability of behavior of a work environment depends on the existence of internal constraints.

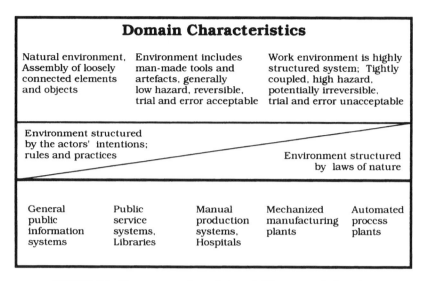

FIGURE 2.9. The sources of regularity of different work domains.

Automated Systems Governed by Laws of Nature. In technical equipment and physical systems (process plants, chemical production, etc.), the source of regularity of behavior can be traced to the laws of nature. The underlying processes are confined and connected by the physical construction of the plant while their functional behavior is dictated by these laws. Thus, predicting this behavior in response to human actions can be inferred bottom-up from knowledge about the involved physical processes. The objective functions—that is, the intentional structure or reasons for the desired functions (as shown in Fig. 2.2)—are "hard-wired" in the form of a set of formal operating procedures combined with an often complex automatic control and safety system. These maintain plant state and operation in accordance with the high level, stable design goals—such as to produce power as requested by customers and to do it as economically and safely as possible throughout the lifetime of the system. The task of the operating staff is basically to ensure that the functioning of the automatic control system actually reflects the intentionality of the original design while their own personal goals and preferences have little significance.

It follows that the contents of the information for the operators with regards to functionality must be based on these physical laws as applied to the productive processes of the particular system including the limiting conditions set by the confinement. However, the significance of providing intentional information (the reasons for the design) is well illustrated by this example. In order to understand the functions and the behavior of the automated control and safety system, the operators must be familiar with the intended control *strategies* underlying this system. The reason for this is that the implementation of the automatic control system is only a medium for processing this intentional information and, consequently, has little significance except for the maintenance crew. Unfortunately, in process plant control rooms, little attention has, so far, been given by designers of support systems to communicating information intentionality to the operating staff. The reason for this is that the rationale for most design choices has been embedded in the minutes of meetings, in professional and company practice, and in industry standards. It is very difficult during some critical situation in the control room to identify and make explicit the original reasons for a particular system design feature. This is another example of the problem of analyzing adaptive systems. When the operating staff at a later point in time is required to reconfigure a system because of changes in requirements or major disturbances, the lack of intentional information can hinder an effective response. Blueprints and operating instructions only communicate *what* and *how*, not important information about *why*. In order to provide effective support, the analysis and deliberate consideration of the path of propagation of both functional and intentional information through the different organizations involved in design and operation are important issues.

Mechanized Systems Governed by Instructed Rules of Conduct. Other work domains, such as most manufacturing systems, are more loosely coupled, even though they have a complex "technical core." Their productive functionality is based on mechanized processes as constrained by the technical equipment. When such systems are organized according to the "scientific management" paradigm, operations are centrally planned and the intentionality, that is, the objectives of activities, are embedded in operating instructions propagating top-down through the levels of the organization. In more modern "just-in-time" type plants, this central planning with its top-down directive flow is replaced to a large degree by on-the-floor improvisations and adaptations to the immediate situation so that the intentionality becomes more decentralized. In a way, the high level goals form an umbrella under which the daily operational decisions can be made (see also the following category).

Thus the individual decision makers confronting their part of the work domain will have to shape their activities under the influence of different sources of regularity. First, of course, are the laws of nature governing the technical side. Then, depending on the application, the intentional direction will come from some kind of combination of preplanned schedules and formal or informal rules of conduct that shape our own behavior as well as the behavior of the other actors.

Loosely Coupled Systems Governed by Actors' Intentions. An important extension of the "just-in-time" approach is found in a wide variety of loosely coupled domains, such as hospitals, offices, and manufacturing plants based on manual work. In such systems, the productive functionality is ***not*** constrained by a technical core but depends mainly on the activities of the staff itself. In other words, there is no concrete physical system in back of the user interfaces upon which the actors' attention is focused.

Coordination and control of activities depend on the communication of company–institutional objectives. The intentionality originating from the interpretation of environmental conditions and constraints by the management of a hospital or a library propagates dynamically downward and becomes implemented in more detailed policies and practices by members of the staff. Making intentions operational and explicit during this process requires an interpretation considering a multitude of details dictated by the local context. Therefore, many degrees of freedom remain to be resolved by situational and subjective criteria by the staff at the intermediate levels of an organization. This in turn implies that the individual actor faces a work environment in which the regularity to a considerable degree depends on the intentionality brought to bear by colleagues. The intentional structure is, therefore, much more complex and dynamic than for the industrial process systems.

Systems Governed by Actors' Personal Objectives. Another kind of work system located further to the left on the map of Figure 2.9 includes the systems that are driven by the actors' personal objectives. Settings include research institutes and universities. Another example is the public library described in Chapter 9. In this category, the intentionality governing the use of the system is brought to bear by the individual user within the envelope defined by the institutional policies. The productive use of system resources depends on the user's own working practices and the coordinating intentionality depends entirely on the actor's current, personal objectives.

However, system design is facilitated considerably by the fact that typical "work situations" can be identified in terms of user problems and needs, and their work practices and intentions can be identified by field studies.

Systems for the Autonomous, Casual User. For some information systems, such as lexicographic data base access, information services for the general public, and so on the task situation and intentions of the casual users are considered to be unknown. These systems normally include a set of generally useful information sources and data bases generated by many different public services, institutions, and commercial companies. No universally relevant intentional structure or task situation can be formulated as a design basis for structuring system functioning or interface design. The intentionality of the task situation is entirely defined on occasion by the actual user. Not even the higher level means–ends structure of such a system can be formulated except in very general terms of making information available to the public and profit making.

Concluding Remarks on System Categories. The five classes described represent a continuum from systems where the intentional properties of the work domain are embedded in the functions of a control system, through systems where intentionality is represented by laws, regulations, social rules and values, to systems in which the entire intentional structure depends on the actual users' subjective preferences (see Fig. 2.10).

To an actor, coping with a complex system, the properties to consider for controlling its state of affairs will be a varying combination of functional and intentional relations. Sometimes, actions will be aimed directly at its functional state. However, control often requires an influencing of other actors' intentional states or, in technical systems, entails modifications of the intentional structure embedded in the control system. Whether control is to be exercised on functional properties directly or

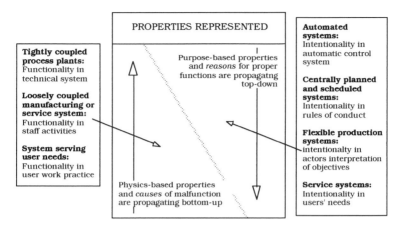

FIGURE 2.10. Different combinations of the implementation of productive functionality and coordinating intentionality serves to characterize work domains.

through mediating intentional properties depends on the level in the means–ends hierarchy where the problem is perceived. Therefore, as stated above, when the interaction with a work system is mediated through an information system, constraints originating in the intentional structures, as well as those based on functional, causal relations, must be represented. This is a central topic for the discussion of interface design in Chapters 5 and 7.

The stratified structure of the work domain description can be related to similar conceptualizations that have been found in management and organizational theories (e.g., by Parsons, 1960; Thompson, 1967; see Chapter 4). The use of all of the means–ends levels is important because neither administrative decision making nor organizational aspects can be modeled without considering the basic system purposes (process plant properties, manufacturing processes, surgery, etc.), which the decision making is intended to fulfill–maintain. Studies (e.g., of office systems) without any consideration of the inherent subject matter appear to give inadequate results for information system design. A similar argument has also been presented in reports from organization and management research (Parsons; and Thompson; op. cit.).

In a recent book, Mitroff and Linstone (1993) reviewed organizational models of work systems with reference to Churchman's (1971) categories of "inquiring systems." The topic of their discussion is the problem of decision making in highly integrated and dynamic sociotechnical systems. The main thesis of the book is not to use bounded thinking for unbounded problems. It stresses the cross-disciplinary and complex nature of problems in modern systems. They advocate the need for "unbounded thinking" based on "multiple perspectives" that are intimately interacting. No science (natural science or physics) or perspective has a priority over the others. This situation leads to a request for systems thinking which, according to the authors, has been expounded by Singer (1959) and Churchman (1971). One basic proposition by Singer is that no discipline can offer an adequate set of elementary or simple acts into which a complex situation can be decomposed. Three different perspectives must normally be taken into account: The *technical* (science and technology), the *organizational* (social entities, small to large, formal to informal), and the *personal* (individuation, the self). The "technical perspective" can be based on several sets of models and data interpretations, "realities." Within a "reality" and the related model, the system considered is normally broken down to subproblems and treated separately. In the unbounded mode, the system is normally treated as an integrated whole. Each perspective reveals insight which, in principle, cannot be obtained from any other perspective. The perspectives, therefore, often interact in a dialectic way. The emphasis on the different perspectives is a matter of one's ethical values and judgments. Some basic characteristics of the different world views and perspectives are shown in Figure 2.11.

The interesting feature of Mitroff and Linstone's discussion in the present context is the emphasis placed on the need to have an analysis including multiple realities and perspectives, which are very analogous to our emphasis on a simultaneous analysis of the propagation of functionality and intentionality through the levels of a means–ends hierarchy. In addition, their pinpointing of the need for considerations of the subjective process and situation criteria is close to our focus on criteria for choice among action alternatives, based on subjective preferences.

Causal Representations Revisited

The distinction made in the means–ends network of the domain representation reflects back on the Aristotelian concepts of causality. Aristotle distinguished between four kinds of causes illustrated by the process of building a house: (1) the *material cause* (i.e., what it is made of); (2) the *formal cause* (i.e., the architects conception of its shape or form); (3) the *efficient cause* (i.e., the process of building the house); and (4) the *final cause* (i.e., the purpose of the whole operation). This latter concept is often taken to imply that Aristotle had in mind a cause subsequent to its effect. His concepts, therefore, in general have discredited teleological approaches. This position, however, according to Bambrough (1963) is based on a misinterpretation of Aristotle:

	Technical	**Organizational**	**Personal**
Goal	Problem solving	Action, stability, process	Power, influence prestige
Mode of inquiry	Data, modeling, analysis	Consensual, adversary	Intuition, learning, experience
Ethical basis	Logic, rationality	Justice, fairness	Individual values, morality
Planning horizon	Far	Intermediate	Short

FIGURE 2.11. The perspectives of analysis proposed by Mitroff and Linstone.

Because of modern developments in the physical sciences, we have come to think of causation primarily as a relation between events; but for Aristotle, the four causes were primarily causes of things or substances. The doctrine is intended as an account of how particular substances originate or "come to be" and why they have those properties that we recognize in them. When we think in these terms, and escape from our customary preoccupation with events, we can see at once how natural and how closely interconnected are the four questions to which Aristotle's four causes indicate the relevant types of answer: What is it? What is it made of? How was it made? Why was it made?

From this point of view, Aristotle's different causes are more closely related to the relationships among concepts within the means–ends relations discussed in this chapter, rather than to modern causality in terms of relations between events or, in other words, his concept of cause is related to the creation of systems, not to their subsequent behavior. Hence:

> it is no accident, but an essential feature of his causal theory, that the four causes should be illustrated so often and so effectively in terms of examples drawn from biology and from human skills of manufacture (op. cit.).

That is, from systems evolving through selection by a designer according to some intended purpose, or by nature, according to some "survival value."

HINTS FOR FIELD ANALYSIS

The data for the analysis of the work domain can be collected from various sources. Before field studies are initiated, a good introduction to the domain should be obtained from textbooks, activity reports from the particular work system including material, such as commercial brochures and reports to the shareholders. Sometimes a careful perusal of newspapers can give useful insights into policies, management intentions, political disagreements, and relevant changes in society's attitudes and opinions. At a more detailed level, formal procedures, and instructions will be useful for learning about the various means–ends relationships in the system. However, the typical detailed procedural body concerning "what and how to" is unreliable as a data source for activity analysis.

Interview Techniques

Unstructured interviews at all levels of an organization are very fruitful even though a body of structured information is the ultimate aim. Unstructured interviews using detailed notes or tape recorders have turned out to be the only realistic approach. Attempts to use an interview form structured according to the analytic model repeatedly cause interruptions in the interviewee's train of thought. Similarly, attempts to make recordings of the information gathered in prestructured records during the conversation tend to distract both the interviewer and the interviewee.

In the first interviews, it is important to identify the various personnel categories and to clarify the tasks that the different persons carry out together with the competence boundaries that exist in relation to the work domain. In addition, task areas common to different personnel categories are of interest for identification of the flexibility and adaptivity of the division of work.

As the basis for more focused activity analyses, the interviews should attempt to identify typical situations and tasks. These will assist in understanding the interaction between domain and task structures as prerequisites to the detailed activity studies. However, the information collected about "typical situations and task" will not be reliable for the activity analysis itself (see Chapter 3).

A discussion of specific work situations and processes is most important for collecting information about goals and intentions as well as both formal and informal constraints. A useful rule of thumb in order to gain a good overview of the total work domain is to formulate questions based on movements in the means-end/part–whole problem space. Thus, an employee can be asked to name all the important functions with which he/she is involved and thereafter the means–ends relationships can be explored with questions about what is done, what is the purpose in doing it and, lastly, how is it done. In this effort it is important to explore alternatives explicitly in order to represent the many-to-many mapping of the means–ends map. Another possibility for achieving a systematic insight is to have an employee describe the functions he/she participates in during the course of a day or, alternately, the functions that form part of a given work process. More means–ends related information will be accumulated in later phases but, as mentioned above, it is important at the beginning to gain an overview of the situation.

Chapter **3**

Activity Analysis

INTRODUCTION

The work domain representation discussed in Chapter 2 is a stationary, situation-independent description in terms of the functional properties of the work system and the intentional structure embedded in it. That is, the representation gives a map of the inventory of the functional resources and the objectives that the actors will need to deal with in the various work situations that arise. In general this map will be very complex, simply because all of the means–ends relationships relevant for coping with any situation (i.e., the possibilities) have to be included—even those that were only implicitly present in the established work practice at the time of analysis. In any particular situation, the relevant activity space constituting the basis for the activities of the involved staff will include only a limited set of "actualities."

To study features of the work domain that shape (or did shape in the past) the behavior of the human actors, it is necessary to focus on the parts of the domain representation that are relevant for each particular activity. This delimitation will select or "instantiate" a subset of the elements and relations of the work domain at a certain slice of time, which then will have to be explored in more detail. The aim of a work domain analysis is to identify the basic sources of regularity (the invariants) underlying the responses of the domain to human actions, that is, the behavior-shaping constraints forming the activities of the work staff. Within the work domain, many degrees of freedom usually exist for choosing means for certain ends, and their elimination will depend on the process criteria brought to bear by the actors together with their interpretation of the relevant intentional structure.

To be able to relate the requirements posed by the work environment to the cognitive resources and subjective performance criteria of the members of the staff, the activity analysis goes through three stages of decomposition and shift of conceptual language as mentioned in Chapter 1. (1) An analysis of activities in *work domain terms* serves to define prototypical work situations and functions. Also, the objectives that are active and the resources that are available to actors for such recurrent activities will be identified. (2) Next, an identification is made of the activity in *decision making terms*, that is, the control functions that the actors carry out during these situations. (3) Finally, the activity is described in *cognitive terms* by identifying the mental strategies that *can* be used by the actors in excercising control. This latter phase is necessary to relate activities to the cognitive resource profile of the actors (see a discussion in Chapter 4).

ACTIVITY IN WORK DOMAIN TERMS

The pattern of interaction between people and their work environment, as well as their mutual cooperation, is more or less continuous and dynamic. In order to analyze this interaction, it is necessary to decompose the pattern into meaningful elements that are manageable for separate analysis. In essence, this analysis will be a causal analysis in terms of the relationships among events (see the implications of this in Chapter 1). Traditionally, the decomposition applied takes the form of a task analysis to produce a description in terms of sequences of actions. In Chapter 1, arguments were presented as to why this is no longer an adequate approach because tasks have become discretionary and involve flexible cognitive processes for which the strategies applied depend on subjective and situational characteristics.

Instead, activities must be decomposed and analyzed in terms of a set of problems to solve or a set of task situations to cope with. In hospitals, for instance, a convenient decomposition is to analyze "prototypical task *situations*" because many tasks arise out of scheduled meetings and, therefore, are well defined in time. In other cases, activities cannot be clearly delimited in time and the elements can be more adequately defined in terms of "prototypical task *functions*," which have to be attended to. In this case, an activity element is characterized by its content independently of its temporal characteristics as in "monitoring production" or "assisting users during information retrieval." In short, work in general can be pragmatically decomposed into segments with respect to time and location or with respect to functional content, whichever is convenient for analysis.

The outcome of this analysis will be a set of "prototypical" activity elements which, in varying combinations, can serve to characterize the activity within the work system through their relational structure. See Figure 3.1 for the hospital example, Figure 3.2 for a manufacturing example and Figures 9.1 and 9.3 for a library example. The description of the individual prototypical task situations or functions serves to decompose the activities for a subsequent identification of the decision making processes and, subsequently, the cognitive resources required by each prototypical activity, while their mutual interaction determines the communication required for proper coordination of these activities.

The decomposition of activities into prototypical task situations or functions can be done at different levels of detail. For each of the activity elements shown in Figures 3.1 and 3.2, several, more detailed task situations or functions can be distinguished,

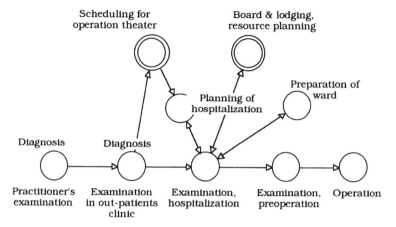

FIGURE 3.1. Prototypical work situations in a hospital.

and the decomposition should be continued until work functions are identified for which the control functions (i.e., the decision making processes) of the actors can be meaningfully described.

Prototypical Work Situations

This first decomposition of the general activity corresponds to the labels used by the staff. Work situations have names, and the decomposition of the activity represented by these names identifies elements that are recurrent, natural islands of activity with reasonably well-defined boundaries. The names of the situations will usually be framed at the level of general functions of the means–ends map. Two examples are discussed here. One is patient treatment in a general surgery wing of a hospital, based on field data collected by Hovde (1990). Another is production management in a manufacturing company, based on field data collected by Kaavé (1990) and Rindom (1990). The definition of "prototypical work situations–functions" in these two examples is quite different.

It is important to consider the difference between the definition of *prototypical* and *typical* work situations. Often, during interviews and discussions in a work place, people tend to describe what they find to be the normal, or usual way of doing things. That is, the data they present are related to an intuitive averaging across cases—the *typical* case. In contrast, to define the *prototypical* case, one has to collect data related to a set of actual, detailed cases and then, by analysis and categorization, define a prototype that can faithfully represent the properties of a category.

Work Situations in a Hospital. A decomposition of the activities required to prepare a patient for an operation in a wing for general surgery is illustrated in Figure

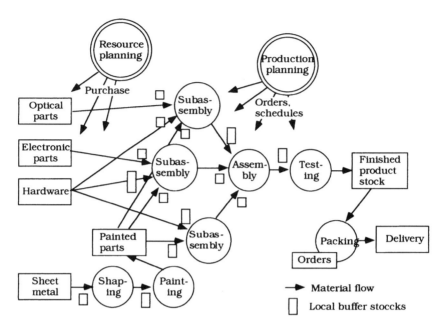

FIGURE 3.2. Feed-forward planning and control of production in a company based on a centralized resource management program.

3.1. The figure shows a sequence of prototypical work situations involved in the preoperative handling of a particular patient through the hospital. To repeat an important point—this representation cannot be based on an analysis of a "typical" sequence. With respect to the decision making and the information needs, no two situations are identical and any descriptions of typical situations will strip away all the context guiding the actual use of heuristics and know-how and the only remnant will often be material reflecting textbooks and formal procedures. The relevant prototypical situations should be identified by an analysis across a set of particular work sequences to identify proper prototypes. Figure 3.1 represents two different kinds of task situations. One includes those where a task force treats a particular patient (single line circles); the other covers the resource planning situations across the patient population (double line circles).

Work Functions in a Manufacturing Company. Similar analyses have been made in manufacturing companies. A particularly illustrating example has been found in the company that responded to the fast change of pace illustrated by Figure 1.1. As discussed in Chapter 1, many manufacturing companies are contemplating a change from preplanned production for stable market conditions to a "just-in-time" production strategy. In the first case, production is planned from a prognosis based on last year's sales in order to achieve optimal efficiency in the production system. For such a company, the production flow and the prototypical task situations are illustrated in a schematic way in Figure 3.2. In this preplanned mode, the local control of activities depends only on the plan and the local state of affairs. That is, the different prototypical work situations are rather well decoupled by buffer stocks. Schedules specifying "what to do" and "when to do it" replace the need for information about company goals and criteria (i.e., why do it? -> just do as you are told).

In a "just-in-time" strategy, production is controlled dynamically so as to meet the requests of individual customers. No orders, no production. Stock piling as well as parts-in-process are minimized and, consequently, the functions of production and sales interact dynamically—also in the short term. In this situation, production management becomes a very dynamic and distributed control process, which leaves little time to consult the decision making hierarchy found in a traditional company structure. The production structure and flow (and prototypical tasks) found in this type of plant are illustrated in a schematic way in Figure 3.3. In this particular system, products are pulled through the system according to demand. In the "kanban" arrangement shown, each work station requests the delivery of material by sending a card (Japanese: kanban) to the upstream stations. In this way, inventory (material in process) can be minimal. Planning is focused on supplying adequate resources to the individual actors who then will take care of the local planning—also during disturbances. The important message in this discussion is the very different relational structure connecting the work situations. For details, see Rasmussen, 1994a.

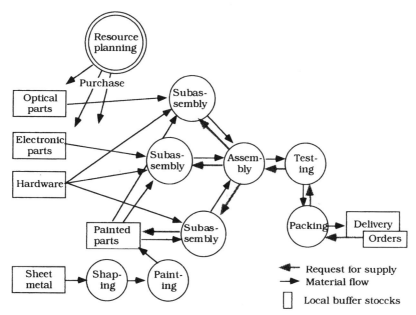

FIGURE 3.3. The "customer controlled" production strategy. The downstream flow of parts is controlled by requests (called "kanban cards" in a Japanese "just-in-time" organization) sent upstream when a new supply is required.

Communication Among Work Situations

The communication structure connecting work situations and the actors is an important aspect for the design of integrated information systems. This structure is very different in the hospital and the manufacturing cases. In the prototypical work situations of the hospital, several actors work as a team whose members are identified with reference to roles and professional background, not to persons. Due to shift-work requirements, most roles in the work of a hospital are not assigned to particular individuals. Therefore, a stable configuration is linked to work situations and roles.

The communication within the team in a familiar situation is very complex.

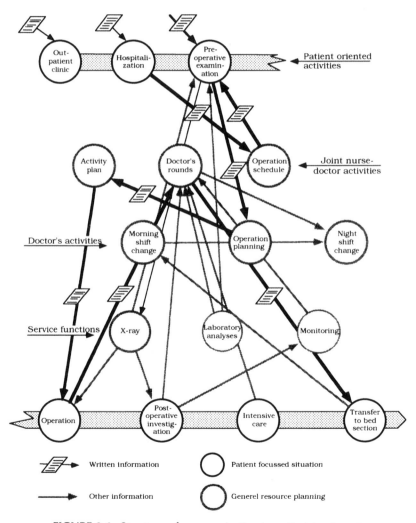

FIGURE 3.4. Structure of communication in patient treatment.

Identification of the communication networks among work situations, such as those shown in Figures 3.1, is an important part of the analysis. The communication utilizes several media, including paper forms, telephone calls, formal and informal meetings, casual talks in the cafeteria and corridors, and computer files, any of which can be decisive. An example of the communication links among the situations involved in a patient's passage through a hospital is shown schematically in Figure 3.4.

In our manufacturing example, no shift work was involved and the roles were closely related to particular persons (for a detailed illustration, see Dutton and Starbuck, 1971). At the same time, the prototypical work functions were not concentrated in time to particular situations, but spread over time. Therefore, a stable configuration of activities was related to particular individuals and their work functions. This has a significant influence on the communication pattern. An example of such a pattern was found for the informal production scheduling system shown in Figure 3.5. Here production management is based on the passing of kanban cards calling for supplies upstream in the production flow. This can involve a delay because of the propagation of messages along the production path. In the actual system, however, the shop floor foremen did not wait for the messages. They created a bird's-eye view through their informal contact network and in this way established a predictive feedback management by bilateral "contract negotiations." See the figure that illustrates the attention horizon of a shop floor foreman in his/her informal network for monitoring the adequacy of his/her resources for oncoming requests. The

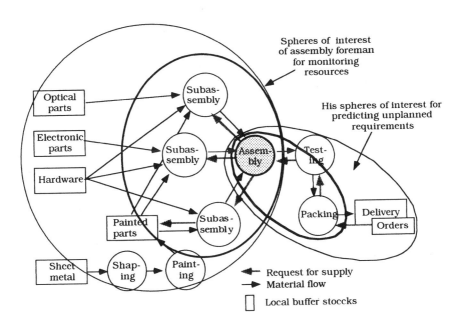

FIGURE 3.5. The varying attention horizons of a shop foreman's informal information network, which has been established to monitor the adequacy of the flow of resources to his/her work station and the likely future demands.

more important they perceive a particular product order to be, the farther up- and downstream the production flow they will tend to monitor the preparation of the resources for their next activity. The bird's-eye view serves to give timely predictions of likely disturbances, rush orders, and inadequate resources. The range of the horizon changes dynamically in response to the perception of urgency. This arrangement is much more responsive to rapid disturbances than the formal system. An illustrating and detailed account of this kind of production control performance is found in Dutton and Starbuck, 1971.

For examples of the identification of prototypical decision situations in the library environments, see Chapter 9.

ACTIVITY IN DECISION MAKING TERMS

So far, the activities have been expressed in work domain terms, such as operation on a patient in a hospital and production planning in a manufacturing company. As mentioned earlier, at a certain level of decomposition of an activity, attention will be focused on the *control of activities* by the actors as they affect the decision processes required to cope with the work situation. This is necessary in order to obtain an activity representation that can serve the formulation of the cognitive task and the required cognitive resources of the individual actors.

This part of the analysis will describe the decision functions found in the various prototypical task situations and functions in order to interrelate the various recurrent "states of knowledge" used by the actors to represent the state of affairs in the work domain and the elements of their plans to control such states. Several different categories of information processing are necessary for these control functions. The actual state of the environment must be identified; this involves information collection, situation analysis, and diagnosis, which are all analytical processes. Next, the implications of the actual state with reference to the current objectives must be evaluated. This process involves prediction, value judgment, and choice. Finally, the proper sequence of control actions must be selected by a process of planning and scheduling. To be able to relate "states of knowledge" regarding observed data, current state, goals, intended state, and required action, basically different information processes are required (such as induction, deduction, value judgment, and choice). Each of these processes requires different kinds of support.

Normally, the individual decision functions of a task will not be clearly separated; their mutual relationships will depend on the work domain, the task situation, and the level of competence of the actor. This complicates the analysis, but will not change the need for a consistent framework. The formulation of a generic set of decision tasks will facilitate the design and evaluation of decision support systems significantly.

Structure of a Decision Sequence

Even though the decision processes involved in actual work are not nicely structured in rational sequences according to formal decision theories, we need a framework to represent the various states of knowledge and the information processes required to go from one state to another during reasoning.

Decision making and communication among decision makers are more manageable when the process is composed of standard subroutines connected by more or less standardized key nodes; that is, "states of knowledge" about the environment, goals, and plans as mentioned in the previous section. Such standardized states of knowl-

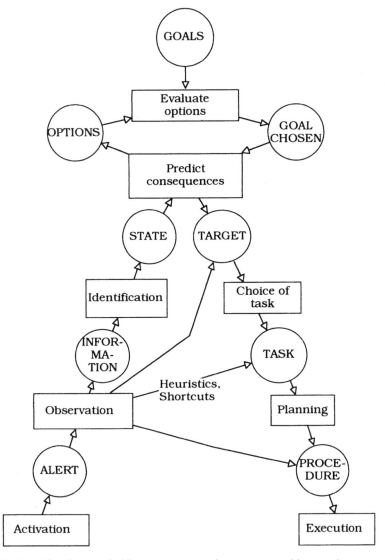

FIGURE 3.6. The decision ladder. It represents the sequence of basic information processes in a decision task together with a number of heuristic short-cut paths.

edge about different features of a task are very useful for linking different processes, for bringing results from previously successful subroutines into use in new situations, and for communicating with other actors. This is important since a complex decision task may be shared, not only by a number of cooperating team members, but also across time with procedure designers, system programmers, and computers.

The decision ladder illustrated in Figure 3.6 represents the set of generic subtasks involved in decision making. It is based on analyses of verbal protocols taken during actual work in different work domains that identified the different categories of statements made by the actors about their knowledge, their questions regarding their task, and their past and intended acts. In addition, the relative frequency of temporal order of the different categories of knowledge statements was analyzed to identify the decision process. The result is shown in Figure 3.6 with the "states of knowledge" arranged in a sequence. This result reflects a kind of normative, rational sequence, together with the heuristic shortcut connections that are typical of natural, not formalized, decision making. The directions indicated by the connecting arrows are only indicative of the normative sequence used to format the figure. Very often, paths in the opposite directions are found. For instance, the actor states that the time has come to enter a task. Then the question is asked whether the system is ready for this; some observations are made to verify the required state and then actions are taken without further planning. In other words, the framework is a kind of sketch pad that can be used to represent observed decision paths and the communication required between different situations. See Figure 3.7, which illustrates the decision processes involved in different diagnostic sessions during a patient's hospitalization together with the exchange of information among the sessions.

As this figure indicates, it becomes difficult to separate the individual "decisions" in the continuous flow of work that normally will involve frequent shifts between the individual decision functions, and thus lead to a very complex path. Since the sequence of elements in a cognitive task is much less paced than the required sequence of acts in the work environment, it will be difficult to structure the results of a cognitive task analysis in terms of a stable procedure, even in a particular decision situation. Instead, such situations can be characterized with respect to the *set* of decision functions that are required. It should be noted that the figure does not include any representation of the control of the sequence in which the different subtasks take place. The actual sequence depends very much on the context in which the decisions are made and the tacit knowledge of the decision maker. This will be discussed in a subsequent section.

As mentioned earlier, the various information processes involved in the situation analysis along the upward leg of the decision ladder, the value judgment at its top, and the downward planning leg require different kind of support systems. Therefore, the decision processes found in the decision ladder are well suited to distinguish among different interface designs. Now, together with the characteristics of the work domain that were identified in Chapter 2 (Fig. 2.9), which specify the sources of regularity of behavior of the environment, they can serve to characterize the decision processes to be supported by the interface, thus shaping a map of suitable interface designs. This is illustrated by Figure 3.8.

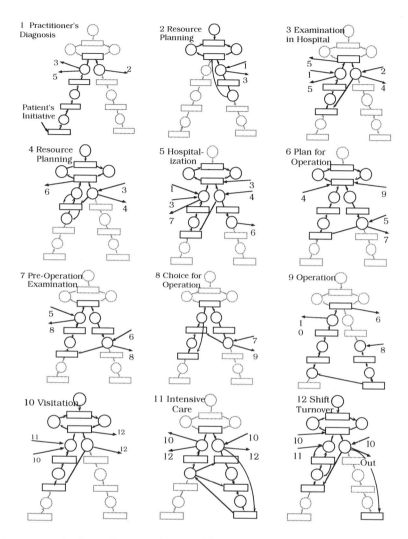

FIGURE 3.7. This figure illustrates the use of the decision ladder as a sketch pad to record the actual decision paths during different diagnostic sessions in a hospital and the exchange of information taking place. The numbers indicate communication links among diagnostic sessions.

Formal Decision Making

The rational, knowledge-based decision process used when no heuristic short-cuts are available (e.g., by novices and by experts facing unfamiliar situations) forms the basic framework for the sequence illustrated in Figure 3.6. It depends on knowledge about the internal, functional, and intentional properties of the work system. Subroutines connect different states of knowledge about the state of affairs in the work system and

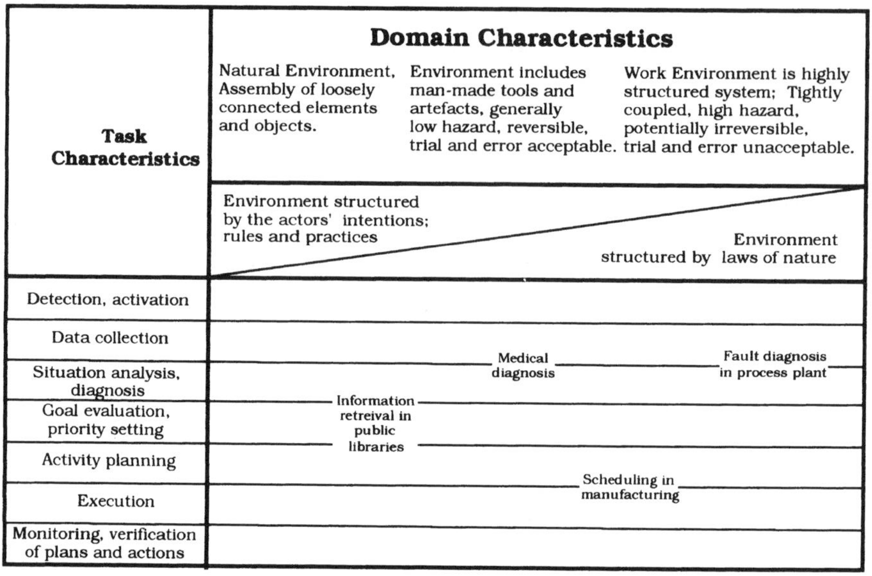

Task Characteristics	Domain Characteristics		
	Natural Environment, Assembly of loosely connected elements and objects.	Environment includes man-made tools and artefacts, generally low hazard, reversible, trial and error acceptable.	Work Environment is highly structured system; Tightly coupled, high hazard, potentially irreversible, trial and error unacceptable.
	Environment structured by the actors' intentions; rules and practices		Environment structured by laws of nature
Detection, activation			
Data collection			
Situation analysis, diagnosis		Medical diagnosis	Fault diagnosis in process plant
Goal evaluation, priority setting	Information retreival in public libraries		
Activity planning			
Execution		Scheduling in manufacturing	
Monitoring, verification of plans and actions			

FIGURE 3.8. A map for prototypical task situations based on the domain characteristics and the typical decision tasks. Several typical tasks are indicated as illustrations.

this, according to the discussion in Chapter 2, can be done in terms of various levels of abstraction and decomposition. Information processing then involves the transformation of one state of knowledge to another (again not necessarily in the directions of the arrows shown). The transformations are based on mental models representing the constraints of the work environment. These comprise the invariant properties that govern its normal regularity of behavior and that determine the internal relationships in the sets of data provided by the environment.

A particular characteristic of the normative decision making is that "decisions" can be separated in time. Thus a problem is encountered; the necessary evidence is collected; a decision is made. Natural or heuristic decision making has more of the character of a continuous control of an activity where it is difficult to identify separate decision making encounters.

Heuristic Decision Making

The heuristic, rule-based mode of decision making is applied by skilled actors and depends on shortcuts in the basic decision sequence based on "know-how." Heuristic decision making builds on induction where familiar states of the environment are associated to actions that have been effective in previous encounters (Rasmussen, 1993).

Humans are *not* passive receivers of information; instead they continually ask questions of the environment based on the context of their activity. Given that humans are highly adaptive and tend to minimize cognitive strain (and follow the path of least effort), only those cues that are necessary to allow them to discriminate between the action alternatives perceived in the actual context will be consulted; know-how depends on empirical cue–action correlations in familiar scenarios of activity.

In this mode of activity, it is actually a question of whether one can even talk about "decisions." In a familiar environment, an actor will read off cues from the environment corresponding to several levels of goals, categories of situations requiring certain familiar activities, and cues for specific actions and movements. In this way, control of activities can be structured as being a cue–action hierarchy in that cues at a high level activate the consideration of a particular goal, while other cues define the relevant task and its action alternatives. At a more detailed level, cues activate the individual action sequence (see Figure 3.9).

Normally, it is difficult to identify the action alternatives perceived by an expert actor because they are only intuitively perceived. However, since the relevant action alternatives actually reflect the control requirements of the work domain in a given situation, an analysis of the domain representation can support an identification. In other words, even though this mode of control of activities does not explicitly depend on a representation of the means–ends network, a means–ends map will be useful as a consistent representation of the system constraints indirectly shaping the cue–action mapping. In contrast, the cues adopted to guide actions reflect situational, subjective

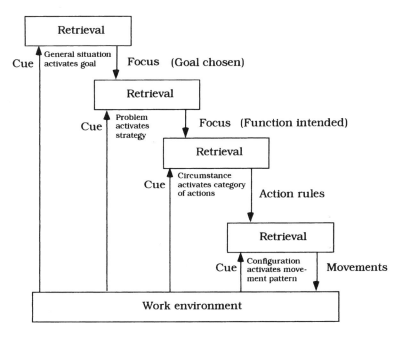

FIGURE 3.9. The structure of natural decision making in terms of cue–action correlations at several levels of abstraction.

experiences of the actor and are difficult to find except by observation during the actual situations (e.g., by eye movement records).

This formulation of heuristic decision making as a continuous control of actions cued by signs from the environment relates closely to Gibson's direct perception of affordances. Chapter 5 discusses this formulation in more detail. It also relates closely to Hammond's distinction between rational and intuitive judgment in quasirational thinking (Hammond et al. 1984). The rational part takes care of both the conscious cue analysis that is sometimes necessary in rule-based behavior and in the functional analysis underlying knowledge-based reasoning, while intuitive behavior reflects the cue–action responses. Hammond conceives the intuitive–rational distinction as a continuum. In the present framework, the distinction between the behavioral categories is quite distinct while the mixture brought to bear in a given situation will reflect a continuum.

The activity analysis in decision making terms is very important for designing information and decision support systems since it identifies the sources of information that are relevant. Also, an identification of the mode of decision making is important in order to determine the format of information required. For formal decision making, information in terms of the means–ends network is necessary more or less independently of the actors experience, while heuristic decision making requires information that matches subject and situation dependent aspects. This aspect will be discussed below in more detail with reference to the mental strategies.

ACTIVITY ANALYSIS IN TERMS OF INFORMATION PROCESSING STRATEGIES

Analysis of the characteristics of the mental strategies that *can* be applied by an actor for the various decision tasks is of course important. Equally so is an identification of the work situation as well as the actors' subjective performance criteria governing the choice among alternative strategies in order to relate task requirements to cognitive resources.

Mental Strategies: A Definition

A clarification of the present use of the concept of strategy is important. A "strategy" is taken here to be an idealized category of cognitive processes, which can serve one of the decision processes discussed in the previous section (such as, situation analysis, diagnosis, goal evaluation, or planning). The actual implementations of a given strategy during a work situation will be different each time, but they will share important characteristics, such as a particular kind of mental model, a certain mode of interpretation of the observed evidence, and a coherent set of tactical planning rules. As a consequence of this definition, different strategies will demand very different resource profiles of an actor with respect to knowledge, cognitive processing capacity, time, and so on. From this point of view, a "strategy" is an abstraction in terms of an idealized structure of a class of cognitive processes. Performance in an actual work situation, in contrast, reflects "natural strategies," which can be best described as a string of frequent shifts among segments of idealized strategies. As discussed later, shifts of strategy appear frequently because they serve to resolve local demand–resource conflicts. The formulation of a set of coherent, idealized strategies is necessary to have a framework for a formal description of reasoning in a complex work context. Such a framework should include: (1) a set of idealized strategies, (2) the users' criteria for choosing among such strategies, and (3) the cues initiating a choice of strategy or a shift between strategies in an actual work situation.

The strategies that are relevant for describing cognitive activity depend on many features of the particular work domain, on the decision function considered, and on the level and kind of expertise of the actor involved. It is, however, possible to identify some basic characteristics that can support a transfer of models of strategies among different design problems. In the following sections, some examples are discussed, based on our particular samples of work analyses.

Mental Strategies: Some Examples

At a very early stage of our research (Rasmussen, 1969) we considered the diagnostic task during operational disturbances of complex work systems to be a critical task. Mental strategies in process control and maintenance have, therefore, been studied and described in detail elsewhere (Rasmussen and Jensen, 1974; Rasmussen, 1981). Also, mental strategies for information retrieval in libraries have been investigated (Pejtersen, 1979a) and are discussed in detail in Chapter 9. Likewise, strategies for medical diagnosis have been analyzed in detail (Wulff, Pedersen, and Rosenberg, 1986; Pedersen and Rasmussen, 1991). In the following sections, a brief review of the mental strategies found in these domains is given as a basis for making more general statements. To our knowledge, the mental strategies actually used during work in a manufacturing domain have not been studied in detail, but a promising start has been presented by Sanderson (1991), Sanderson and Moray (1990), and Moray and Dessouky (1990, 1993). They approach strategies for production scheduling from the normative side, using formal scheduling strategies as a reference structure.

Strategies for Trouble Shooting. In general, the diagnostic task in trouble shooting is a search to identify a change from normal or planned system operation in terms that will refer the diagnostician to locate a failed component to be replaced.

Observations in the form of a set of symptoms reflecting the abnormal state of the system can be used as a search template to find a matching set in a library of symptoms related to different abnormal system conditions. This kind of search will be called *symptomatic search*. On the other hand, the search can be performed in the actual, maloperating system with reference to a template representing normal or planned operation. The change will then be found as a mismatch and identified by its location in the system. Consequently, this kind of search strategy can be called *topographic search*.

These two kinds of search strategy use the observed information in very different ways. Any observation identifies an information source and reads off the content of its message. In a symptomatic search, a reference to the location of the fault is obtained from the content of the message; in a topographic search, a reference is obtained from the topographic location of the information source, whereas its message is subject only to a good–bad judgment, which is used for the tactical control of the search.

The Topographic Search. The topographic search is performed by a good–bad mapping of the system through which the extent of the potentially "bad" field is gradually narrowed down until the location of the change is determined with sufficient resolution to allow the selection of an appropriate corrective action. The search depends on a map of the system that gives information on the location of potential sources of information for which reference information is available for judgments. The information available in observations is used rather uneconomically in a topographic strategy because it depends only on good–bad judgments. Furthermore, the observations do not take into account previously experienced faults and disturbances. Therefore, switching to other strategies may be necessary to reach an acceptable resolution of the search or to acquire good tactical guidance during the search. However, the topographic search is advantageous because of its dependence on a model of normal system operation, which can be derived during design or obtained by data collection during normal operation.

Symptomatic Search. Symptomatic search strategies are based on the information content of observations to obtain an identification of the system state instead of the location of the information source. The search decisions are derived from the internal relationships in data sets and not from the topological structure of system properties. In principle, a search is made through a library of abnormal data sets (symptoms) to find the set that matches the actual observed pattern of system behavior. The reference patterns can be collected empirically from system maloperation or derived by analysis or simulation of the response of the system to postulated disturbances. Furthermore, reference patterns can be generated "on the job" if the diagnostician has a functional model available that can be modified to match a current hypothesis about the disturbance.

The symptomatic search can be a parallel, data-driven *pattern recognition*, or a sequential *decision table search*. If a search is based on reference patterns generated at the time by modification of a functional model in correspondence with a hypothetical disturbance, the strategy can be called search by *hypothesis and test*. Symptomatic search is advantageous from the point of view of information economy and a precise identification can frequently be obtained in a one-shot decision.

A very important characteristic of the different strategies for this diagnostic task is the representation of the functional properties of the system that is required. *Direct recognition* depends on an intuitive induction from the observed symptom pattern; the *topographic strategy* relies on a search guided by a topographic map of the system that gives reference information defining normal states at different locations; search by *hypothesis and test* depends on a functional model of the system that can be used for a deduction of the effect of component faults; search by decision *table look-up* depends on stored symptom–cause relations developed from an analysis of the system or stored empirically from past experience. All of this relates closely to the continuing discussion of mental representations in Chapter 4. However, it is clear that the different possible strategies for the same task can have very different resource requirements—for instance, in terms of data, knowledge of basic system functions, processing, and memory capacity.

Strategies for Information Retrieval in Libraries. From an analysis of actual user–librarian conversations in public libraries, several different strategies for identifying books to match user needs have been identified (Pejtersen, 1979a). These strategies are discussed in detail in Chapter 9, but a brief review here will allow a comparison to be made with the technical strategies and indicate any potential for generalization.

Bibliographical Search. The user is able to identify reading needs by author and title; they explore the book stock and compare needs with book contents. The librarian acts as an assistant and explains the arrangement of the books, equipment, and the use of auxiliary tools. This strategy appears to be a kind of decision table search using catalogs and card indexes.

Analytical Search. The user explores reading needs systematically and compares them with the relevant aspects of the available books. This strategy is the rational, problem solving strategy.

Search by Analogy. The user identifies their reading needs by mentioning a previously read book and asking for "something similar."

Empirical Strategy. This strategy represents the use of shortcuts by the skilled librarian. It is based on the librarians' prototypical classification of users and books. Titles to suggest are selected on the basis of the correlation experienced between user characteristics and typical reading habits. Thus, in addition to considering their expressed wishes, users are classified according to a number of informal features (such as visual appearance, verbal style, dress, and age) and books are split up into simple genre classes.

Browsing Strategy. Finally, an information seeker in a library may have a need that is so ambiguous that a specification of a search template is avoided and, instead, the contents of a shelf or a data base are scanned to find a match with the intuitively present need.

A comparison with the strategies discussed for technical trouble shooting indicates that they both are formulated at the same level of generality and can be used to specify the mental model and the categories of information that need to be considered for the design of appropriate support systems. Again, the representations of the properties of the work domain required for the strategies are very different; The *analytical search* requires a faithful model of the users' problem domain and reading needs; the *bibliographic search* depends on having tools with titles and authors with reference to location of books; and an *empirical search* uses induction from past encounters.

These two sets of examples represent strategies for situation analysis at the upward leg of the decision ladder and have been discussed without reference to the downward planning of the corresponding action task. This is only possible because the basic properties of the task in these two examples are independent of the diagnostic result. In the trouble shooting situation, the task is always to replace the failed equipment; in the library context, the task is to locate a book in the library stock. This clear delimitation of strategies for situation analysis is not normally possible. In most work situations, the task and action to consider depend on the result of the situation analysis which, in turn, may depend strongly on the tasks that are relevant to consider and the alternative actions available to the actor. Some additional examples will serve to illustrate this circular relation between situation analysis and choice of responses.

Strategies for Disturbance Control in Process Plants. An example related to trouble shooting is the diagnostic task of a process plant operator monitoring plant operation (Rasmussen, 1984). Faced with an operational disturbance, their immediate goal and task is not clear: Is it to protect production and continue operation by means of a control action that compensates for the disturbance; or is it to protect the plant and its environment from an accident by activating protective actions that will shut down the plant? The context of the diagnosis depends very much on the answer to this question and, therefore, the diagnostic reasoning becomes circular: The goal to adopt depends on the diagnosis, but the diagnostic strategy to follow depends on the goal adopted. To maintain operation, the diagnostic task is focused on the identification of the productive function that has been disturbed and on the possible compensating control actions. To protect the plant, the concern is with the prediction of the potential consequences of the disturbance.

This decision problem is an extension of the cue–action problem discussed in the previous section; an expert decision maker will primarily be looking for the information serving to distinguish among the action alternatives he/she has.

To sum up, the context in which a diagnostic search is performed depends strongly on the repertoire of action alternatives available to the diagnostician.

Strategies for Medical Diagnosis. Another setting for using mental strategies along the upward leg of the decision ladder should be mentioned. The technical examples discussed so far are related to a well-structured system designed for particular physical processes. Diagnostic strategies can then be developed with reference to the normal "as designed" processes; that is, the strategies can be based on causal relations among events defined by their functional relationships (laws of nature).

This is not always the case and a comparison of the typical diagnostic reasoning found in medicine and in engineering will serve to illustrate how the diagnostic context depends on the domain and especially on its internal source of regularity. For diagnostic strategies, there are rather great differences between cases where it is possible to give a consistent description of the underlying functionality and those where we do not know enough to rely on functional reasoning. In the first case it is possible to apply a method we can call *functional, variationist,* whereas in the second case we must rely on an *empirical, historical* approach based on categorization and correlation.

In the variationist approach, diagnostic strategies depend on the identification of a change with reference to normal. That is, the hypothesis and test strategy is a search for a variation with reference to a model of normal function, while the decision table search strategy is a search in a decision tree constituted by variations from normal states, labeled by causes. This decision tree is developed by deducing relevant symptom sets from variations of the model of normal (as designed) function.

In the empirical, historical approach, it is not possible to give a satisfactory functional description of the system—basically because its design and working mechanisms are unknown or at least only partially known. Therefore, it is only possible to construct a discrete tree of past events that are logically related as causes and effects. Such logically structured trees depend on factorial analyses, correlation analyses, and partial knowledge about the underlying mechanisms.

Utilizing engineering and medical diagnosis as examples for these two approaches is very simplistic and should only be taken to be an illustrative demonstration of some general features. The functional diagnostic approach is—to an increasing degree—being employed in medicine and the empirical approach to diagnosis is, of course, used by all technical experts within their own domain. The distinction is, however, clearly visible in the theoretical treatment of diagnosis in the engineering and natural sciences as compared with the more empirical sciences, such as medicine, psychology, and sociology. (For a detailed discussion of diagnostic reasoning in medicine and technology, see Pedersen and Rasmussen, 1991.)

The Context of Mental Strategies. In connection with an analysis of the task behavior of the actor, we can conclude that a description of the contextual field in which the mental strategies unfold varies widely. Yet, it is equally as important as the description of the structure of the strategies themselves, since it can point to the mental representation to be applied for the strategy. This is also an important point in the design of support systems.

To distinguish the different types of fields from each other, some generalizations will be useful. The following categories can be identified from the examples in the previous sections.

Inductive Strategies. Direct inductive judgment depends on the individual actor's pool of episodical experience. The experience that is activated will be biased by the action alternatives relevant to the situation which corresponds to the intuitive cue–action hierarchy in Figure 3.9. An example in this category is the recognition strategy in diagnosis.

Hypothetico-Deductive Strategies. Such strategies are based on a hypothesis about the solution followed by a deduction of its effect based on a representation of the internal (functional) sources of regularity of the domain. Examples are the hypothesis and test strategies in engineering, which utilize a (design) model of functionality and in medicine, based on models of normal physiological or anatomical functioning.

Deduction and Search Strategies. In this case, the characteristics of the solution to look for are deduced from a representation of the problem and a search for a match is performed among the available solution alternatives. An example here is the analytical strategy for information retrieval.

Search and Comparison Strategies. This category includes sequential searching in a field that can represent empirically or analytically derived data representing different aspects of the domain. The search can be carried out in the physical or functional topography of the domain, such as in a topographic search for doing a technical repair and when browsing through shelves for books in a library. Or a search can be done in a decision tree bringing order to a set of symptom–cause or cue–action correlations, which were derived empirically from past encounters or deductively from a functional model. Examples are the decision table search strategies in technical and medical diagnosis and the empirical strategy in libraries.

This discussion shows that even for one particular decision task, such as diagnosis, representations of many different aspects of the actual work domain are required for the repertoire of mental strategies that can be chosen by an actor, depending on the situation and level of expertise. As mentioned earlier, this fact has important implications when selecting the kind of mental models of the users to consider when designing support systems. The complexity of the repertoire of mental strategies that is relevant for the situation analyses found at the upward leg of the decision ladder is summarized in Figure 3.10.

The description of the idealized strategies serves to formalize their processes, their information requirements, and their cognitive loading. The cognitive processes applied in actual work situations will for several reasons involve frequent shifts among these idealized strategies and among the different perspectives—according to very subjective and situation-dependent factors. An example of shifting strategies is shown in Figure 3.11, which shows the trajectory in the work space taken by a computer maintenance engineer. The engineer starts (1) by guessing about a familiar fault from mere recognition. When proved wrong, he/she gets some hints from a passing colleague who reported on an odd experience with the system the day before and he/she (5) continues from that episodic evidence without success. Finally, the engineer (13) enters a sequence composed of pieces of strategies from topographic and hypothetical searches. In this way, each particular footprint of a strategy becomes unique and complex.

One reason to shift between strategies is because of their very different resource requirements with respect to time taken, information needs, and necessary background. Shifts of mental strategy are a very effective way to circumvent local difficulties along the decision making path.

Category of strategies	Field in which they unfold	Information processes	Examples
Induction	Pool of episodic experience	Recognition, biased by action alternatives	Recognition of case in techncal diagnosis
Hypothetico-deduction	Functional structure	Deduction of symptoms from hypothesis of cause	Analytical diagnosis in technology and medicine
Deduction and search	Relationl structure of problem space	Deduction,search in solution space	Analytical search in libraries
Search and comparison	Search in physical or functional topography	Search and comparison with template	Topographic search in technical dignosis; Browsing in libraries
Same	Search in decision trees	Same	Decision table search in mediacl and technical dignosis Empirical strategies in libraries

FIGURE 3.10. This figure illustrates the complexity of the repertoire of mental strategies that is relevant for situation analysis along the upward leg of the decision ladder.

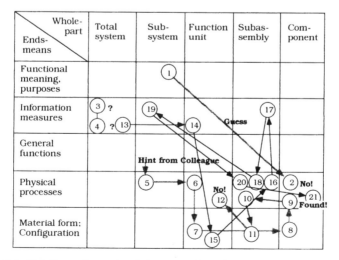

FIGURE 3.11. This figure shows the shifts among formal strategies in an actual case. Shifts of strategy take place in the intervals between behavior segments Nos. 2/3, 4/5, and 12/13.

The abstraction–decompositon map of a work domain is very useful to represent the trajectories of strategies during a task sequence as shown in Figure 3.11. Similar use of the map to analyze behavioral traces during experimental tasks has been discussed by Bizantz and Vicente (1994) and by Christoffersen, Pereklita and Vicente (1993).

Implications for System Design

Several important implications can be drawn from the nature of natural strategies. One is that modeling and simulation of diagnostic performance in actual work should take into account the repertoire of possible strategies. A second important modeling problem is to identify the performance criteria and the (subtle) cues in the work setting that control the transitions from one strategy to another.

Conflict Resolution. The fact that several cognitive strategies are available for most information processing tasks implies that resource conflicts can be resolved and mental work load relieved by a shift of strategy. Momentary difficulties can normally be circumvented by such a shift. To enhance the use of information systems and ensure their proper acceptance by users, all of the strategies that are likely to be chosen by a user should be supported. It should also be remembered that new information tools normally influence the demand–resource match between strategies and users and, therefore, changes in their preferences will be likely. See the discussion of the library system in Chapter 9.

Explanation and Trust. The different strategies have very different resource requirements. This difference has important implications when humans and computers cooperate in a task. The information processing resources of computers and humans are very different with respect to the capacity of short-term memory, volume and consistency in data processing, and so on. Different strategies will be effective for a computer and for a human in connection with the same decision function. This must be considered carefully for information system design, since the user's understanding and acceptance of advice given by a computer can be a critical issue when the computer finds its results by means of a strategy different from the one preferred by the user. A replay and explanation of the computer strategy may not be an acceptable explanation. Instead, a mapping of intermediate results from the computer's strategy onto a representation of the strategy applied by the user may be necessary. For this purpose, a structuring of the reasoning of the computer into a sequence of natural "states of knowledge" can be useful for exchanging results.

METHODOLOGICAL HINTS FOR ACTIVITY ANALYSIS

Identification of "Prototypical Task Situations"

The work domain description represents the general situation-independent properties of the work system from which a selected subset is relevant for each work situation.

Activity analysis serves to identify (1) the part of the work domain representation that is relevant for particular situations, (2) the mode of representation most suited for the activities of the involved actors, and (3) the decision processes applied by the actors. Since, as mentioned previously, procedural descriptions are unsatisfactory, it is necessary to identify a set of prototypical decision and action situations that can be combined in various ways to form the total response of the actors. The description of the prototypical decision situations should include an identification of the involved people, the resources necessary for the activity, and the information sources used.

As mentioned above, information about typical work situations is unreliable for designing information systems. The generalization necessary for design cannot take place reliably from the typical case. No two work situations are identical since the context of a decision situation and the heuristics brought to bear depend on minute differences in situational and personal characteristics. Consequently, data collections should aim at detailed records from a number of particular work scenarios and the analysis should reflect a detailed understanding of the structure, decision processes, and information exchanges for each individual scenario.

Then generalization should be based on an identification of recurrent "prototypical" decision situations using a categorization along and across the representative set of records. The prototypical decision situations identified by such a generalization can then be combined into the representation of the work situations on which to base a system design and predict the effects of the resulting changes in the resources available to the actors.

It follows that data acquisition must be based on detailed descriptions of an adequate number of specific situations—for example, how a particular item is produced in a factory or how a particular patient is processed through a hospital—in order to obtain reliable information on the various options for action and the criteria of choice. In addition, an identification is necessary of the heuristics and rules of thumb, which characterize the expertise of the actors.

Thus, in this phase, the analysis delves into more detail about the prototypical work situations based on the broader interview sessions. It must aim at selecting a characteristic set of concrete cases (concerning a product, patient, or case). At this level, an appropriate set of work sequences underlying these cases should be described, which can be used to generalize and to identify representative decision making situations. Such descriptions are formulated in the terminology of the work place. The sequences are established with the help of interviews concerning specific and concrete examples.

A very useful form for data acquisition is based on anecdotal and episodic descriptions that are easier to remember, formulate, and capture accurately than more abstract formulations. However, the interviewer must keep the interviewees "on course" with their recounts without permitting them to revert to citations from textbooks and/or other formal rules and regulations which, in principle, are intended to prescribe how the given sequence should be handled. This normative material falls outside of a discussion of specific work processes but can be a useful source of information about the domain, goals, resources, constraints, and so on. If the interviewee refers often to procedures in describing a given task, then the sequences referred to will not become sufficiently specific so that the individual steps or subtasks can be identified with reasonable certainty together with the criteria underlying choice in a particular situation.

At any rate, descriptions of the work are collected from interviews of as many people as is necessary to ensure the elicitation of information about the total process and to get a representative collection of heuristics. The people do not necessarily have had to carry out the functions they report about themselves, but they must have been in close contact with the situation. Later, when the decision making and the underlying strategies are uncovered, it will be important to talk to the people who were directly involved. When several detailed and concrete cases have been gathered, they have to be analyzed in order to identify the distinctive segments–situations that can make up the elements in the work description. In Figure 3.1 the prototypical decision situations are distinct and reasonably well delimited. In hospitals, there is a clear situational distinction that typically can be characterized through the formal and informal meetings that take place among the cooperating parties in order to carry out a given set of activities. This is not always the case since an activity can be spread out over time and location and require that one person move around, talk to others, and initiate certain actions in various places in the course of a single task. Here the descriptor "prototypical work function" is more suitable than the "situation."

This analysis uncovers the functional relationships and helps to identify characteristic prototypical work and decision making situations. Normally, it should be possible (and desirable) to distinguish between "transactional" activities (i. e., the handling of individual cases) and "resource planning," which encompasses all individual cases (see Fig. 3.1 again). Interviews with the people involved in concrete examples are a necessary start.

As a supplement to the above, a systematic analysis of the normal planned meetings and conference activity is necessary. This analysis should be based on detailed minutes and notes (if possible with the help of a tape recorder, when acceptable). This will give us an independent chance to identify significant prototypical decision making situations and supplement those gleaned from the more transactional analyses. Figure 3.12 illustrates a structure of work and decision making situations that aggregates the information from activity analyses and the results of a single day of meetings in the surgical wing.

Still another independent check of the coordination and communications analyses can be achieved by following selected people through the course of a work day. This can identify the same and possibly additional activities in the coordination and communication areas.

Activity Analysis in Work Domain Terms

Starting with all the identified and relevant work and decision making situations and functions, the *participants* and *tasks* for each must be found and described in work terms together with a denoting of information needs, information sources, and communications media. A direct participation in these work situations is necessary so as

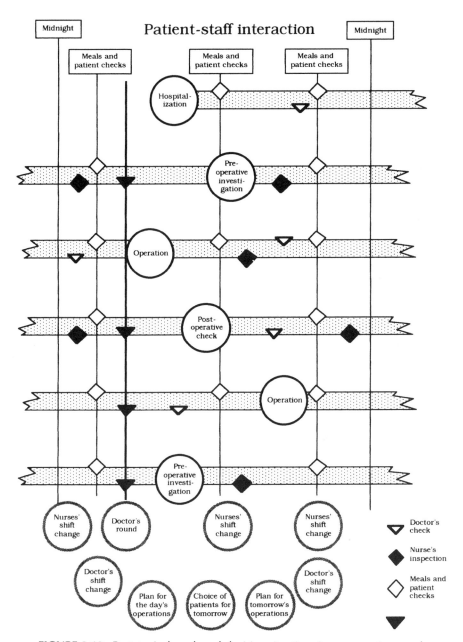

FIGURE 3.12. Prototypical work and decision situations in an operation ward.

to ensure that all the tasks, which actually are carried out, are included. Very often, "left over" or "miscellaneous" tasks are carried out that have no direct connection with the main on-going effort. These ad hoc decisions can become important in their effects on the total coordination and communication. The analysis of meetings that cut across individual cases can give an independent check of the coordination activities and the related communication channels.

For each of the prototypical work situations considered, the total activity is analyzed for decomposition into more elementary activity elements (such as the following from an operation theater planning meeting in a hospital: patient treatment planning; personnel planning; evaluation of treatment and methods; allocation of doctors to operations theaters; purchasing, budgets). In addition, the allocation of roles to the individual actors is identified with reference to the domain description, and the communication paths and content within and external to the group are analyzed. An important part of this analysis in work terms is the identification of the information used with respect to its content, its source, and the communication medium. Figure 3.13 illustrates the information sources used for different considerations in an operating theater planning session, with reference to a representation of the patient, as well as the hospital domains.

The analysis consists of working through all the material accumulated from

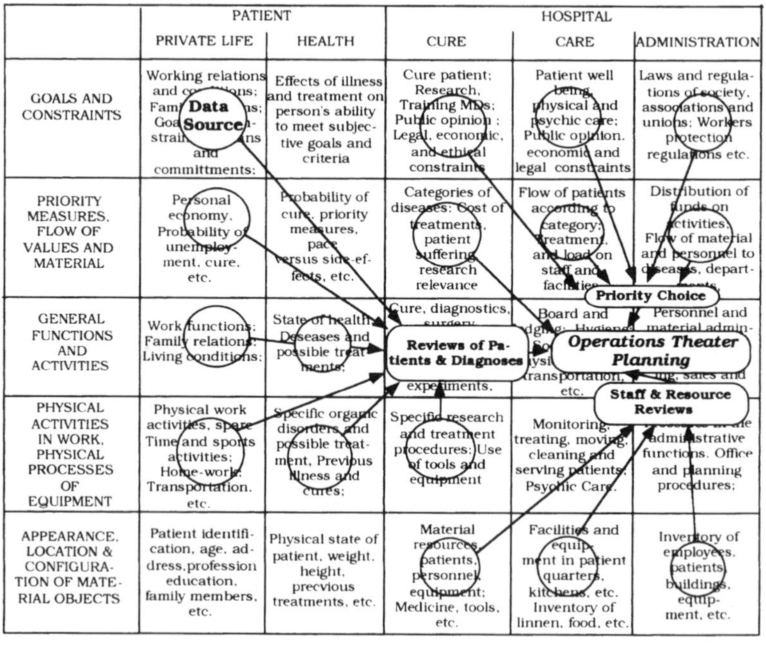

	PATIENT		HOSPITAL		
	PRIVATE LIFE	HEALTH	CURE	CARE	ADMINISTRATION
GOALS AND CONSTRAINTS	Working relations and connections; Fam **Data** us; Goal **Source** ns strai and committments;	Effects of illness and treatment on person's ability to meet subjective goals and criteria	Cure patient; Research, Training MDs; Public opinion ; Legal, economic, and ethical constraints	Patient well being, physical and psychic care; Public opinion. economic and legal constraints	Laws and regulations of society, associations and unions; Workers protection regulations etc.
PRIORITY MEASURES, FLOW OF VALUES AND MATERIAL	Personal economy. Probability of unemployment. cure. etc.	Probability of cure, priority measures, pace versus side-effects, etc.	Categories of diseases; Cost of treatments, patient suffering, research relevance	Flow of patients according to category; treatment and load on staff and facilities. **Priority Choice**	Distribution of funds on activities. Flow of material and diseases, departments.
GENERAL FUNCTIONS AND ACTIVITIES	Work functions; Family relations; Living conditions;	State of health; Diseases and possible treatments;	Cure, diagnostics, surgery; **Reviews of Patients & Diagnoses** experiments.	Board and ... Hygiene; **Operations Theater Planning** transportation, etc.	Personnel and material administration; ... sales and **Staff & Resource Reviews**
PHYSICAL ACTIVITIES IN WORK, PHYSICAL PROCESSES OF EQUIPMENT	Physical work activities, sports. Time and spare activities; Home-work; Transportation. etc.	Specific organic disorders and possible treatment. Previous illness and cures;	Specific research and treatment procedures; Use of tools and equipment	Monitoring, treating, moving, cleaning and serving patients; Psychic Care.	Administrative functions. Office and planning procedures;
APPEARANCE. LOCATION & CONFIGURA-TION OF MATE-RIAL OBJECTS	Patient identification, age, address, profession education, family members. etc.	Physical state of patient, weight, height, precvious treatments, etc.	Material resources, patients, personnel, equipment; Medicine, tools, etc.	Facilities and equipment in patient quarters, kitchens, etc. Inventory of linnen, food, etc.	Inventory of employees, patients, buildings, equipment, etc.

FIGURE 3.13. The information sources in oprating theater planning with reference to the work domain shown in Figure 2.7.

interviews, observations, and tape recordings in order to identify and classify the tasks in terms related to the contents of the associated work. The classes will often be retrievable directly from the participants' use of professional terminology and other phrases which, in most cases, reflect a division of the ordinary daily activities into suitable subtasks around which planning, discussion, and cooperation can take place. A more coherent division and classification can first take place after the next phase, where the decision making functions are dealt with. As in the previous case, the analysis of the described prototypical work situations and functions should aim at identifying a set of typical tasks as culled from some kind of overlaying of the analyzed individual cases, which can lead to a prototype that is defining for the set.

Activity in Decision Making Terms

At this phase, the focus will be on an identification of prototypical decision making tasks expressed in information processing terms that characterize the selected work situations. This is based on a determination of the underlying decision tasks, including situation analysis, the establishment of goals and priorities, as well as the planning of the necessary action. In other words, the intention must be to define typical decision tasks within the work domain with particular weight placed on which means–ends relations are involved.

While the analysis in domain terms to some extent could be based on interviews after the fact, the analysis in decision making terms requires the analyst to participate in the process, preferably with a tape recorder. The analysis will serve to structure and categorize the activities into a combination of the decision function elements defined by the decision ladder in Figure 3.6. In this way, it is possible to identify the information requirements of the different typical, recurrent activity elements, and to find all the sections of the activity records belonging within the same category of mental activity in order to have material for detailed analyses of the mental strategies. For this analysis, the use of the decision ladder as a sketch pad giving a visual overview of the data material can be very helpful. See this use of the ladder in Figure 3.7.

Mental Strategies: Analysis of Behavioral Traces

When the relevant decision functions have been identified, as described in the previous paragraphs, a further analysis of the behavioral trace through time is necessary in order to decompose and categorize the mental activity and to identify the various information processing strategies that *can* be used for the decision functions.

A basic principle for a proper analysis is that the behavior fragments must be analyzed individually and understood in their context by someone who is familiar with the work domain. Performance in work is governed by individual subjective criteria. Therefore, averaging across samples before the material has been understood, structured, and categorized—as discussed in connection with the analysis of the verbal protocols later on in this section—will be misleading. To average across samples at a too early stage of analysis will be to compound data of different kinds only to have to sort them out again later by statistical analysis. It is evident that we, like Eddington's (1939) ichthyologist, will be able to obtain results more readily by a conceptual analysis defining the categories in which a behavioral trace should be decomposed and analyzed than by a statistical analysis of data across categories. The crucial point is not to have statistically significant, average data, but to *understand* the individual traces and to find prototypically significant categories.

A detailed description of the analysis of event traces is found in Hollnagel et al. (1981). For a recent and comprehensive analysis of methods for analysis of behavioral traces, see Sanderson and Fisher, 1994. The various domains of analysis are the following:

Raw Data. Raw data are the records obtained from interviews, observations, and performance measurements in the field or from a simulator, and so on. These data all comprise performance fragments of one kind or another.

Intermediate Data. This represents the first stage of processing of the raw data. In this stage, the data are combined and ordered along a time line in order to provide a coherent description of an actual, particular "occurrence." A description of the actual performance results, using the original terminology and language of the work domain—and it is thus a description by a work professional rather than an outside expert. This step is relatively simple—being basically a rearrangement rather than an interpretation. Hence, special translation or transformation aids should not be required.

Analyzed Event Data. At this stage, the intermediate data format is transformed into a description of the task performance using formal terms and concepts that reflect the theoretical basis for the analysis. In our framework, this would consist of a combination of information processing and decision making. The description is still ordered along a time line specific to the given situation. However, the transformation has changed the description to a formal description of the performance during the specific event. This analytic step may be quite elaborate since it implies a theoretical analysis of the actual performance, which can support a transformation from user task terms to formal terms.

The analysis basically involves a decomposition of the behavioral traces into uniform and recurrent segments, which can be categorized meaningfully. This stage is difficult and labor-intensive and frequently, special analytical tools will be needed, for instance in the form of graphic representations for fast replays of data streams to facilitate visual recognition of similar segments, or categorization tools as developed by Sanderson, James, and Seidler (1989), Sanderson et al. (1991), and James and Sanderson (1991). In some cases, the analysis of task requirements can result in structures to use for visualization. For instance, in experiments to study strategy formation in computer games, an analysis of the rule structure and the related choice trees gave ideas for graphic representations that supported visual decomposition and categorization. An important complication of this stage of analysis is that the emphasis is changing from being descriptive to explanatory, and thorough domain expertise and understanding, as well as the conceptual tools, are mandatory.

Conceptual Description. At this stage, the description is no longer a set of specific behavioral traces but a set of categories presenting common features from a number of recurrent, uniform segments. By overlapping and/or combining formal descriptions of the individual segments, a description of the generic or prototypical instance of a category is found. The performance under a specific event can be considered as an example of, or a variation on, the prototypical performance. The validity of a prototypical performance can be tested either by determining whether a given formal description of an actual performance can be subsumed under the prototypical description—or by comparing this formal description with predictions of typical performances based on the prototypical description. The step from formal to prototypical is again quite elaborate. It requires a number of special translation aids but also an experienced analyst. They have to be able to generalize from the specific into prototypes that can predict typical performance in specific tasks.

Competence Description. This description is the final level of analysis and is concerned with the cognitive control of the behavioral trajectories represented by the material. Categorizing fragments across samples instead of averaging across samples allows the analyst to obtain several variants of fragments within the same category of behavior when relating the conceptual descriptions with the modes of cognitive control. The description of competence is concerned with basic elements, such as mental models, decision strategies, performance criteria, and preferences, which, in a given performance, are combined to produce the result. The description at this level is thus context-free; it has to do with the behavioral repertoire of the user, independently of any particular situation within the given class of situations, that is in task-independent terms of the generic strategies, models, and performance criteria that underlie the performance. Later, for a design scenario, when a new context is given, the description of competence can lead to a description of the prototypical performance and, given further information, a description of the typical performance. In this way, a competence description essentially gives a basis for performance prediction during system design–evaluation.

The competence analysis depends on inference with reference to a particular theoretical model framework. Competence aspects cannot be observed in the data but depends on a theory-based interpretation. Therefore, the analyst must have a good background in cognitive psychology as well as considerable methodological experience and, at the same time, familiarity with the actual work domain.

An Example: Analysis of Mental Strategies From Verbal Protocols. * This section describes the analysis of the mental strategies of service technicians in electronic

---

*The figures accompanying this section are deliberately reproduced in their original form as they were created during the analyses in order to give the reader a visual impression of the outcomes of the various codings that are described.

trouble shooting to illustrate a detailed analysis of behavioral traces. This example is chosen because diagnosis in maintenance can be reliably separated from the overall work sequence since the objective is clear: Replace the failed component. Therefore, no repertoire of diagnostic objectives and action alternatives complicates the analysis.

Raw Data. In our study, maintenance technicians were asked to verbalize their work procedures and tape recordings were made. It was made clear to the technicians that we did not want them to do an introspective analysis of their thought processes but merely tell what they were trying to accomplish, what they were doing and/or observing, and with which instruments. In order to obtain qualitative formulations of the basic features of their diagnostic procedures rather than quantitative data from standardized repetitive experiments, recordings were taken of several individuals locating faults in different instruments during their normal work day. This approach also enhanced the confidence of the group because it made comparisons of the individual technician's performance irrelevant. In order to enable them to familiarize themselves with the experimental procedure and the tape recorder, the instruments used for the initial cases were selected by the technicians themselves when they found it convenient to make a recording. However, it soon turned out that stereotyped procedures were used for the simpler instruments, such as amplifiers and power supplies. Therefore, the cases to be recorded were later selected by the planners of the experiment.

More complex systems were then chosen which gave a greater variety in the external display of symptoms. Instruments such as multichannel analyzers, oscilloscopes, TV-receivers, and analog-to-digital converters were included. Analyses indicated no use of a detailed functional analysis of the observed behavioral symptom patterns, even when it was obvious that such an analysis could lead directly to the fault. Therefore, in later phases of the experiment, simulated faults were introduced to explore more carefully why functional reasoning was not used.

The technicians were asked to relax and relate what they were thinking, feeling and doing, to express themselves in everyday terms, and to use only short hints during fast work sequences. A record was immediately typed and the technician was asked to read it, while remaining in his/her working position in front of the instrument, in order to correct mistakes and supply supplementary information when he/she felt something was missing. At the same time, the analyst had the first review of the record and could have a short discussion with the technician to clarify weak passages in the verbalization.

Analyzed and Intermediate Data. The initial systematic analysis was based upon the definition of a set of elementary events describing the microstructure of the sequences. The records were coded and a computer print-out made giving a graphical picture of the sequence, as well as a connectivity matrix (Fig. 3.14) describing each case. The graphical read-out (see Fig. 3.15) turned out to be very effective for a visual identification of recurrent passages in the protocols. Because of the large number of parameters, attempts to identify recurrent sequences from the connectivity matrices

	A	B	C	D	E	F	G	H	I	J	K	L	M	N	O	P	Q	R	S	T	U	V	W
	22	14	10	-3	22	-4	-9	11	62	27	16	15	16	-9	--	21	-4	19	20	56	42	-5	48
A	-3	--	--	--	-4	--	--	--	-6	-2	--	--	-1	--	--	--	-1	-1	-4	--	--	--	--
B	--	--	-2	--	--	--	--	--	-9	-1	--	--	-1	--	--	-1	--	--	--	--	--	--	--
C	--	--	--	--	--	--	--	--	-8	--	--	--	-1	-1	--	--	--	--	--	--	--	--	--
D	--	-1	--	--	--	--	--	--	--	--	--	--	--	--	--	-1	--	-1	--	--	--	--	--
E	-2	--	--	--	--	-1	--	--	-6	-3	--	--	-1	--	--	-3	--	-6	--	-1	11	--	10
F	--	--	--	--	-1	-1	--	--	-1	--	--	--	--	-1	--	--	--	--	--	--	-3	-1	--
G	--	-1	-1	--	-1	--	-1	-2	--	--	--	--	--	--	--	-1	--	-1	-1	-1	-3	-1	-2
H	-1	-1	--	--	-1	--	-1	--	-1	-3	--	--	--	--	--	-1	--	-2	--	-2	-6	-2	--
I	--	--	-3	--	--	--	-1	-1	-4	-5	16	14	-9	-6	--	-1	-1	--	-1	21	-6	--	22
J	-3	-1	--	--	--	--	-1	-2	-8	-6	--	-1	-1	--	--	-2	--	-1	-1	-1	--	--	-1
K	--	-2	-1	--	--	--	--	--	-6	--	--	--	-1	--	--	-2	--	--	-4	-9	-2	--	-5
L	--	-2	-2	-1	-3	-1	--	-1	--	--	--	--	--	--	--	-1	--	--	-4	-8	-1	--	-6
M	-1	-3	--	--	-2	--	-1	--	-2	-1	--	--	--	--	--	-3	--	--	-2	-7	-7	--	-1
N	--	-2	--	--	--	--	-1	--	-2	--	--	--	--	--	--	-2	-1	--	-1	-5	-2	-1	--
O	--	--	--	--	--	--	--	--	--	--	--	--	--	--	--	--	--	--	--	--	--	--	--
P	-1	-1	--	-1	-4	--	--	-2	-5	-2	--	--	--	-1	--	--	--	-3	-1	-1	-1	--	--
Q	--	--	--	--	-2	--	-1	--	-1	--	--	--	--	--	--	--	--	--	--	--	--	--	-1
R	-6	--	--	--	-1	-1	--	-1	-3	-4	--	--	--	--	--	--	-1	-1	-1	--	--	--	--
S	-5	--	-1	--	-3	--	-2	-2	--	--	--	--	-1	--	--	-3	--	-3	--	--	--	--	--

*

FIGURE 3.14. A connectivity matrix for visual pattern identification of the most frequent microaction patterns in an action sequence. Letters A, B, C, . . . represent coded "microactions."

gave only a very few hints about the general pattern. Based upon the graphic read-out, sequences identified as recurrent routines were reanalyzed from the original records and classified according to the characteristics of the information processing used.

Conceptual Analysis. The task of finding the structure in verbal records generated by several persons trying to locate individual faults in different types of systems makes one realize how colorful actual work performance is and how much effort is required for an analysis.

An almost immediate experience is the danger that the analyst himself/herself will develop fixed routines in his/her analysis in order to manage the classification of the multitude of situations. It proves imperative to have long breaks in the analyzing effort to reduce inflexibility and to be able to return to the original material with an open mind. It also seems important to have several analysts criticizing each other's models of the structure in order to break fixations and to decide whether differences in classification are caused by ambiguities in the definition of categories or by different interpretations of the records.

The analyses started by extracting the most obvious and frequent routines, leaving for later analysis the complicated and more individual parts of the records. For each collection of such typical routines, the original data in the records were analyzed across protocols to identify the information processes used, the mental models of the instruments that could explain the stream of action, and the interpretation of observations used. In this way, the protocols can be recoded at a higher level as a sequence of elements representing different mental strategies (see Fig. 3.16). For each

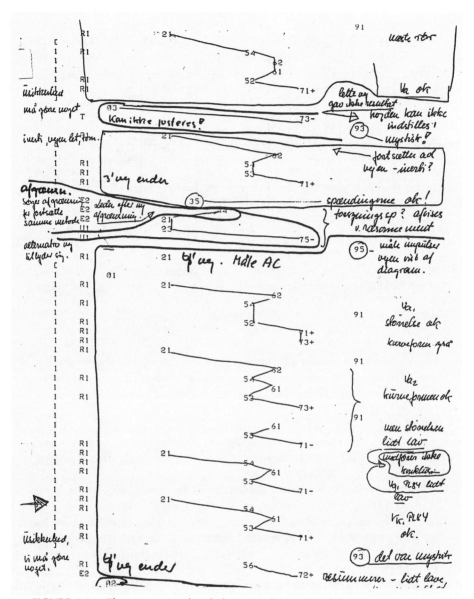

FIGURE 3.15. The sequence of coded microtasks is printed for visual identification of (recurrent) strategy patterns. This figure and the following figure 3.16 are reproduced as they emerge during the analysis to illustrate how sketches and annotation are used to facilitate visual analysis.

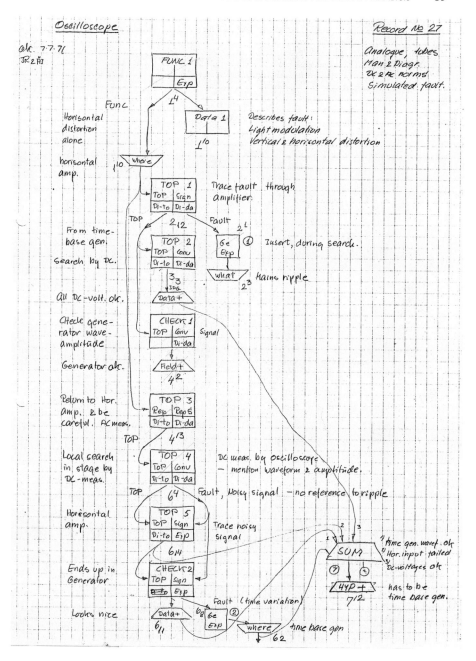

FIGURE 3.16. Coded protocol sequence.

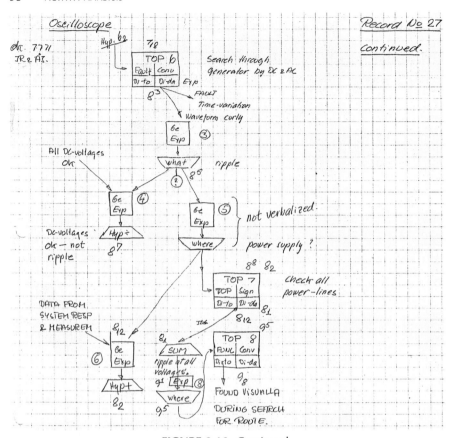

FIGURE 3.16. Continued.

of the strategies identified, a graphic symbol was designed, identifying the type of strategy, with codes indicating important search parameters, such as the information used to guide the search path, the type of information observed, and the source of reference for judgment.

The necessity for a highly iterative classification contributed greatly to the amount of work; each new category introduced made it necessary to review all those already used. For each subroutine a graphic symbol was chosen, with a set of codes indicating important features. The records were finally coded as a graphical pattern showing the interconnections of the subroutines and with comments to facilitate later reviews. For this phase of the analysis, the tools "SHAPA" and "MacSHAPA" developed by Sanderson, James and Seedler (1989) and Sanderson et al. (1991) can provide effective support.

In formulating the generalized subroutines, the analysts had to cut away most of the details related to the individual cases. However, in order to find why a specific routine was chosen, they had to return to the details in the original record and to the features of the specific instrument. It is important for the analysts to be thoroughly familiar with the work domain in order to be able to imagine themselves in the task situation and thus have a clear understanding of the significance of the manipulations and observations. They can then better interpret the protocol and find the structure in the information processing.

During the analysis it was found that there are certain cases of interference from the verbalizing tasks that can cause uncertainty during the classifications of sub-routines. For example, it may be more attractive for the technician to do something physically, to manipulate, than to use more complex mental processes, such as functional inference, because an action is more readily explainable on the tape record and fits better with the pace of speech. Sometimes the impression is that activities reported sequentially would normally be part of a parallel data processing, that is, a set of automated routines. Thus a routine may be disturbed because it is forced into consciousness and, therefore, the record can be erroneous and incomplete. Some of the records indicate that the technicians subconsciously collect information con-currently with the reported activities and that such information supports "bright ideas," which are difficult to explain later in the sequence.

However, careful discussions with the technicians after the analysis of the records does not indicate any serious misinterpretations of the general structure of the proce-dures. After getting accustomed to the recorder, they found the execution of the task, as well as the time spent, to be normal. Actually, attempts later in the experiments to change the behavior of the technicians by training them in the use of more "rational" strategies in more difficult cases had no observable effect on their behavior. When engaged in the actual task, they are so deeply embedded in the task that it takes over control of their strategy.

Chapter **4**

Analysis of Work Organization and System Users

INTRODUCTION

In the previous chapters, we considered a work domain as a coupled assembly of processes and functions that is controlled and coordinated through the activities of human actors. In addition, the decision functions and information processing strategies relevant for this control were identified, without referring to the allocation of functions to individual actors or teams. In this chapter, the topic is the identification of the structure of the organization coordinating activities and the allocation of functions to the individuals involved.

First, the work organization is considered as a distributed, self-organizing control system, where the aim of analysis is to identify the mechanisms that shape this organization and govern the allocation of functions to individuals.

Next, the competency of the individual actors, their cognitive control mechanisms, and their cognitive resource profiles are discussed together with their subjective preferences. This is a prerequisite to discussing the work strategies they are likely to adopt and the work interfaces that are suitable.

Together, these aspects represent the lower path of analysis shown in Figure 1.9.

Organizational Modeling

While we recognize that characterizations of current organizational structures and processes can range from stable and formalized to flexible and dynamic, our analyses of such organizations—especially in connection with accidents—have pointed to the need for a framework for modeling the division and coordination of work that is capable of capturing the mechanisms behind organizational adaptation to change. In accordance with the conception of a system of work as an adaptive system, organizational analysis will basically apply an open systems perspective. At the moment, this is a fruitful paradigm in organizational theory and has been further explored and developed since the idea was shaped by the work of Woodward (1965), Thompson (1967), and later by Mitroff and Linstone, (1993).

Many work domains will be rather loosely coupled so that it is only in special work situations (assembly line work, process plant control) that the requirements of the work domain will control the pace of activities of the actors directly in a coordinated way. In most cases, an extensive communication among actors will be necessary for coordination. In the present context, the term *organization* does not refer to stable groups of people as they appear in organizational charts but instead refers to the relational structure necessary to coordinate the work activities of individuals. In contrast to the formal organization, which reflects the allocation of authority and financial and legal responsibility to groups and individuals, the actual organization of work and social relations depend on very dynamic relationships.

At the outset, it is important to develop a framework for discussing organizational and management aspects that, on one hand, is compatible with the cognitive, information processing point of view and can be used for designing information systems. On the other hand, it should map well onto the frameworks available from the social and organizational sciences, and thereby facilitate the transfer of research results from this source. The following framework has evolved from this point of view.

A system of cooperative work is an extremely complex organizational phenomenon involving many forms of social interaction. To cope with this complexity, it is useful to distinguish (1) the system of work as a functional organization for coordinating activities in a loosely coupled work domain from (2) the social organization of the relationships among actors. This distinction defines two basic perspectives for the social system of work: a "work organization" perspective and a "social organization" perspective. If we follow this point of view, we have to consider organizational aspects at two different levels of analysis:

1. The functional work organization required to coordinate activities will be determined by the interaction of the control requirements of the work domain with the competency of the staff members. This functional organization will determine the allocation of roles to the individual actors and the *contents* of the communication required for coordination.

2. On the other hand, the social role configuration chosen by or imposed upon the individuals and/or teams depends on the management style which, in turn, influences the *form* chosen for the communication and social interaction within the functional team or organization. The form depends on the conventions and constraints chosen for how this communication takes place and is expressed.

The control requirements of a work domain change over time as will the functional work organization. A particular division of activities and, consequently, work organization, will evolve for each situation depending on the competencies of the actors, the "technology" of the work domain, and on the external environment of the organization. Studies show that, even in traditionally tightly controlled organizations, such as the military and high hazard process plants, the actual cooperative structure changes dynamically to match the actual circumstances and, therefore, a framework for modeling must be able to capture this feature. Likewise, the social organization may change independently of this (e.g., under the influence of new managers, alternate management styles as well as from a changing perception of general human values).

FUNCTIONAL WORK ORGANIZATION

The focus of the functional work perspective is on the cooperative patterns as they evolve in the interaction of functional requirements of the work environment with the cognitive resources and subjective preferences of the actors. As discussed in Chapter 2, any work system is a complex network of many-to-many relations among ends and means and its internal, more or less loose coupling requires human intervention in order to remove ambiguities and to control its functional state. In this way, the perspective is congruent with the "rational systems" perspective as defined by Scott: "From the standpoint of the rational system perspective, the behavior of organizations is viewed as actions performed by purposeful and coordinated actors" (Scott, 1987).

The work organization perspective conceives the system of work as a rational system in the sense that it, by and large, is functional with respect to the environment by producing a product or providing a service under the specific conditions and constraints characterizing the environment. The configuration and allocation of work roles can be specified in absolute terms (depending on topographic location, on access to information, and on regulations such as union agreements) and in terms of the criteria governing the dynamic allocation employed by the actors themselves (such as preference for particular social patterns, and mutual concern about equal work load). In the actual work situation, alternate ways to allocate work roles to the individual actors will depend on different criteria and these are often competing. This conflict turns out to be an important contributor to accidents as discussed in Chapter 6. Therefore, a framework to sort out the different architectures of organizational structures, as well as the criteria governing their adaptive changes under varying conditions, will not only be useful for designing information systems but also for the control of hazards involved in the work activities.

Different perspectives turn out to be useful in a discussion of forms of cooperative work:

1. *The Architecture of Cooperative Work.* The first question is *What is partitioned* and allocated to the individual staff members? That is, what kind of representation of the activities is used as a map on which to draw the boundaries between the different actors' activities.

2. *Criteria for Division of Work.* The next question is *How are the boundaries defined?* The choice of boundaries between the activities that are adopted by or allocated to different individuals can be guided by several different criteria related to

the characteristics of the work domain as well as to actors' competency and preferences.

So far we have only considered activities in the context of the work requirements or, in other words, from a functional point of view. Another perspective having to do with the perception by the individual actor is necessary in order to analyze other factors (e.g., work load).

3. *Task Design.* This perspective leads to a consideration of the set of activities faced by an individual actor within a particular time frame. "Tasks" frequently represent the overlapping and interleaved activities with which an actor has to cope in a particular situation, irrespective of whether they are directly and functionally connected.

4. *Job Design.* Finally, we have to consider the aggregated activities of an actor across time in order to evaluate the combination of competencies required by this individual. The tasks and activities that are taken on will go together into a job design that also will be subject to certain constraints requiring attention in the task allocation—particularly with reference to competence as well as the general "quality of working life."

In the following sections, we will have a closer look at these aspects of cooperative work.

Architecture of Cooperative Work

Before the cooperative structures are discussed, a closer look at different degrees of cooperation will be useful. Very often, distinction is made between activities of individuals, groups, teams, and organizations, each of which has some very typical characteristics.

Individual Work. An individual actor can be assigned to cope with the entire work space for a particular work scenario, irrespective of the decision functions and information processes the work implies. This is often the case for shorter term work situations during which an actor works at a separate work station without direct interaction with colleagues. Over longer time horizons, individual work is found, for example, in research and other creative activities such as design, which, for this reason, will take on a special character.

Collective Work in Groups. In other cases, a group of people acts together, pooling their resources and knowledge in a kind of collective activity. Multiple actors are allocated to the same task or type of task so as to resolve limitations in physical resources or information processing capacity. A typical case occurs when the requirements cannot be met by one individual and a combined effort becomes necessary (activity in separate places, lifting heavy loads). Another example is group problem solving during committee meetings and brain-storming sessions during a design task when no structured cooperation is found. Instead, the "collective mind" of the staff seeks to find a creative solution.

Collective work in groups is characterized by not being structured by task requirements, but by social dynamics depending on the personal characteristics of the individuals. We will return to this issue later.

Collective work in this sense is normally only found for certain kinds of tasks and can take place locally in any organization, often for a limited period of time.

Cooperative Work in Teams and Organizations. In general, a functional work organization is a complex, cooperative structure involving several individuals. Different activities are allocated to specialized roles and actors. We distinguish between teams and organizations. *Teams* are sets of actors cooperating in a structured way (in contrast to the functionally unstructured *groups*) and interacting dynamically during a task situation. *Organizations* include more teams that cooperate over a longer time span and are more loosely coupled than the members of a team.

Organizations and teams can be considered distributed control systems with each unit taking care of one particular segment of the work space. The internal functional structure and tightness of coupling among elements of a work system depend on the type of work domain according to its location in the map of Figure 2.9, its operational state, and the related tasks faced by the team. Since the state of a system will change dynamically with time, so will the cooperative structure of an organization and its teams (i.e., the active work organization).

The allocation of roles to individual members of a work team can be resolved in different ways, resulting in different architecture of the work. Two aspects are important. The first is the *division of work* itself, *What* is divided among staff members? This question relates to the kind of representation that is used as a map to draw the boundaries between the activities of the individual. The other aspect is the criteria used for setting the boundaries in the map, *How* is work divided. We consider "what is divided" first.

Division of Work: What Is Divided

Domain-Oriented Architecture. A work place typically involves a very complex network of means–ends relations constituting a work domain of the type discussed in Chapter 2. In general, such a domain includes a large number of work functions that must be taken care of by a given number of specialized actors.

The cooperative architecture of a manufacturing company in a stable market traditionally has the form of a functional hierarchy. The activities are divided vertically according to the various levels of the work domain representation and allocated to different categories of employees. In private companies, high level executives are oriented mainly toward financial operations and are typically graduates from business or law schools with little competence in the "bottom" level manufacturing domain. In other branches, such as hospitals and universities involved in research and teaching, the professionals will be involved in strategic planning and the mediation of outward-directed policies at the same time as they pursue their substance matter activities. In this case, the division of activities cuts across the levels of the means–ends domain map. When manufacturing companies become involved in a dynamic, turbulent market, an empirical shaping of their strategies is no longer adequate and a knowledge-based management is necessary. This means that the top executives will have to be able to explore both the functional user domain at which their products are aimed, as well as the potential of the company's production technology. Drucker (1988) suggests from a similar line of thinking that the management style of present day hospitals or universities is the paradigm for future commercial organizations and Dertouzos (1988) concludes that:

> In the postwar years American managers took the production process largely for granted and ranked it below finance and planning in the hierarchy of managerial concerns. The management profession must now reassess its priorities. Managers can no longer afford to be detached from details of production; otherwise they will lose the competitive battle to managers who know their business intimately.

These examples illustrate how the functional requirements of the work domain influence the cooperative architecture.

Actor Function and Process Oriented Architecture. Cooperative work can be further specialized by function or process. In this case, selected functions or processes are allocated to actors who are specialized or who have access to the information needed. In some cases, for instance, for certain management staff functions, for consultants, advisers, and other forms of specialized services, the cooperative architecture is independent of the domain related architecture.

The next question is, How is work divided among actors?

Criteria for Division of Work. The boundaries between the activities of different actors, groups, or teams will be chosen during a work situation according to various criteria that result in different cooperative structures. Since control requirements are defined with reference to particular work scenarios and situations, this architecture will change dynamically over time. Within a given system of work, several different, often competing and conflicting criteria can exist and be determining factors for the decomposition with respect to the individual actor's functions.

In every cooperative activity, the informal allocation will reflect an immediate effort to achieve a balance between current requirements and the resources available. Consequently, the cooperative structure can become very dynamic. Also over longer periods of time, changes will be made as knowledge and skills are acquired "on the job."

In the following paragraphs, some illustrating criteria for the division of work will be presented. The list is only meant to be illustrative and should be revised in connection with a detailed analysis of a particular case.

Norms and Practice. In stable systems with a long prehistory, the formal role configuration is often closely related to the actual and frequently hierarchical organizational structure and the corresponding social status rankings. Frequently, this formal structure poses very strict constraints on the actual work allocation, in particular when strict boundaries between professions are established through union agreements. These kinds of fixed criteria regarding "who may do what" can lead to very unproductive limitations in periods with rapid technological developments and this, in turn, can have consequences for the work domain's control conditions. In the case of rapid innovations with their effects on a system's functional character and its technological basis, these implications need to be carefully studied.

For a discussion of the requirements when changes in the system environment and the technological basis have to be considered, some additional functional criteria are relevant.

Load-Sharing. Frequently, a division of work is made from considerations of work load—both formally during work planning and informally and dynamically during the work process itself.

Functional Decoupling. This criterion will serve to minimize the necessary exchange of information among actors. From this, one can identify separate sets of work functions that can be controlled as units with a minimum of interaction and, therefore, effectively allocated to actors or teams.

Competency. The competence required for different decision tasks often determines the division of work. When basic education and profession are considered, this criteria is closely related to formal, union boundary criteria but over time, know-how and practice evolve, and the boundaries between formal competency erodes. Nurses, for instance, may take over activities from medical doctors in many work situations before the formal rules are actually changed.

Information Access. In stable, preplanned work domains, the boundaries around the work segment under the control of an actor and the segment from which information is accessible are more or less the same. For a dynamic work domain, this is not the case. The potential for discretionary problem solving depends heavily on the width of the information window available to the decision maker (see the wide boundaries of the attention envelopes of the foreman in a flexible manufacturing plant, Fig. 3.5.). A horizontally wide window is necessary for dynamic coordination and a vertically wide window is necessary for selecting proper means–ends relations. The effectiveness and the ease with which practical reasoning can serve problem solving depend very much on the access not only to factual information but also to intentional information. This is necessary in order to be able to draw on analogies, to judge whether solutions are reasonable, and to interpret ambiguous messages from other actors on the basis of a perception of their motives and intentions. All of this depends greatly on having access to information from several levels of the means–ends hierarchy (see Rasmussen, 1985). This alone is a conclusive argument against the effectiveness of hierarchical task allocation in a system that has to survive in a changing environment and to adapt to rapid internal technological changes (Drucker, 1988).

Safety and Reliability. For work in a domain posing a hazard to the staff, the investment, or the environment, the criterion of redundancy is often used to plan cooperation. In this case, critical functions are allocated to more than one individual or team so that different individuals independently test or verify the performance in a particular function.

Team Structure Paced by Task Requirements: An Example.
When the cooperative architecture and the boundaries around actor roles have been identified with reference to the work domain, the communication required among actors can be identified from an analysis of the control requirements of the work functions. An explicit analysis of the internal work system coupling is important because of its considerable influence on the amount and types of communication among the actors involved in the work system and, therefore, on the actual work organization. It is important to remember that several configuration architectures will compete, and the governing model will change with time and work conditions.

The need for a team to dynamically restructure in response to changes within the work system depends on the type of work domain (i.e., its horizontal location on the map of Fig. 2.9). For tightly coupled technical system, such as process or power plants, the team restructuring can be paced forcefully by the system. For instance, consider a power plant in normal, automatic operation with a control room staff comprising three individuals. The elements of the plant are effectively coupled and coordinated automatically and monitored by the alarm system. All measured variables of the system are relatively closely correlated and one operator can monitor the production with reference to current demands while the rest of the staff is busy with logging, reporting, and other administrative activities. In addition, during this period an important role of the members of the control room staff is to monitor the activities of the test and maintenance teams working on the plant. This monitoring includes authorizing reconfigurations of the system for test, calibration, and repair.

If, however, a fault occurs and the automatic coordination of plant functions becomes ineffective, the system breaks down into separate functional sections that each need the attention of an operator to reestablish a safe operational state. Thus the group has to dynamically restructure itself into a distributed decision team according to one of the criteria discussed previously. When a safe state has been established, the team will again restructure, possibly into a kind of collective decision making in order to identify the cause of the disturbance. In this way, the division of work, the communication network necessary for coordination, and the management structure is paced dynamically by the work system. The communication channels include direct verbal communication, nonverbal communication of intention and focus of attention (skilled operators in a traditional control room are able to see the panel locations of attention and actions of other staff members' activities), and communication from the systems themselves in terms of their responses to the other crew members' control actions.

In other cases, changes of cooperative structures over a longer time span are caused by conflicts among criteria. For example, as discussed in detail in Chapter 6, a conflict may arise during stressed periods because a criterion by which work load can better be distributed takes over control from a formal criterion intended to ensure safety of work. The result is an increased risk for an accident. In fact, organizational adaptations to competitive pressure and the resulting conflicts among cooperative coordination criteria actually appear to be important contributors to accidents (see the discussion of the influence of organizational factors on accident causation in Chapter 6).

Task Design

In the previous discussion, the activities have been considered from the functional work requirement point of view. However, it is also important to structure the description as seen from the individual. A particular individual will, at any point in time, attend in a time sharing mode to different activities that are not necessarily functionally related. This set of activities, which may only be related haphazardly in time and place, will together shape the actual task of the person, determine the work load, and influence the actual division of labor. As a result, the division of work will, in addition to the criteria discussed above, depend on factors external to an activity's functional structure. This result should be explicitly considered in any work analysis.

Job Design

Another aspect that has not been given much weight in the current discussion is the aggregation of activities across time for a particular individual in order to determine the total job content as the basis for a proper design of job structures as well as training and education schemes. As important as this may be, it is considered to be outside the scope of the presented treatment of the design of information systems.

SOCIAL ORGANIZATION: MANAGEMENT STYLE

The functional work organization has been represented from the perspective of the work system as a purposive instrument performing a set of functions within its environment. The social organization will be studied from the perspective of the work system as a complex of social interactions between individuals or coalitions of individuals with diverging–discordant interests and motives. The social organization perspective essentially corresponds to the "natural system" perspective on organizations as defined by Scott (1987). In recent years, this perspective has been explored and elaborated in a large number of organizational studies and approaches.

As discussed in the previous sections, the functional work organization is determined by the control requirements of the work domain and by the criteria for setting boundaries adopted by the staff. The required coordination of the cooperating individuals following from this architecture will specify the *contents* of the communication. On the other hand, the architecture of the social management will determine the *form* of the communication (i.e., the conventions chosen for social interaction). Various structures of social organization are possible and they are more or less independent of the task and the role configuration principle adopted, as well as the characteristics of the work domain.

Architecture of Social Interaction

The different patterns for the social organization adopted depend on the social style and values of the persons involved and on the management style installed from the top (as well as cultural ties). These social constraints on the form of communication reflect back on the cooperative structure and influence the flexibility and speed of response of a work organization. In the following paragraphs, a number of organizational architectures are briefly described.

Autocracy. In principle, one decision maker can be responsible for the coordination of the activities of all other actors in a team or organization. In a work team, one individual may take the initiative and control the social interaction. In an organization, this mode of meta-control of coordination is, of course, most applicable when one person controls the mode of coordination in a smaller, private company. That is, one manager acts as a roving coordinator and ensures that the priorities are appropriate in the working network. Also, in a group or team, a dominating character may take the lead in the cooperation within the constraints given in the task situation.

Authority Related Hierarchy. Frequently, the control of coordination is stratified in a team or an organization such that one level of decision makers evaluates and plans the activities at the next lower level. This hierarchical structure has a predecessor in the American military command, control, and coordination paradigm. It is based on the fact that, in a large organization, the information traffic and the time requirements regarding response lags can be arranged in several distinct levels—this is the case for a system of very specialized actors, such as those involved in a military mission. Different coordination, or management styles, are possible within this structure depending on whether the communication downward through the system is based on the communication of specific goals (the military model), on the communication of procedures (the bureaucratic model), or on passing down overall objectives for local interpretation (the adaptive model).

In modern organizations, the hierarchical structure is normally maintained for formal purposes, such as allocating legal responsibility and financial competence, whereas more flexible (and often informal) management structures are accepted for the more dynamic work coordination. This can also be the case in military organizations, where the formal rank structure is abandoned if a high-tempo coordination of activities across levels and units is required.

Function Related Heterarchy. In a general, dynamic work situation, the formal hierarchical structure disappears, even in military organizations. The work of Rochlin et al. (1987) demonstrates the pronounced ability of the organization on an aircraft carrier to shift between (1) a formal rank organization, (2) a flexible, self-organizing "high-tempo" work coordination across ranks and organizational units, and (3) a flexible emergency organization responsive to the immediate requirements of the actual situation. Similar organizational forms can become necessary in modern commercial enterprises that wish to adopt a flexible customer driven policy in a dynamic and competitive market (see the various discussions on this elsewhere in this volume).

For the coordination of activities in these flexible organizations, several meta-strategies can be adopted to shape the coordinating activities.

Negotiation. The individual decision makers negotiate with their immediate cooperators and the necessary communication is planned locally. This mode is typically found when coordination planning has to be arranged ad hoc, that is, when high-tempo performance is required in a very flexible and dynamic situation for which particular extemporaneous patterns of "contracts" evolve (Flores et al. 1988).

A similar mode of coordination has been called debative cooperation. In complex decision tasks, one individual decision maker may have neither the necessary conceptual background nor the required information on the state of affairs. In addition, conflicting goals and criteria may have to be resolved. The cooperation, therefore, involves debate and negotiation among actors with different heuristics and perspectives.

Anarchy. Each actor plans their own activity without interaction with other decision makers on the meta-level. Communication is entirely through the work content. This mode is sometimes seen in academic research departments.

Democracy. Coordination involves interaction and negotiation among all members of a team or an organization (worker participation committees). This mode is frequently found in companies when there is a reasonable time horizon for planning and where special meetings and frameworks for the planning can be arranged.

Generally, it should be evident that these different architectures imply, or evolve from, different forms of communication among actors (i.e., whether information is passed as neutral facts, advice, instructions, or orders). The effective way of influencing the social organization independently of the work organization will be through constraints and conventions for communication.

Team Structures Shaped by Social Style: An Example. As mentioned previously, the structure of coordination within a team operating a technical system (e.g., a process plant) will normally be forced by the internal coupling and control requirements of the system. In less constrained cooperative task situations, team cooperation will be influenced more by the social dynamics within the team, but still the cooperative structure can change dynamically depending on the characteristics of the different phases of a task. An illustrative example is described by Sonnenwald (1993) who analyzed the cooperation within various industrial design teams. Sonnenwald identified several characteristic phases of a design project (similar to the phases discussed in Chapter 7) and found characteristic social patterns of cooperation for each phase. For each, different social roles were adopted by team members, some similar to those guided by the criteria for division of work discussed above and related to the domain and interests they represent (consumer, technician, backer), some related to their social role and initiatives in the form of communication (facilitating actor, intergroup star, intragroup star, or gate keeper).

For example, in design situations that took place in less constrained environments (where team members could structure their patterns of behavior somewhat independently of the environment) a "facilitating agent" emerged. The facilitating agent fostered communication among team members and helped them negotiate conflict. This enabled the team to complete tasks that required a consensus among team members who had different, and conflicting, perspectives. In this way, a social role, a facilitating agent, aided task completion.

Concluding Remarks

To sum up, a work organization is considered to be a distributed control system serving the coordination and control of a loosely coupled work system. The cooperative pattern evolves in the interaction of the control requirements of the work domain from below specifying the necessary coordination and the social management style propagating from above, as illustrated Figure 4.1. The figure illustrates four actors or decision makers each of whom is allocated particular, but overlapping,

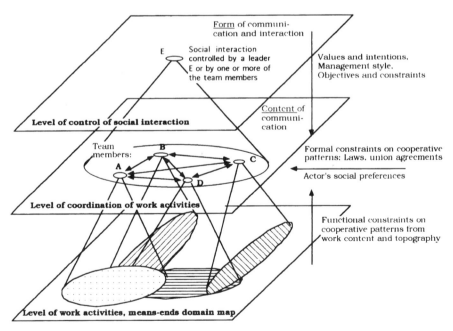

FIGURE 4.1. Work organizations as emerging in the interaction between the bottom-up propagating control requirements of the work domain and the top-down propagation of social practice and management style.

"activity windows" giving access to a part of the overall work domain. The concerted action within this, normally rather loosely coupled, work domain requires coordination and communication among actors. The *structure* of the communication net and *content* of the communication and, therefore, the functional work organization, are determined by the control requirements of the work domain. In this way, the information content of the data bases of a support system and the necessary retrieval attributes are identified by analyzing the control requirements of the work domain. This is the objective driving a cognitive work analysis.

The social organization, in contrast, is determined by the conventions chosen for the *form* of the communication and this depends on the "management school" or "culture." The management culture, in turn determines whether the decision making at the level of social control involves another actor E or one particular actor of the group ABCD in a hierarchical organization, or all of them in a democratic coordinated process. In other words, the locus of control, the coordinating initiative, and the communication format are determined by an analysis of the social norms and practices of the organization. This falls in the domains of cognitive sociology and Computer Systems for Cooperative Work (CSCW) research.

COGNITIVE RESOURCES AND PREFERENCES OF THE ACTORS AND USERS

In the previous sections, the work organization and the allocation of functions to the individual members of the work staff have been described. The analyses in the following section serve to identify the characteristics of the individual actors, their levels of expertise, their competence, and the subjective preferences and performance criteria that they may bring into play. The aim is an identification of the cognitive control structure and resources that can be applied to the task, depending on background and expertise.

Demographic Characteristics of the Actors and Users

Very different categories of system operators and users are found in the different systems. We will see that basically different categories of people are involved and, furthermore, by considering the different categories of work domains illustrated in Figure 2.9, that the categories to consider are correlated with the location within the continuum of the work domain characteristics (see Fig. 4.2).

At the far right end of the spectrum we find tightly coupled systems, such as industrial process plants, and nuclear power plants which are served by dedicated operators. Depending on the particular system, their complexity, and the hazards involved, operators are graduates from technical universities, skilled professionals, or unskilled workers, all subject to specialized education and training and, for some systems, subject to formal authorization before released for operation. Training is based on elaborate systems of textbooks, manuals, and working instructions and often, for instance, in nuclear power and aviation, high fidelity training simulators are used for training and certification. Furthermore, operators serve the same system or

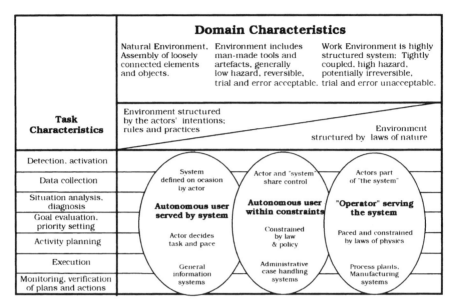

FIGURE 4.2. The figure illustrates the correlation between different categories of system operators–users and the work domain characteristics.

kind of system for very long periods of time, and they have the opportunity to develop very extensive know-how and expertise.

As described in Chapter 2, the intentionality of the work domain is normally determined during system design and operation planning. The personal goals and preferences of the individual is of less importance. However, the ability to handle critical situations is important, and, for this, specialized medical and psychological selection tests are in use. The design of this kind of system has been discussed in some detail elsewhere (Goodstein and Rasmussen, 1988; Rasmussen and Goodstein, 1988).

To sum up, for this category of work systems operators are carefully selected and trained to have the cognitive resources required by the system.

At the opposite end of the spectrum, the systems are designed to serve causal users. Consequently, the population to be considered for system design consists of a very wide set of categories with respect to basic education. In this case, systems have to be designed to match those characteristics of users that are relevant for the functions the system is intended to support. This requires a careful analysis of the intended user groups and their subjective needs and preferences.

In the middle part of the spectrum, identification of user characteristics is a complex problem. We are dealing with systems that, on one hand, require certain functions attended by system users with a particular educational background but, on the other hand, leave many degrees of freedom open to the discretionary choice of the individual. In this case, system design is a very complex effort to match system characteristics to the characteristics and preferences of certain categories of staff members and the users they serve. The library system discussed in Chapter 9 is a typical example of this kind of system.

It is important to realize that in any work organization, activities are found that relate to several locations in the map of Figure 4.2. Administrative work is found in power plants served by clerks and in offices, while tightly coupled technical systems (such as complex reproduction equipment) may call for dedicated operators. Therefore, the demographic characteristics must be looked upon for each prototypical work situation or function.

Modes of Cognitive Control of Activities

Humans have different modes of controlling their interaction with the environment and the cognitive resources as well as the subjective preferences, which depend very much on a proper match between the features of the work domain and the requirements of the various cognitive control modes (Rasmussen, 1983).

In the present framework, the cognitive control of human activities is considered to be the function of a hierarchically organized control system, that is, a control function organized in several levels of perception, action, and information processing.

The lowest level performs a continuous control of the movements required by the interaction with a work environment, directly and through a computer system. This is called the skill-based level and, technically, corresponds to a real time, continuous control system. The next higher level of control is taking care of the organization of the routine patterns of movements (i.e., the individual acts) into the proper procedural sequences. In technical terms, this constitutes a sign-driven sequence controller. This is the rule-based control level. Finally, the highest level is concerned with the generation of plans to be used by the sequence controller for new situations. This is the knowledge-based, problem solving level.

In control technical terms, the levels can be described as the continuous, signal processing level; the adaptive, information processing level, and the self-organizing, problem solving level (cf. Mesarovic, 1970).

Skill-Based Control. Interaction at the lowest level, the sensory-motor skill level, is based on a real-time, multivariable, and synchronous coordination of physical movements with a dynamic environment. (see Fig. 4.3). Quantitative, time–space signals continuously control movements, and cues are perceived to be signs adjusting the internal world model and maintaining its synchronism with the environment. The dynamic control of the movement patterns depends on high capacity signal processing governed by the internal world model. High speed performance is possible due to an open-loop, feed-forward generation of movements. During the run-off of sensori-motor routines, the conscious mind is free to attend to other matters on a time sharing basis.

The flexibility of skilled performance is due to human abilities to compose and adjust from a large repertoire of prototypical movement patterns the sets suited for specific purposes. The particular patterns are activated and chained by cues perceived as signs and no choosing among alternatives is required. During skill-based perfor-

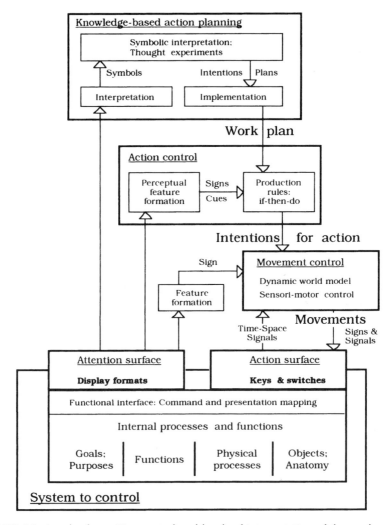

FIGURE 4.3. Levels of cognitive control and levels of interpretation of the work system content.

mance, higher level control may be active, consciously anticipating upcoming demands, and thus updating the state of the dynamic world model, preparing it for the proper response when the time comes.

This anticipatory updating of the internal model is equivalent to Gibson's (1966) atunement of the neural system underlying the direct perception of invariants of the environment in terms of "affordances." Performance at this level is typical of the master, the expert, whose smoothness and harmony have fascinated philosophers and artists throughout the ages.

To sum up, at this level control of activities require on-line, real time control based on tacit knowledge that cannot be described by the actor. It depends on interaction with the temporal–spatial configuration of objects that can be real material objects or configural representations of concepts.

Rule-Based Control. At the rule-based level, conscious attention may run ahead of the skilled performance, preparing rules for coming requirements. It may be necessary to memorize rules, to rehearse their application, and to update more general rules with the details of the present environment. Stored rules will frequently be formulated at a general level and, therefore, will need to be reformulated and supplemented with details from the current physical context. In other cases, rules are not ready, and successful coping with a similar situation in the past is memorized to establish a transfer. In general, control at the rule-based level requires a conscious preparation of the sequence beforehand, or else a break in the smooth performance will take place. The conscious mind operates infrequently in synchronism with interaction with the environment. Attention will look ahead to identify the need for rules for actions in the near future, and it will look back to recollect useful rules from past encounters. The composition of a sequence of stored rules can be derived empirically from previous occasions, or instructed by another person (e.g., a work procedure or cookbook recipe). If this is not the case, rules must be prepared on occasion by conscious problem solving and planning based on declarative knowledge (i.e., a functional model of the environment).

To sum up, at this sequence control level, generation of proper organization of patterns of movement (i.e., acts) into plans depend on access to stored rules and to experience from past work scenarios. The planning is done ahead of the action, that is, it is not synchronized with the interaction and is based on recall of past experiences and imagination of future encounters. It depends on the availability of convenient cues to release acts, cues that are only conventional signs with no functional significance.

Knowledge-Based Control. During unfamiliar situations for which no know-how or rule for control is available from previous encounters, control must move to a higher conceptual level in which performance is goal-controlled and knowledge-based. In this case knowledge is taken in a rather restricted sense as the possession of a conceptual, structural model or, in Artificial Intelligence (AI) terminology, of deep knowledge. Therefore, the level might also be called "model based." In this situation, the goal is explicitly formulated and based on an analysis of the environment and the overall aims of the person. Then a useful plan is developed—by selection. Different plans are considered and their effects tested against the goal— either physically by trial and error or conceptually by means of "thought experiments." At this level of functional reasoning, the internal structure of the system is explicitly represented by a "mental model" that may take on several different forms. A major task in knowledge-based action planning is to transfer those properties of the environment that are related to the perceived problem to a proper symbolic representation, since the information observed in the environment is perceived as "symbols" with reference to the mental model being utilized.

The means–ends scheme is a very important framework to represent "mental models" required for knowledge-based reasoning in that it represents the entire repertoire of means and ends (i.e., objectives, priorities, functions, processes, people, and parts) from which the relevant subset has to be drawn for a particular situation. In other words, it maps the field or "territory" in which an actor (decision maker) has to navigate in order to comply with their work requirements. It is important to emphasize that in a dynamic environment, the opportunity for an effective exploration of options in the means–ends space is very important and should be effectively supported by any information system. In Chapter 5, new approaches to make these options visible within the means–ends structure are discussed in detail.

The role of a multilevel abstraction hierarchy in problem solving is very clearly seen in Duncker's (1945) research on practical problem solving related to physical, causal systems (radioactive tumor treatment and functioning of a temperature-compensated pendulum). Based on verbal protocols, Duncker describes how subjects go from the problem to a solution by a sequence of considerations where the items proposed can be characterized by a "functional value" feature pointing upwards to the problem and a "by means of which" feature pointing downwards to the implementation of a solution. The relation to the means–ends hierarchy is clear.

Another observation on the role of an abstraction hierarchy that was used in understanding a mechanical device has been reported by Rubin (1920), who recounts an analysis of his own efforts to understand the functioning of the mechanical shutter of a camera. He finds that a consideration of purpose or reason plays a major role in the course of arguments. He conceived all the elements of the shutter in light of their function as a whole. He did not perceive the task to be an explanation of how the individual parts worked but rather what their overall functions were. How they worked was immediately clear when their function was known. He mentions that he finds it an analytical task to identify the function of parts, the direction of thought being from overall purpose to the individual function (top-down considerations). The hypothesis necessary to control the direction is then readily available. This approach was found to have additional advantages: solutions of subproblems have their place in the whole picture, and it is immediately possible to judge whether a solution is correct or not. In contrast, arguing from the parts to the "way they work" is much more difficult because of being a synthesis. In this approach, solutions of subproblems must be remembered in isolation and their correctness is not immediately apparent.

Indeed, having access to information at several levels of abstraction along the means–ends dimension is important for effective problem solving. Shifts in the level of abstraction during problem solving have proved to support very useful heuristics, as has been convincingly demonstrated by the experiments of Wason and Johnson-Laird (1972). Such shifts are the basis for analogical reasoning (Rasmussen, 1986).

To sum up, at this level, control is based on a symbolic, mental model representing the deep, internal sources of regularity of the behavior of the work environment, and information is interpreted symbolically with reference to this model.

Shifts Between Levels of Control. One important feature of this complex interaction is the incessant changes in the control and its distribution over the different levels that take place as high skill evolves. Control moves from level to level and the complexity of behavioral patterns, rules, and models within levels will grow with training.

Such shifts between the levels of cognitive control occur when less familiar situations are met. Then, concurrent shifts are necessary in the internal control structure from procedural to declarative knowledge and in the interpretation of information from stereotype signs to symbols. Such shifts are difficult to initiate (cf. the psychological research on functional fixation; Luchins, 1942). In addition, emotional aspects add to the difficulties. Lanir, Fischoff, and Johnson (1988) recently pointed to this fact in a discussion of the "boldness" required to quit a familiar rule set or a standing order and to enter a rule-breaking problem solving mode in unusual situations.

To sum up, tasks are frequently analyzed in terms of sequences of separate acts. In general, however, several functions are active at the same time (see Fig. 4.4), and at the level of skilled sensori-motor control, activity is more like a continuous, dynamic interaction with the environment. Attention, on the other hand, is scanning across time and activities in order to analyze past performance, monitor current activity, and plan for foreseeable future requirements. In this way, the internal, dynamic world model is prepared for oncoming demands and the related cues and rules are rehearsed and modified to match predicted requirements, and/or symbolic reasoning is used to understand responses from the environment and to prepare rules for foreseeable but unfamiliar situations. Attention may not always be focused on current activities, and different levels may simultaneously be involved in the control

Mode of cognitive control	Mental functions	Temporal characteristics	Related real world
Knowledge based	Planning in terms of functional reasoning by means of symbolic model.	Achronic, that is, temporal scale is not maintained in causal reasoning	As can be
Rule based	Planning in terms of recall of past and rehearsal of future, predicted scenarios.	Diachronic, that is, temporal scale is maintained but not synchronized	As has been and may be
	Attention on cue classification and choice of action alternatives	Synchronic, that is, operation in the actual time slot, but not synchronous	As is
Skill based	Data-driven chaining of sub-routines with interrupt to conscious, rule-based choice in case of ambiguity or deviation from current state of the internal world model.	Synchronous with real world, operation in 'real time'	As is

FIGURE 4.4. The interaction between the levels of cognitive control mode.

of different tasks, related to different time slots, in a time sharing or in a parallel processing mode.

Evolution of Expertise. When expertise evolves, cognitive control shifts levels. An important point is that the behavioral patterns of the higher cognitive levels do not become automated skills. Automated patterns of movements evolve while the activity is controlled and supervised from the higher levels. When a state of expertise is attained, the basis for the higher level control (i.e., conscious declarative and procedural knowledge) very likely will deteriorate. In fact, the period when this is happening may be error prone due to interferences between a not fully developed sensori-motor skill and a gradually deteriorating rule system. This kind of interference is familiar to highly skilled musicians when they occasionally start to analyze their performance during fast passages and run into trouble. It also seems plausible that this effect can play a role for pilots with about 100-h of flying experience, which is known to be an error-prone period.

Following the lines of reasoning suggested above, the transfer of control to new mental representations is a very complex process involving changes along several different orthogonal dimensions. When trained responses evolve, the structure of the underlying representation shifts from a set of separate component models toward a more holistic representation. This was discussed by Bartlett (1943) in relation to pilot fatigue. Moray (1987) found that model aggregation can lead process operators into trouble during plant disturbances because the process is irreversible; that is, the regeneration of partial models needed for causal reasoning in unfamiliar situations is not possible from the aggregated model. The learning model implied in the skill, rule, and knowledge framework indicates that skill acquisition involves more than an aggregation of mental models. Typically, control by a structural, functional model will also be replaced by empirical procedural knowledge concurrently with a shift from a symbolic to a stereotypical sign interpretation of observations. Thus the acquisition of expertise involves at least three concurrent and structurally independent changes—in terms of model aggregation, in terms of shifts from declarative to procedural knowledge, and in terms of a "symbol to sign" shift in the interpretation of information. These qualitative changes in the level and mode of control reflects itself in the errors made during the process.

Cognitive Control and Natural Decision Making

The distinction between "cognitive control" as opposed to "decision making" is ambiguous and basically a question of the context. When considering an actor's cognitive planning and control of actions in a dynamic work context, the cognitive control perspective is natural. When coming from the study of decision making for separately formulated problems, the decision theoretic perspective is natural. However, the different types of decision making form a continuum and reflections about distinctions, similar to those made in the SRK model are often found when studying "dynamic decision making" (Brehmer, 1992) or "naturalistic decision making"

(Klein, 1989), which has to do with decision making as a continuous process, integrated in the interaction with the environment.

In normal work contexts, decision making is not usually an effort to resolve separate conflicts, but is more like a continuous activity to control a continuously changing state of affairs in the work environment. In other words, decision making is embedded in a more general cognitive control and depends very much on tacit knowledge related to skilled actors' intuitive reactions in a familiar context.

In hospitals, we observed a kind of "naturalistic" decision making in very familiar contexts, which is closely related to the skill-based control of actions. Operation theater planning is done during planning conferences or meetings including doctors and nurses. A typical feature of the hospital system seems to be a kind of collective memory. No one individual has available all the relevant information about the individual patients, but a kind of collective memory stores this information. When the treatment of an individual patient is planned, the context from previous considerations defines an elaborate background knowledge. If, at a meeting, an action is proposed that is not supported by the knowledge possessed by a member of the group, this will be voiced promptly. If the situation is ambiguous, a member of the group will very likely offer comments updating the context. This goes on until the context is properly established and a decision can be taken by the surgeon in charge without any alternatives being explicitly mentioned or negotiated.

Similarly, Amalberti et al. (1992) found the level of expertise of fighter pilots to depend on the size of the repertoire of possible flight mission scenarios they anticipated during prebriefing and for which they prepared proper cue–action sets.

In other words, decisions emerge when the landscape (the representation of work context) is well-enough shaped so as to let the water (behavior) flow in only one direction. One important aspect of this cooperative conditioning mode of decision making is the built-in redundancy. Several persons with different perspectives on the patient's situation accept the result of the negotiation. Another important aspect in this evolutionary aggregation of the context underlying the conclusion is that information is provided for resolving ambiguities that could not be retrieved from a data base by an explicit question because the question could not be phrased. In addition, this important piece of information would not be rendered outside the actual face-to-face encounter (e.g., entered into a data base) because only the specific context makes it worth mentioning.

When this intuitive reaction to the context is no longer effective, a mismatch can be experienced between the state of affairs in the environment and the intuitive expectations of the actor. In this case, a number of alternatives for action will normally be perceived by a skilled professional, and the efforts will be focused on searching for information that can resolve the ambiguity. The important point to consider here is that an expert will need no more information than is necessary to resolve the choice between the perceived action alternatives. This mode has several important implications for interface design. One implication is that decision makers are not subject to "information input" from the environment that has to be analyzed. Instead, they ask the environment very specific questions. They are, in a Gibsonian sense, able to consult "invariants" in the environment by direct perception if the environment is well structured and transparent.

Direct Perception and Cue–Action Hierarchies

In Chapter 3, on heuristic decision making, the performance in a familiar environment is represented by a cue–response hierarchy, based on human abilities to read-off cues at several levels of abstraction in the means–ends representation. This point of view relates closely to Gibson's theory of direct perception as being a basic precondition for effective interaction with a natural environment.

Gibson on Affordances. The challenge that Gibson (1979, 1970, 1966, 1950) took up was to try to explain how perception of the natural environment is possible. One of the key notions that he developed was that of an *affordance* (Gibson, 1979). Unlike some traditional approaches, which assumed that perception is a reconstructive process based on elemental sensations, Gibson proposed that higher order properties of the environment can be perceived directly through a phenomenon referred to as resonance (Gibson, 1979, 1966). In this way, the affordances of the environment—the possibilities that it offers the organism—can be directly perceived by the atunement of the perceptual system to the invariant structure in the ambient optic array. Gibson claimed that in normal cases, the meaning and value of the objects are directly perceived, and not just the individual characteristics of these objects.

It is useful to consider in greater detail the hierarchy of levels at which direct perception of the environment can take place. Figure 4.5 illustrates how some of the

Value Properties: Purpose, Goal			
Survival	Pleasure	Altruism	
Priorities: Abstract Functions			
Reward	Danger	Nutrition	Manufacture
Cooperation	Nurturing	Copulation	Privacy
Comfort	Pain		
Context: General Functions			
Communicating	Warmth	Drinking	Eating
Washing	Bathing	Injury	Support
Fighting	Shelter	Aiding	Punishment
Locomotion			
Movement: Physical Processes			
Sit-on	Bump-into	Fall-off	Get-underneath
Climb-on	Sink-into	Swim-over	Walk-on
Stand-on	Grasp-able	Barrier	Obstacle
Breathing	Pouring	Cutting	Lifting
Throwing	Piercing	Carrying	
Objects and Background: Physical Form			
Layouts	Objects	Surfaces	Substances

FIGURE 4.5. Gibson's levels of affordances.

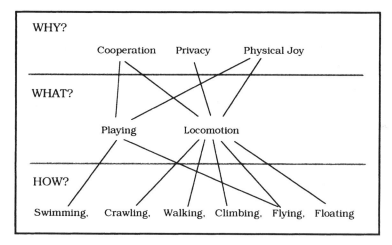

FIGURE 4.6. The many-to-many mapping across levels of abstraction in Gibson's affordances.

examples provided by Gibson (1979) to explain the concept of affordances can be described with the means–ends structure used to represent a work domain. This table emphasizes the fact that the world is actually composed of a hierarchy of affordances, a point that Gibson briefly alludes to (cf. Gibson, 1979, p. 137) but does not address in any great detail. An interesting property of this hierarchy is that the relation between levels is one of means–ends relations. This has important implications for the design of "ecological interfaces" discussed in Chapter 5. Perception of affordances at the various levels plays different roles in the control of human activity. Selection of goals to pursue are related to the perception of value features at the highest level, the planning of activities to perceptions at the middle levels, while the detailed control of movements depends on perceptions at the lowest level of physical objects and background.

In Figure 4.6, several relations between levels are represented in order to show how there is a many-to-many mapping between affordances at various levels. For instance, the possibility of locomotion can be fulfilled by a number of different alternatives: swimming, crawling, walking, and so on. In addition, a single affordance can be relevant to more than one value-related affordance. For example, swimming may be relevant when the situational context affords either playing or locomotion. Similarly, locomotion will be relevant when there is a desire for privacy (to escape from a crowd), or for cooperation (if one's partner is on the other side of the pool). An important implication is that the perception of the situational context in terms of higher order, value-related affordances can help people deal with the complexity of the environment by constraining the number of meaningful lower level affordances that are relevant, given the current context.

Direct perception of an artifact's affordances will generally be possible if the artifact has been adapted to human needs and capabilities. This is accomplished either by the natural evolution of cyclical redesign (Alexander, 1964), or by forethought on the part of the designer. An artifact that is well designed should, through the appropriate use of invariant features make obvious what it is for and how it should be used (cf. Norman, 1988). In other words, its affordances should be easy to perceive.

While Gibson was concerned with the natural environment, the focus of the present discussion is modern work domains. Such domains can be viewed as a man-made ecology. It is argued that the theory of direct perception is equally as applicable in high-technology work domains as it is to the natural environment and to high level problem solving mediated through a computer screen.

Natural and Man-Made Environments. In his discussion on affordances, Gibson (1979) anticipates this issue by addressing the implications of the human alteration of the natural environment (pp. 129–130). In ecological terms, man-made artifacts can be seen as efforts to change and expand the environment's affordances. In changing the substances and shapes of the environment, humans have "made more available what benefits him and less pressing what injures him" (Gibson, 1979, p. 130).

In spite of these changes, Gibson makes it very clear that he does not view this trend as involving the creation of a new kind of environment:

> This is not a new environment—an artificial environment distinct from the natural environment—but the same old environment modified by man. It is a mistake to separate the natural from the artificial as if there were two environments; artifacts have to be manufactured from natural substances (Gibson, 1979, p. 130).

This argument against two qualitatively different types of environments is based on a belief in the primacy of the basic components of the natural environment:

> The fundamentals of the environment—the substances, the medium, and the surfaces—are the same for all animals. No matter how powerful men become they are not going to alter the fact of earth, air, and water. . . . For terrestrial animals like us, the earth and sky are a basic structure on which all lesser structures depend. We cannot change it (Gibson, 1979, p. 130).

Apparently, Gibson did not have high technology work domains in mind, when he presented these arguments; we will return to this issue in Chapter 5, which presents a detailed discussion of the coupling between a work domain and its actors.

Semantic Interpretation and Mental Models

The direct perception–action mode of cognitive control depends on a human actively engaging in a goal-directed activity in contrast to a human passively analyzing and making judgments about a given environmental situation, as is the case in knowledge-based semantic interpretations (e.g., of a work of art). In a flexible and changing work environment, the interaction between direct perception and semantic analysis becomes important. For the design of complex interfaces, therefore, it is necessary to consider the interpretation of information as related both to the syntactic control of actions and to a semantic interpretation, with reference to a mental model of the sources of regularity of the environment, that is, its invariants.

Michaels and Carello (1981) espouse Gibson's ecological approach with great zeal. As they put it, much of the notion of "information-about" is expressible via the concept of invariant. Psychologically, invariants are those higher order patterns of stimulation that underlie perceptual constancies (i.e., the persistent properties of the environment that we are said to know). From the perspective of ecological physics, invariants come from the lawful relations between our activities and the objects, locations, events in the environment and the structure or manner of change in the corresponding patterns of light, sound, and so on. Finally, geometry serves to describe invariants more explicitly. Structures and transformations can both be invariant. Structural invariants are properties that are constant with respect to certain trans-formations, while transformational invariants are those styles of change common to a class of transformations that leave certain structures invariant. Affordances ("in-formation for" in their terminology) are perceived as the acts or behavior permitted by objects, places, and events. Two comments from Gibson (1988) are relevant— "affordances link perception to action" and "learning about affordances entails ex-ploratory activities." The action alternatives we perceive relate to our potentially purposive behavior and will affect the affordances that we will detect. Thus informa-tion about affordances is subjective and situation dependent. All of this illustrates the interplay between perception and action that underlies the ecological approach; indeed, the authors note that "the theory of affordances claims that perceptions are written in the language of actions."

For semantic interpretation, the sources of regularity (the invariants) must be read off with reference to a symbolic mental model. Such a model can be formed at all the levels of abstraction discussed with reference to the means–ends relations in Chapter 2. This complicates the design of interfaces (see Chapter 5). In general, the concept of a mental model is used to characterize features of the resident knowledge base that represents the properties of the task environment to support the planning of activities and the control of acts when instantiated and activated by observations of the actual state of affairs. In the present framework, a distinction is made between the term "mental model" and the more general term "mental representation." Mental *models* here refer to declarative representations of the relational network that conceptualizes the invariant structure of the environment and the constraints governing the regularity of its behavior. In addition, mental *representations* include the implicit representa-tions of such constraints in terms of procedural and episodic knowledge, and tacit background knowledge.

Categories of Mental Representations

The point of view of the following discussion is taken from analyses of performance in complex work situations and from the requirements met when designing computer-based interfaces. For such purposes to be satisfied, it is necessary to study the interactions of a wide variety of mental strategies and models. In particular, a study of the interaction and interference between different modes of cognitive control appear to be important for the understanding of human errors (see Chapter 6).

In this discussion, a typology of mental representations will be proposed. As mentioned earlier, we restrict the term mental model to the representation of the "relational structure" of the environment, following Craik's definition of a mental "model" (Craik, 1943). This means that the mental model is a representation of the fundamental constraints determining the possible behavior of the environment. Therefore, it is useful in anticipating the response of the environment to acts or events when instantiated by information about the current boundary conditions. The study of errors has made it clear that a taxonomy of representations should not only consider the higher level knowledge-based functions related to inference and reasoning, since the role of the body in the control of sensori-motor performance is also an integrated part of the system. In the following section, the representations related to actual working performance according to the skill, rule, and knowledge distinctions are briefly reviewed.

Representations at the Skill-Based Level. Performance at the skill-based level depends on a dynamic world model which, like Johnson-Laird's mental model (Johnson-Laird, 1983), has a perceptual basis. The model is activated by patterns of sensory data acting as signs, and synchronized by spatio-temporal signals.

This dynamic world model can be seen as a structure of hierarchically defined representations of objects and their behavior in a variety of familiar scenarios (i.e., their functional properties, what they can be used for, their potentials for interaction, or what can be done to them). These elements of a generic analog simulation of the behavior of the environment are updated and aligned according to sensory information acquired while interacting with the environment. The model is structured with reference to the space in which the person acts and is controlled by direct perception of the features of relevance to the person's immediate needs and goals.

This conception is similar to Minsky's (1975) "frames." The main—and fundamental—difference is that Minsky's frames depend on a sequential scene analysis; they are structured as networks of nodes and relations, and they are basically static. Minsky defines frames as a data structure for representing stereotyped situations that are organized as a network of nodes and relations. On the other hand, Gibson's (1966) ideas on "direct perception" are far more convincing, when viewed as a model of the high capacity information-processing mechanisms underlying perception and sensori-motor performance in fast sequences, than is Minsky's symbolic information processing version. The latter is more adequate for higher level information processing based on the manifestations of the "dynamic world model" at the conscious level in terms of natural language representations.

The "dynamic world model" in the present context is very similar to the mechanisms needed for the "atunement of the whole retino-neuro-muscular system to invariant information" (Gibson, 1966, p. 262), which leads to the situation where "the centers of the nervous system, including the brain, resonate to information." (For more detail, see Rasmussen, 1986.)

Representations at the Rule-Based Level. At the rule-based level, system properties are only implicitly represented in the empirical mapping of cue–patterns representing states of the environment or actions or activities relevant in the specific context supplied by the underlying dynamic world model. According to the definition adopted here, this representation does not qualify as a mental "model" since it does not support the anticipation of responses to new acts or events that have not been previously encountered, and it will not support an explanation or understanding except in the form of references to prior experience.

However, in order to prepare for rule-based control of activities, conceptual relations may be important. Descriptive relations are useful in assigning attributes to categories and, therefore, in labeling scenarios and contexts for the identification of items to retrieve from memory. Descriptive labels are the basis for updating the focus for intuitive judgments and for establishing the proper "background" for action and communication. As could be expected from research on memory (Bartlett, 1932; Tulving, 1983), episodic relations are important for the structuring of memory. Episodic relations appear to be important for labeling prototypical situations to serve as tacit "frames" or contexts for intuitive judgments and skilled performance.

Representations at the Knowledge-Based Level. In the present context, representations at the knowledge-based level constitute proper "mental models" of the relational structure of the causal environment and work content. Many different kinds of relationships are put to work during reasoning and inference at this level—all depending on the circumstances (e.g., whether the task is to diagnose a new situation, to evaluate different aspects of possible goals, or to plan appropriate actions).

Mental Models in Action

When we refer back to the discussions of mental strategies, we will realize that the different strategy categories relate directly to the three levels of cognitive control and, therefore, are related to the corresponding types of mental representations.

Strategies depending on recognition and intuitive classification are based on skill-based control and, therefore, on parallel processing and feature identification (recognition primed decision). Empirical strategies based on decision table look-up depend on rules and cues for their selection. Finally, analytical strategies based on conceptual situation analysis and planning depend on symbolic mental models of system functions. Therefore, mental models applied for cognitive control during an actual work situation will be a kind of a semantic network that evolves and changes as the situation progresses. Since shifts among strategies are frequent, the mental model used will consist of many fragments replacing each other repeatedly. The items of work that are the objects of reasoning are selected from the means–ends work space representation. However, the properties of the items retrieved from long-term memory into working memory depend on the immediate line of reasoning while the resulting semantic net and the path chosen depend on the particular task circumstances and the actor's performance criteria. Consequently, a stable representation of the mental model of a particular task and strategy cannot be found. It exists only in the form of a work space representation of potential means–ends relations and a representation of the set of relevant properties of each item. The actual mental model can then be generated as the strategy in a particular situation unfolds.

In addition to the part–whole and means–end relations, a number of other conceptual relations will be useful for operating on a problem representation in the knowledge-based domain. When means for action has been chosen from perceived means–end relations in a particular work context, causal relations are used to judge the effects of the actions. Value aspects are important for choice and for assignments of priority in decision situations when the constraints given by goal specifications and functional requirements leave some freedom for optimizing, for instance, cost, reliability, effort required, and emotional aspects. A choice among possible strategies in a work situation will depend on performance criteria (i.e., value aspects assigned to the work process, as well as its product). In addition, generic relations that define a concept as a member of a set or category in a classical Aristotelian classification can be used to label a part of the environment and assign it to a category for which functional properties are readily available. The generic relations are, in particular, useful for drawing formal logical inferences (syllogistic reasoning).

The complexity in concepts and representational forms, which are involved in the creation of the mental model during an actual task, is best illustrated by the discussion of the fields that frame the context of reasoning in an actual task (see Fig. 3.10).

For system design, the problem of relating the information presentation to the mental models of the system operators and users depends very much on the kind of system considered. For tightly coupled, technical systems where operators are serving system needs and have to cope with rare events, the mental model required for control must be identified by studying the system processes and their causal structure. From here, interface formats have to be designed that will lead the operators to adopt the proper mental model. In contrast, for systems designed to serve the casual user, interfaces have to be designed to match the mental models the users have acquired by their activities in their normal context (see Figs. 7.11 and 7.12).

HINTS FOR ANALYSIS OF WORK ORGANIZATION AND USER CHARACTERISTICS

Analysis of Functional Work Organization

The material collected from the analysis of the prototypical work situations must be structured to give an overview of the persons participating in the various situations, as well as the communication paths between them. For each of these channels, the information content and form should be identified. In addition, a "connectivity matrix" can be developed as shown in Figure 4.7. By means of column–row manipulations, a matrix with the elements concentrated around the diagonal can be found. (This technique has been formalized for identification of tightly coupled, functional units for control system design, see Himmelblau, 1973). This representation of the communication structure identifies the groups needing close cooperation for a particular task situation. The connectivity matrix analysis can be a very effective tool for matching a formal organizational structure to the natural group formation.

In Figure 4.7 actors B–E form a group as do actors E–I. The two groups are interconnected by actor E. It appears that actor J acts as a kind of secretary receiving messages from the group E–I, and passing on the reports to actor K and to actor B.

The result of the analysis is (1) the identification of the actual work organization as evolving from the adopted criteria for division of work, (2) the interaction between the work organization and the existing technological underlay in the lower levels of the work domain, and (3) the interaction between the external surroundings and the prioritization and planning activities in the upper ends of the work domain. We have repeatedly and successfully used this tool for reorganization of an electronic service organization of a national laboratory in response to the requirements posed by changing technology (computerization).

Analysis of Social Organization

The analysis of the social organization has not been the central focus of the field studies we have made. Therefore, only a few reflections of our experiences can be given.

The analysis of the social organization, which is more or less independent on

Information Receiver

Actor	A	B	C	D	E	F	G	H	I	J	K
A	O										
B		O	v	v	v						
C		v	O	v	v						
D		v	v	O	v						
E		v	v	v	O	v	v	v	v	v	
F					v	O	v	v	v	v	
G					v	v	O	v	v	v	
H					v	v	v	O	v	v	
I					v	v	v	v	O	v	
J		v								O	v
K											O

(Information Source is the vertical axis label for the rows.)

FIGURE 4.7. The actual work organization can be identified by means of a communication matrix.

functionality issues, serves to identify the persons who are in charge and see to it that communication regarding coordination is effective. This organization is often closely related to the social status structure. The social organization that manages the allocation of work and the form for coordination has to be analyzed and described on the basis of the earlier acquired interview results. As mentioned earlier, the analysis should concentrate on the *form* of the coordination while the work analyses determine the *content*.

An important topic of analysis is the propagation of criteria functions. The importance of explicit knowledge about goals and performance criteria for the flexible and fast response of operations management was discussed in Chapter 2. The analysis naturally will include the goals that are derived from market conditions and company policy and are then communicated "downwards" through an organization where they are decomposed and transformed to match the choices among the local action alternatives. However, careful attention should be given the many options that are open for local choice as well as the many influential, local, and subjective performance criteria that are active.

Another feature of the informal work organization of a company in a transitory phase from preplanning to customer-driven operations is the use of the formal, centralized planning system as a vehicle for informal, feedback management, as mentioned in Chapter 2. This implies that the analysis of informal violations of the formal systems can serve to uncover the actual criteria functions at the local scenes.

Further relevant points to consider include:

- Customary privileges and prejudices connected with task allocations.
- Institutional forms of manifestation and regulation of conflicts of interest (e.g., labor organizational influences).
- Systems of incentives and penalties.
- Forms of social control (i.e., the degree to which information flow about actors is regulated, inhibited, or impeded by formal hierarchic lines of communication).
- Forms of allocation of power and authority (e.g., on the basis of a recognition of accomplishment or as an arbitrary bestowal).

Expressed in another way, the form of the necessary communication and its control (i.e., the way in which coordination takes place) are, like institutional goals, "top-down" influences originating in company policy and management culture that condition the social organization. On the other hand, the communication contents propagate "bottom-up" and are based on the substance of the work being carried out.

Chapter 5

User-Work Coupling

INTRODUCTION

The point of view taken in this book is that the role of an actor is to control the state of affairs in a work environment that is dynamically changing in response to external conditions and to the activities of other actors. The work organization is considered to be an adaptive, distributed decision and control system interacting with the productive functions and processes. To a large degree, the decision makers and actors are coupled with the productive work processes through an interface of tools, equipment, and computerized information systems. As discussed in previous chapters, an actor can direct his/her activity towards any of the levels of the means–ends network; the related decision task can be decomposed into subroutines represented by the decision ladder and, depending on the level of expertise of the actor, different modes of cognitive control can be applied for each of these subroutines. Furthermore, when computers are used as mediators between the work environment and the actors, the various subroutines of the control task can be allocated to the actor and to the computer in different ways. Consequently, the functional coupling between work and actors is a complex issue. This coupling is discussed in this chapter to help us to identify some important functional requirements that must be considered for the design of effective interfaces.

A schematic illustration of the functional coupling is shown in Figure 5.1, which illustrates different role allocations between the actor and the interface functions by means of a simplified version of the decision ladder of Figure 3.6. The situation analysis of the upward leg of the decision ladder can be considered as the input function to the decision making process. This process involves judgment with respect to the consequences of the observed state followed by goal evaluation and selection of a proper target state. The downward planning functions of the ladder are the output functions, involving decomposition of the task and addition of the context within which the elementary actions are to be taken.

The basic role of human actors in modern work systems is to act as a flexible

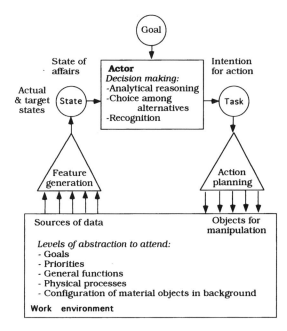

FIGURE 5.1. Cognitive user-system coupling. The figure shows a simplified representation of the decision ladder (Fig. 3.6) coupled to a means-end representation of the work environment.

analyst and decision maker. The central function of Figure 5.1 is normally allocated to these human actors. Direct support can be given to analytical decision making at the knowledge-based level, when problems can be explicitly formulated and separated in time and from their context. Then, support can be based on normative strategies derived from formal decision theories. Such aids are very important in many work situations, but they are often tuned to particular domains and problems where these approaches are suitable. As such, the topic falls somewhat outside the main line of the discussion in this chapter. Good reviews are found elsewhere (Goodstein, 1990; Sage, 1981a,b; 1987a,b).

The coupling of this decision function to the work environment can be implemented in different ways, depending on the way the situation analysis at the input and the planning function at the output are shared by the actor and the computer-based interface. When the actor and the work environment interact dynamically in a closed-loop control function, the situation analysis or feature formation at the input in Figure 5.1 forms the "measuring" function of the controller and the effectiveness and reliability of work performance then depend entirely on the quality of this function (see the discussion of feedback systems in Chapter 1). When measurements are reliable and the goal is properly chosen, any discrepancy originating from an inadequate decision or ineffective action will be detected and corrected if the control loop is intact. Consequently, a very important issue is the design of an effective and reliable support of the interaction at the input end of decision making.

However, if the effect of an improper decision takes the control function outside its capability limits and thus breaks the closed loop, then proper control is lost, and the ultimate effect of actions including those by other actors (such as colleagues, supervisors, or authorities) depends entirely on the properties of the work system. This raises another important design issue for many work domains—the control of performance when the closed-loop control allocated the individual staff member exceeds their local capability limits. In that case, a mismatch occurs in the work–actor coupling, an event frequently judged to be a "human error."

In case of such a coupling mismatch, either human behavior has changed from being normally successful to a point that brings the work system out of control or the work requirements have changed in a way that makes the usual human behavior unacceptable. In both cases, we are faced with a human-system coupling that is too narrowly adapted to the normally successful conditions—and this can be due to features in the system design, work planning, or the particular actor's individual experience. Since adaptation to a close match between work requirements and behavioral patterns is the hallmark of expertise, the fundamental design problem considering the work interface is that the actor will typically demonstrate a tendency towards tight adaptation. Therefore, the design must seek to arrange for adequate flexibility so as to widen the tolerance band when required. This requirement implies that a design for error tolerance is an important interface issue. Therefore, Chapter 6 presents a detailed discussion of human error mechanisms and their relations to the capability limits of adaptive systems.

The Cognitive Coupling

As mentioned earlier, the design of the coupling and the allocation of roles with respect to the input function in Figure 5.1 is critical. When, as in many traditional work places, the actor has to perform the diagnostic input transformation by a conscious analysis of raw or unfiltered data (such as several instrument readings, verbal messages, and other observations) considerable cognitive strain is added to the decision task and the reliability of performance can be low under stressed conditions. Consequently, many attempts have been made to support this task through the use of computer aids as exemplified by diagnostic algorithms or heuristic programs (expert systems). The success of such solutions depends very much on the nature of the work domain. For example, it is only realistic to devise automatic situation analysis (diagnosis) tools for well-structured technical systems. Unfortunately, such systems often lie in the high hazard category, where an unreliable diagnosis can be critical (for a detailed discussion of this problem see, e.g., Rasmussen and Rouse, 1981). Therefore, an allocation of this function to rational analysis by an actor or to a computer preprogrammed by the designer has some serious drawbacks. An alternate approach is advocated here; that is, to rely on human abilities to directly perceive the state of affairs in the environment, given the information is present in a proper format. For this, it is necessary to arrange for interfaces that are transparent in the sense that the deep structure of the work domain is accessible to direct perception in the Gibsonian sense.

Accordingly, we call this approach the ecological approach to information system design (Rasmussen and Vicente, 1990; Vicente and Rasmussen, 1988a,b, 1990, 1992). We consider it to be important for several reasons. First, it supports activities independent of any predefinition of problems and preplanning of responses. Second, it is focused on the support of the "measuring function" and the perception of objectives that are so important for closed-loop feedback control in a dynamic environment. Finally, in relieving a decision maker from the strain involved in situation analysis and diagnosis by being able to directly see the problem at the conceptual level, resources are saved for the decision process itself during unfamiliar situations. Therefore, the discussion in the following sections is focused on the features of *ecological information systems*.

At the output side of decision making (Fig. 5.1 again), interface functions can help an actor in dealing with the resource management problem of selecting the proper means to carry out an intended task or action and to plan the proper sequence of acts in a work procedure. In complex, tightly coupled systems, dedicated automatic sequence controllers are an example; they are provided for this very purpose and cover a wide variety of normal task sequences.

Such an automatic sequence controller is in fact an implementation of "direct manipulation." The operator decides to start the plant and presses the start button, from here the controller generates the necessary control actions at the physical configuration level.

This kind of manipulation directly from a higher functional level is, however, difficult to implement in most complex work systems due to the many options present at the lower levels of the means–ends network. In most cases, local, time dependent, and often subjective criteria will be used to choose among options, see Figure 2.6, and an automatic generation of action sequences is impractical. Therefore, even in process plants, less structured tasks such as disturbance control, require interactive resource management tools depending on users' decisions. Such tools, Disturbance Analysis and Surveillance Systems (DASS), have been designed for nuclear power plants (Gallagher et al., 1982; Johnson, 1984).

For less structured work environments, such tools take on the form of separate, generally applicable computer programs, which serve to replace well-established manual work routines with more automated functions. These include word processors, spreadsheet programs, CAD programs, scheduling and planning tools, and so on. Such programs can play an important role in the design of integrated ecological information systems by supporting many of the directly manipulative requirements. But again, the design of interfaces should give priority to proper visualization to support direct perception of "*affordances*" for action in the Gibsonian sense, as an extension of the direct manipulation interfaces discussed below.

ECOLOGICAL INFORMATION SYSTEMS

The ecological information system concept is closely related to the direct manipulation interface concept developed, in particular, by Shneiderman (1983) and Hutchins,

Hollan and Norman (1986). A brief description of this approach will serve to bring the ecological concepts into context.

Direct Manipulation Interfaces

The *direct manipulation interface* (DMI) was originally described by Shneiderman (1983). Many attempts have been made to describe the idea, for example, "a representation of reality that can be manipulated" or "the user is able to apply intellect directly to the task—the tool itself seems to disappear" (both citations in Shneiderman's paper). There exist two theories of DMI. The first of these is the syntactic-semantic model (SSM) of Shneiderman (1983). The second is the description based on the gulfs of evaluation and execution (Hutchins, Hollan, and Norman, 1986). Vicente and Rasmussen (1988b) discuss DMI and these underlying theories and their relation to the framework underlying this book. A brief overview of DMI will be useful in order to indicate its relevance to the ecological approach.

Norman (1986) states that the general problem in human–computer interaction is that "the person's goals are expressed in terms relevant to the person—in psychological terms, and the system's mechanisms, and states are expressed in terms relative to it—in physical terms." Hutchins et al. (1986) characterize this mismatch by two gulfs between the person and the machine (see Fig. 5.2). The gulf of execution refers to the gap between the person's goals and intentions, and the inputs that the computer recognizes. The gulf of evaluation, on the other hand, refers to the gap between the computer's output and the person's conceptual model of the task. Either of these gulfs can be bridged by the computer or by the person. Of course, placing the majority of the burden of bridging these gulfs on the person greatly increases the cognitive demands of the task, or the *distance* introduced by the interface. Each of the gulfs has two types of distance associated with it. *Semantic* distance refers to the disparity

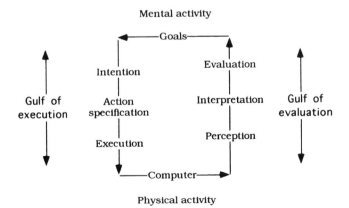

FIGURE 5.2. The gulfs of execution and evaluation (Based on Norman, 1986).

between the user's intentions and the meaning in the interface language, while *articulatory* distance refers to the distance between the physical form of the interface language and its meaning.

On the execution side, semantic distance of the interface is reduced if the user's intentions can be expressed in the command language in a concise manner. The goal should be to match the level of description required by the interface language to the level at which the person thinks of the task. On the evaluation side, semantic distance is reduced if the displayed objects represent the higher level concepts that people naturally adopt when reasoning in the problem domain. In this case, the goal should be to provide a powerful, productive way of thinking about the domain.

The articulatory distance introduced by an interface is also an important factor. On the execution side, the articulatory distance is related to how closely the form of the action is to the meaning of the action. For instance, articulatory distance is reduced if the user can drag a mouse to move an object on the screen, rather than enter a command on the keyboard. On the evaluation side, articulatory distance refers to how closely the form of the displayed objects resembles their meaning. As an example, effective use of icons allows users to infer the meaning of an object from its visual appearance.

Hutchins et al. (1986) go on to say that the success of DMI is related to the feeling of direct engagement that they produce in the user. Thus, the person feels as if he/she is interacting with the concepts of the domain rather than an electronic intermediary. The shorter the semantic and articulatory distances, the greater the feeling of direct engagement. Therefore, the difference between DMI and conventional interfaces is that the gulfs are bridged by the computer, leaving the person to concentrate on the task at hand rather than the interaction process per se.

Compared to the ecological interface design advocated in the subsequent section, some important differences are found. First of all, in our approach we recognize that the user of a computer system operates at several different levels of abstraction and, therefore, the computer cannot simply "bridge a gap" that has several dimensions. Second, in consequence of this, the gap has to be bridged by the user by means of his/her direct perception facilities. We will now have a look at this concept.

Ecological Interface Systems

In a dynamic environment, appropriate support cannot be based on tools suitable for stable, preplanned tasks. An exploration of the ends and means, opportunities and constraints present in a work situation is necessary for adapting to new requirements. Thus an information system that presents users with a complex, rich information context for direct perception, as well as "thought experiments," will be very effective for natural decision making.

The design of ecological information systems attempts to exploit the large capacity of the human perceptual and sensory–motor system as encountered in natural environments by furnishing a complex yet transparent information environment. Through a suitable interface, the attempt is made to stimulate a user's direct perception of information at the means–ends level most appropriate to current needs and, at the same time, support the level of cognitive control at which the user chooses to perform. These interfaces should serve to couple all the means–ends levels of the work environment to the three levels of cognitive control and allow the users to dynamically switch their attentional focus and levels of control.

In Gibson's terms, the designer must create a virtual ecology, which maps the relational invariants of the work system onto the interface in such a way that the user can read the relevant affordances for actions.

Ecological displays will often be complex because the user has to be able to generate a complex cue–action hierarchy at several levels below the currently viewed level of the work environment. In addition, in order to support analytical, functional reasoning, the displays have to present an effective, externalized mental model of the object of interest to the user. By recursion, a perception of this symbolic representation as accessible objects on the display surface will make the symbolic processing amenable to direct perception → manipulation operations.

Decision makers can often be overloaded by the presentation of many separate data, whereas complexity in itself need not be a problem, provided that meaningful information is presented in a coherent, structured way. Such users are not passively receptive to information input; instead they actively ask questions of the environment, based on their perception of the context. Therefore, a rich support of this perception in context is a key design issue. It is our experience that designers and consultants who are not thoroughly familiar with a work domain overestimate the amount of information required for action by a specialist who has effectively adapted to the work and is submerged in its context—while, at the same time, they underestimate the complexity of meaningful displays that are acceptable to such experts selectively seeking cues for action.

While heuristics evolving in a stable work environment can be embedded in a work practice shared by the staff, it is reasonable to expect that the heuristics found in a dynamic work environment are different across members of a staff. In such a case, the adaptive features of advanced information systems could enable them to attempt to match work demands to the individual user and facilitate natural decision making based on personal experience. For example, they could store and replay prior cases, format information according to the user's subjective preferences and strategies, and serve as a blackboard for cooperative work coordination within a group of highly proficient people.

Direct Perception and Types of Work Domains

The ecological interfaces we are advocating are closely related to Gibson's theories of direct perception of "*invariants*" (sources of regularity) and "*affordances*" (cues for action relevance). Therefore, a brief discussion of the opportunity to learn the basis for such perception in different types of environments is useful.

For this purpose and to relate the problem of user coupling to the domain characteristics reviewed in Chapter 2 (Fig. 2.9), three categories of work environments will be briefly discussed with reference to Gibson's direct perception concepts (see Chapter 4).

Natural Environments. Evolution has granted humans an extremely effective perceptual–motor system serving the control of the dynamic interaction with the natural environment. The ability to interact dynamically with the environment and its inhabitants was a key to the survival of primitive people. Thus, the evolutionary process has resulted in the development of capabilities for reading off—by Gibson's direct perception—the functional properties and affordances of the environment at several levels of abstraction (discussed in Chapter 4, Figs. 4.5 and 4.6). Some of the characteristics of this perceptual–motor system are genetic, while some depend on the individual's adaptation to the characteristics of the environment.

When considered as a "display" of relational properties, the natural environment has some very characteristic features that are illustrated by Figure 5.3. The intentions of the person in the bed can be directly perceived without any complicated analysis as can the functional structure of the overall arrangement. If interested, the viewer can also visually manipulate the physical processes involved at a local level. This means that shifts in abstraction, as well as decomposition, are immediately possible at the discretion of the spectator from this single representation of a "complex" system.

This discussion has special implications for the design of ecological interfaces for

FIGURE 5.3. A natural environment. The cartoon illustrates how the human perceptive system is able to read off messages from a natural environment at several levels of abstraction. Storm P. Museum, Copenhagen. Reproduced with permission.

the casual user of systems at the left-hand edge of the continuum shown in Figure 2.9. These untrained users will only have good chances for adapting to the functionality of the system as reflected in the interface if a metaphorical mapping is utilized that incorporates cue–action correlates familiar from other contexts.

Traditional Man-Made Environments. In traditional man-made environments, humans will not be able to rely only on natural evolution. For instance, when new tools are introduced, an additional adaptation is necessary to supplement natural evolution. Learning to use new tools involves two phases. One is the adaptation guided by exploratory behavior. The second is the adapted phase, when behavioral patterns have reached a state of equilibrium. Which of these phases is more important depends on the characteristics of the work domain and the actor.

The first phase is a transient in stable work domains and is eventually overcome with training and experience. The major part of work falls into the stable phase of unchanging stereotypical working patterns, which are governed by surface features of the environment. In this way, activity is guided by empirical cue–action correlations. As a result, for such work situations, users are not generally concerned with the system's internal physical functions—only the surface features. In case of difficulties, a trial-and-error exploration is acceptable because the cost of an error can be expected to be less than the value of the information gained. Consider the example of a radio receiver. Everybody knows how to operate one without knowing anything about its functional properties; if success is not immediately achieved, a few trials will do, based on a perception of the known features of a prototypical radio.

In this case of rather stable work systems, the user can be expected to have an opportunity to learn the invariants and affordances by trial and error. However, easy and fast learning will still benefit from metaphorical relationships to familiar cue–action sets. This category of work environments reflects features found in the left-to-middle portions of the continuum of Figure 2.9.

Modern Work Systems. A number of complications confront an interface designer for many modern work domains. The automation of work processes moves workers from prescribed manual activities to situations of a discretionary nature involving decision making, conscious planning, and choice. As a result, the exploration of opportunities and boundaries of acceptable performance form a part of the daily work activities. In other words, an ultimate stable phase of adaptation is never actually reached. Instead, task demands are in a constant state of flux, requiring continual adaptive efforts on the part of the workers. A simple example from the consumer world involves the contrast between radios (mentioned above) and rapidly changing models of video cassette recorders—now a common household item—which annoy many ordinary users who have little or no feel for their broad repertoire of functional options and interface features. Another important example is office work that has been called "open-ended," consisting as it does of

> . . . the complete set of actions relevant to the organizational world (which is) unknown and unknowable (Barber, DeJong, and Hewitt, 1983).

Office functions have been defined as rule-based state transitions. However, this must be interpreted cautiously and in terms of "practical action" having to do more with

> ... what things should come to, and not necessarily how they should arrive there. . . (Suchman, 1982).

> ... office workers are able to handle unexpected contingencies in their daily work because they know the goals of office work. These goals and actions are often implicit in the office workers' knowledge of their work (Barber, DeJong, and Hewitt, 1983).

This underlying "goal feeling" contributes to a certain degree of built-in "error tolerance" in these systems—especially those where trial and error is feasible. A transition from the middle of Figure 2.9 to the right could, for example, involve banking and other investment-heavy enterprises where the cost of errors begins to be significant.

In such cases, it will be advantageous if the interface is transparent and presents a truthful mapping of the functional goals–constraints and affordances of the work system so as to enable the actor to directly perceive relevant action alternatives without having to learn them all by trial and error.

High-Tech, High-Hazard Systems. For some systems, these problems become greater. Malfunctions in high-tech high-hazard systems can have unacceptable consequences for people, property, and/or the environment. Therefore, learning and adaptation based on trial-and-error explorations are not acceptable, nor will users–operators always be able to rely on patterns of performance derived from an adaptation over time to surface properties. In addition, unlike users of cars and radio receivers, operators in high-tech work domains are also responsible for safely controlling the system after it fails. Because the systems are so complex and are therefore often automated, there is a very large number of possible malfunction modes. Many faults will be unique and cannot be anticipated. As a result, it will not be possible for operators to learn all the appropriate response patterns in advance. Instead, they will often need to select and plan their actions based on an understanding of the internal processes and functional structure of the system. Unfortunately, in many cases, these are not visible. For instance, the internal mass and energy flows of a power plant cannot be observed directly, and yet they are both the object of control of the operator and, at the same time, the basic sources of large accidents.

For this category of work environments it is a technological imperative that the interface is transparent and truthfully maps the deeper functional relationships of the work system for direct perception of system invariants and affordances. These systems relate to the far right side of the map in Figure 2.9.

Functional Requirements for Ecological Information Systems

From this discussion, we can recapitulate some basic requirements for ecological information systems in modern, dynamic work environments.

Support of Direct Perception

- It is highly recommendable that the internal functional and intentional invariants—the sources of regularity—of the system are represented at the surface of the interface—so that they can be directly operated upon under unfamiliar as well as familiar conditions.

- In order to activate the normally very effective and reliable sensori-motor system, the representations used for the interface should support direct perception and manipulation of the "afforded" control objects, as well as analytical reasoning.

Support of Multiple Interpretations. Therefore, an ecological interface should support the different modes of cognitive control that can be applied at different levels of expertise:

- Support of skilled routines: In order to support interaction via time–space signals, the user should be able to act directly on the display, and the structure of the displayed information should be isomorphic to the part–whole structure of movements. Thus an actor should be able to read off features of the interfaces at the different levels of detail matching the structure of the integrated movement patterns found at different levels of training (e.g., the notation in a musical score, see Chapter 7).

- Support of rule-based know-how: Provide a consistent one-to-one mapping between the work domain constraints and the cues or signs provided by the interface. Faithfully represent the internal state of affairs of the work system with reference to its sources of regularity; that is, the functional and intentional "invariants" and the available means for action, the "affordances."

- Support of knowledge-based problem solving: Represent the work domain in the form of a means–ends/part–whole relational network to serve as an externalized mental model that will support knowledge-based problem solving. Since demands for knowledge-based reasoning can be infrequent interruptions of longer stable and familiar periods (as in process plant operation), it is advantageous that this mode of interpretation is made possible within the same display format intended for the skill- and rule-based mode of cognitive control.

Support of Adaptation to Work Requirements. Another basic requirement for the design of interfaces is that they must support adaptation in the cognitive coupling between actors and their work environment. Adaptation takes on many forms, which must be considered explicitly during design. Adaptation to the work situation makes the interaction very dependent on subjective criteria; that is, a personal style of interaction emerges. Therefore, to accommodate these adaptations, the interface must either be "wide" enough to match the styles of all members of the intended user category or the system must adapt to any user's preferred style. A summary of some of the aspects of user adaptation discussed in Chapter 4 will be useful here:

- Evolution of expertise. Any modern information system must be able to support a novice without frustrating the expert. In this respect, it is very important to realize that expertise does not evolve because cognitive control becomes automatic through training but through an evolution of new control structures while the previous structures degrade. Consequently, to conform to the evolution of expertise, an interface must be able to support all the levels of cognitive control at the same time. In Chapter 6, it is argued that adaptation and errors are closely related and that an effective support of the evolution of expertise, therefore, requires error tolerant design, which makes the effects of errors immediately observable and reversible.

- Shift of mental strategy. In most cognitive tasks, several different mental strategies can be used, posing different resource demands for the user, who therefore will be able to shed work load and circumvent local difficulties by shifting strategy. As a result, user acceptance will depend on an information system's capabilities for supporting all relevant strategies or for following users' shifts among strategies.

- Adaptive division of work. Shifts in the boundaries between roles taken up by different individuals can also take place in order to resolve resource-demand conflicts and to match performance to individual preferences. The subjective criteria active in this adaptation will be very situation dependent and directly related to the particular work process, such as perception of differences in work load among colleagues, the amount of communication necessary among actors for coordination, and subjective preferences for certain activities. This adaptation has a significant influence on the work conditions in computer supported cooperation and in the design of interfaces for shared work environments.

Design Competency Required

- In consequence of these requirements, the design of interfaces can no longer be allowed to evolve empirically through trial and error; nor can they be solely based on traditional general purpose human factors design criteria and guidelines. The interface design is an integrated part of the functional systems design since it constitutes the "measuring device" of the controllers and, as for the rest of the system, requires subject matter expertise.

- Designers of systems must have an in-depth knowledge of the cognitive context in which the system is to be implemented in order to choose display formats that (1) will lead users to adopt effective mental models but that (2) do not include display elements that are in conflict with any established stereotypes "belonging to" the given professions or company.

This approach will be expanded upon in Chapter 7.

Chapter **6**

At the Periphery of Effective Coupling: Human Error

INTRODUCTION: THE CONCEPT OF HUMAN ERROR

With current trends toward large, integrated work systems, the consequences of operation beyond the boundaries of normal operation are becoming increasingly hazardous. Analyses of industrial accidents invariably find that around 80% are caused by "human errors" in one form or another. Consequently, human error has become the topic of many research programs.

According to the view promoted in the previous paragraphs, the occurrence of a "human error" signals a mismatch in the coupling to the environment. A definition of a "human error" thus tends to be ambiguous and to depend on the circumstances under which it is observed. Sometimes the event is observed directly and locally within a closed loop as a *deviation from normal* behavior of an individual. In other cases, the event is identified at a more global level, and probably at a later point in time, when inappropriate system performance is observed and an analysis of the chain of events identifies a human error as the *cause of a disturbance*.

Defining errors locally as deviations from normal performance can be useful for activities in well-structured technical systems when a correct action sequence can be derived from the functional requirements of the equipment operated by the staff and a normative work instruction can be formulated. It is then possible to define particular events as being human errors *for a particular interface and task configuration*. This aspect of technical systems has been the basis for most engineering approaches to the study of human error and for the development of predictive human reliability analyses used for system evaluation. This local definition of human errors has also been the basis for some research on errors based on the personal diaries of the researchers (Norman, 1981; Reason and Mycielska, 1982).

The definition of a "correct" sequence of actions for a certain task is difficult, if not impossible for activities in less structured work environments (such as administration, staff management, work planning, or maintenance) and the local definition of an error becomes much more diffuse and less useful. In such cases, errors will very often be identified at a later point in time when the effect of the work procedure chosen has turned out to be unacceptable to somebody "downstream" in the causal path. Then an instance of loss-of-control by an individual actor can be judged to be an error by backtracking along the causal path leading to the observed system failure. In this case, the identification of an error depends entirely on the characteristics of the causal explanations that makes the concept of human error quite ambiguous. This will be discussed in more detail below.

Performance at the Periphery

When an adaptive, closed-loop control function is effective, the performance of the controlled system can be predicted from the objectives and criteria governing its adaptation without considering properties of the forward, productive functions (see the examples in Chapter 1). When the limits of its capability in the forward, productive path, or in the feedback, measuring path, are violated and the closed loop thus broken, drastic and frequently unpredictable performance can result. Therefore, for any work system presenting a hazard to the staff, to the investment, or to the environment, an analysis is necessary of the performance at the periphery of its normal function to determine explicitly the boundary to a loss of control and the effects of its violation.

If the results of the activities of the staff controlling a work system turn out to be unsuccessful, it can be considered an instance of loss-of-control by one or more of its human controllers that results in ineffective closed-loop control. This can happen because of simple human errors or misconceptions of the actual work requirements. On the other hand, the loss of control can be due to the lack of resources to act properly, inadequate time, or because the human actor is trapped by peculiar system features. Such cases can in turn be referred back to managers' resource or work planning. System failures due to equipment failure can reflect an inadequate control in the choice of proper component quality during the design process. In any case, a breakdown of work performance indicates an operation too close to the capability limits and/or a lack of the ability to recover control at the periphery of normal closed-loop functioning. An analysis of system properties at the periphery of normal performance and of the ways to cope with violations of the boundary to loss of control are important interface design issues and are particularly in focus in the design of ecological interfaces, as discussed above.

Studies of cases of system break down and human error are important for another reason. As discussed in Chapter 1, the behavior of a system within the boundaries of normal, adaptive, and closed-loop control largely reflects the control objectives. The internal processes of the system and its controllers only reveal themselves when the loop is broken. Compare this with Herbert Simon's well-known dictum:

> A thinking human being is an adaptive system. . . . To the extent that he is effectively adaptive, his behavior will reflect characteristics largely of the outer environment (in the

light of his goals) and will reveal only a few limiting properties of—the physiological mechanisms that enables him to think (Simon, 1969, pp. 25–26).

From this point of view, an analysis of human behavior when closed-loop control breaks down because of errors is an important source of information about cognitive mechanisms and the strategies used during work. In this connection, a careful study of errors in context is important for improved system design. In the following sections we will discuss in more detail some approaches to studying human errors and the implications for improved designs of the coupling of the staff to a work system.

Causal Explanations and Human Errors

The nature of causal explanations is discussed in Chapter 1, but a brief recapitulation in the present context will be useful due to its implications in treating the concept of human error when analyzing complex systems.

A description of human behavior in terms of a sequence of acts depends on a decomposition that can be done at many levels of detail. Normally, a decomposition will only be carried through to the level that is familiar to the analyst. A description of an event cannot include all details of a complex environment in a particular situation and will normally be focused on the unusual features while the normal context will be taken for granted. This aspect is also observed in most causal representations of accidents, as shown in Figure 6.1. In this way, the causal context within which errors are defined depends on the tacit knowledge of the analyst about the usual state of affairs.

Causal explanations also pose another fundamental difficulty. When the causal path is followed upstream along the flow of events, no objective stop rule can be devised to terminate the search. It is always possible to go one more step backwards

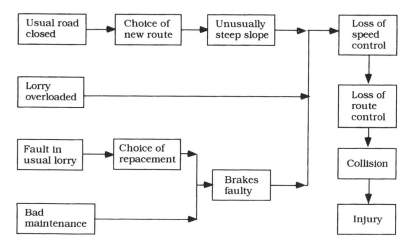

FIGURE 6.1. This figure illustrates the causal tree of an accidental course of events. The events included are all unusual events.

and the decision to stop and to accept an event as the "cause" depends entirely on the discretion of the analyst. The immediate implication is that the choice depends on the perspective adopted by the analyst. Sometimes the search is terminated when an event or act is found that appears to be a familiar explanation; sometimes it stops when a cure to prevent an event is known; and sometimes simply because the causal path disappears due to the lack of information.

Perspectives on Human Errors

It follows from this discussion that the concept of human error is very elusive and, therefore, it will be relevant to discuss in some detail how the nature of causal explanations influences the use of the concept for system design.

Post hoc analyses of accidental courses of events are often used in efforts to improve system performance. In this way, accidents and incidents activate a higher order feedback loop serving to adapt system properties in a better way to the variability of the operating conditions. In this context, different decision makers may be involved and different perspectives of analysis become relevant:

- Explaining an unusual event: The common sense perspective.
- Understanding human behavior: The scientist's perspective.
- Evaluating human performance: The reliability analysts' perspective.
- Improving human performance: The therapist's perspective.
- Finding somebody to punish: The attorney's perspective.
- Improving system configuration: The designer's perspective.

The influences of these various perspectives on the identification of a human error in a particular case will now be reviewed.

The Common Sense Perspective: Explaining the Unusual. The common sense perspective is used for everyday explanations of unusual events. This perspective is important because it illustrates the bias from causal reasoning and will influence the judgment of events as seen also from the other more specialized perspectives as well. When something happens unexpectedly, we normally try to go back along the course of events to find a cause. Typically, during this process only the unusual events will be considered, while the normal context will be taken for granted. The search will stop when an event or act is met that is *familiar to the observer* and is found to be a reasonable explanation. Therefore, the cause identified is likely to be an error made by a person involved in the dynamic flow of events: To err is human and it is easy to find a scapegoat. Furthermore, it is normally difficult to continue backtracking "through" a human being. In all, the nature of causal explanations focuses on people and, in particular, those involved actively in the dynamic flow of unusual events.

The Scientist's Perspective: Understanding Human Behavior. When effective closed-loop control of human activity is broken and the "inner mechanisms" reveal themselves, analysis of the behavior can support the modeling of cognitive mechanisms. In this case, the stop rule in the causal search is to consider any actor in the flow of accidental events that did not manage to maintain control under the unusual conditions, irrespective of whether they were a "root cause" of the flow or not. The involved actors function as causal links between external conditions and the resulting unsuccessful acts. Identification of categories of errors across a number of cases can serve to identify the cognitive processes involved that most likely linked stimulus and response under the various conditions. These inferences basically depend on the psychological theory adopted. From a behavioristic point of view, no assumptions about cognitive processes are accepted. The antecedent to an overt act will be an external stimulus (i.e., a cue presented by the environment) and a theory of "error prone individuals" emerges as a natural explanation of accidental events. From a cognitive point of view, however, human performance can be further decomposed by taking into account some intermediate mental states, events, and processes.

Psychologists interested in errors as being the "windows of the mind" have made a distinction on the basis of an analysis of actors in everyday tasks in terms of the state of the actors' intention. Thus *slips* and *lapses* depend on the wrong execution of a proper intention, whereas *mistakes* occur from a proper execution based on the wrong intention (Reason and Mycielska, 1982; Norman, 1981). A more differentiated distinction of the mental events in control of human acts has been made from analyses of human involvement in failures of complex systems (Rasmussen, 1980; 1982). Here the antecedents of an erroneous act will be (1) a failure in one of the decision functions found in the decision ladder discussed in Chapter 3 (Figure 3.6). From this, the question can be asked, about (2) which psychological mechanisms were the cause of the failing decision function? Next, further backtracking will bring us to (3) the input conditions of the human actor where one looks for external events activating the psychological mechanisms. In this way, a multifaceted description of human errors is obtained that can distinguish a very large number of erroneous overt acts from a very limited number of categories within each facet—a very useful feature for the analysis and design of complex systems. The facets and categories of a classification scheme developed by an OECD (the International Organization for Economic Cooperation and Development) expert group on human errors (Rasmussen et al. 1981) to be used for analyzing event reports from nuclear power plants and the development of a human error data bank are illustrated in Figure 6.2. For the analysis, the events of the causal chain of the figure are followed backwards from the observed accidental events and the mechanisms involved are categorized at each step.

The system did not become operative due to some basic difficulties. These included the need for the analyst to have a cognitive psychology background as well as intimate knowledge about technical functions and operational practice, and to have access to the people actually involved in the event. In addition, it proved difficult (impossible) to characterize the work situation properly with respect to error recovery features, to the frequency of successful performances, and to possible external causes. However, even if the use of the taxonomy for development of a human error data base appears to be less realistic, it has proven very useful for the analysis and understanding of human involvement in accidents (Rasmussen, 1982).

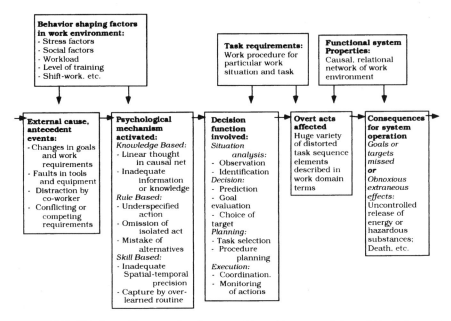

FIGURE 6.2. This figure illustrates the decomposition of human activity to identify internal mental processes and states in the explanation of erroneous behavior.

As stated previously, the categories of error mechanisms identified depend very much on the psychological theory used to infer mental states and processes involved in error production. For a comprehensive and up-to-date review of psychological theories and related error classification schemes, see Reason (1990). In this review, he also describes a human error classification scheme for analysis and design of complex work systems, which is similar to our approach.

Within the research perspective, an iterative process of analyses of error cases and development of models and classification schemes is necessary. In our case, the skill-, rule-, knowledge-based model initially emerged from analyses of error cases under the assumption that errors are not stochastic events but reflect limitations of and interferences among processes within a hierarchically organized set of cognitive control modes (Rasmussen, 1980).

The Reliability Analyst's Perspective: How Reliable Is the Performance? The concept of human error is very much in focus for the development of predictive risk analysis of hazardous industrial installations. In this case, the effects of the likely human error modes during safety related tasks on plant performance are predicted by an analysis tracing the effect of human errors in the individual acts of a task procedure downstream the causal chains toward their ultimate outcome in terms of plant production and safety.

For the effects of human error involved in the direct operation and maintenance of technical equipment, the analysis is formally rather straightforward, since the reference in terms of correct performance can be found in the operating procedure derived from the requirements of the equipment. For each of the steps in a normative procedure, the influence of the different modes of cognitive error mechanisms (as identified from the scientist's perspective in a previous section) is traced downstream the causal path by an error mode and effect analysis.

In practice, however, the fidelity of such predictive analysis is questionable because the action sequence actually used during work often deviates significantly from the normative procedure. In general, designers are not able to foresee all local contingencies of the work context. In particular, a work procedure is often designed separately for a particular task in isolation whereas, in the actual situation, several tasks are active in a time-sharing mode, which pose additional constraints on the actually effective procedures that were not known to the designer or work planner. See, for instance, the analysis of nuclear power operators' modification of procedures by Fujita (1991), who found their violations to be quite rational, given the actual work load and timing constraints. In this case, a basic conflict exists between error as seen as a deviation from the *normative procedure,* and error as seen as a deviation from the rational and *normally used, effective procedure.*

In addition, different problems are faced, depending on whether the predictive analysis is directed toward the *reliability of performance* in a particular task, or toward the *risk of unacceptable side effects* of performance. For human reliability prediction, the result will estimate the probability that the *correct* target has *not* been reached. This calculation depends on cumulative error rates for each step in the normally successful procedure, including all error types, frequent as well as infrequent. Well-developed techniques for human reliability prediction exist (Swain and Guttman, 1983) even if real empirical data on error rates are scarce. Human reliability prediction is an important technique for designing tasks in which actors (plant operators) are in a monitoring and supervisory task because then they are in the feedback control path and their reliability is a crucial parameter.

Prediction of the risk of unacceptable side effects is considerably more difficult because then the effects of particular error types must be identified and, in any well-designed system, hazardous effects are likely to be related to rare error types, for which data cannot be obtained.

Reliability and risk predictions are even more difficult in case of analysis of less structured tasks, such as work planning and management. Predictive risk analysis is then very questionable because the reference for defining errors is next to nonexistent. Again, an important problem has its origin in the human adaptation to the environment and the actors' efforts to optimize performance beyond the prediction of a designer.

The Therapist's Perspective: What Can We Do? In case of significant system failures, therapists of various professions are usually involved in the analysis, each with a well-equipped, professional tool box.

Viewed from the therapist's perspective, when looking for possible improvements in the operation of an existing system, the stop rule for termination of the search depends on the *availability of a cure* and the bias from the nature of causal explanations often focuses the search on persons directly involved in the course of events: One can always ask people to try harder; one can give better training; and one can improve supervision. Only recently has serious attention been drawn toward the role of people in the work system conditioning the paths of unusual events through their normal activities, such as work planning and equipment design. The bias from the availability of cures implies that judges from different therapeutic professions, such as the organizational sociologist, the work psychologist, the safety officer, and the training center representative, will reach different causal explanations of an observed mishap and find different actors to be the best target of their efforts. This condition not only reflects the difficulty in identification of a "root cause" of a system failure and of somebody to blame by causal analysis, but also the fact that the performance of a closed-loop system can be improved at many locations within the loop.

The Attorney's Perspective: Who Is Guilty?   The ambiguity of causal explanation becomes a serious problem when the aim of the retrospective analysis is to find a responsible person to punish. The stopping rule then is likely to be that an erroneous act is performed by a person *who was in control of their behavior* (Fitzgerald, 1961; Feinberg, 1965). Being in control of behavior normally means being *physically* in control, that is, no one is forcing the person and no physical illness (e.g., seizures) prevents the person from "being in control." Once more, focus has traditionally been on the individuals directly involved in the dynamic flow of accidental events. Furthermore, little consideration has been given as to whether the individual was cognitively "in control"; that is, behavior was not forced by subconscious stereotypes leading the person into a trap due to characteristics of the work setting. Analysis of traffic accidents typically identifies the pilot or train driver as being responsible.

A change can, however, be seen in the focus of the analysis of accidents and the allocation of responsibility. For several recent major accidents, such as the Zeebrügge (HMSO, 1987) and the Clapham Junction (HMSO, 1989) accidents, the causal analysis has also included the normal activities of supervisors and managers who actually prepared the work conditions of the people in the dynamic flow of events and, therefore, conditioned them for error. In consequence, significant causes of the accidents in these cases were attributed to higher level management.

The common law liability concepts evolved over centuries without an explicit theoretical basis. In the United States this situation has been changing in response to the fast pace of technological change and a movement toward a rational criterion of liability has been introduced. The stopping rule for causal search suggested by this "liability science" is to find the "cheapest cost avoider," that is, from a social point of view to minimize the cost of accidents (Calabresi, 1970). Ironically, this approach brought with it the present boom of liability cases in American courts involving scientific expert witnesses (Huber, 1991).

Another trend has been reinforced by a recent US court practice. During recent years, a clear trend has been to extend the criminal law to also cover "wishful blindness" on the part of corporate executive officers (CEOs). Previously, criminal law was only in effect for acts of deliberate illegal intentions, whereas civil law took care of careless acts. This situation has changed due to the traditional mismatch between the size of penalty from violation of environmental laws and the cost of environmental protection measures. Difficulties of reinforcing environmental protection laws led court practice to apply criminal law to cases where CEOs delegated the environmental protection measures to lower level staff without effectively monitoring whether the measures were in place and active. Not knowing the risk involved in operation or understanding the implications of management decisions on environmental protection and safety do not excuse top managers even if risk management has been delegated to specialists. That CEOs have the power to be "in control" is sufficient reason for them to be imprisoned after a violation of environmental laws as a result of decisions by anyone in the company (Addison and Mack, 1991).

The System Designer's Perspective: How to Improve Safety. Faced with the evidence from error reports, the system designer's perspective is the search for potential changes of the work system that will improve its performance. "System designer" in this context should be taken in a wide sense, including law makers, authorities, engineering designers, and operations planners. From their perspective, the only relevant causes of system malfunction are acts of God or inappropriate human acts; that is, the only possible alternatives are (1) to accept the occurrence as being the manifestation of a risk to be accepted or (2) to identify some person within the wider system context who can be influenced so as to make safer decisions in the future, being it an operator, a work planner, a company manager, or an equipment designer.

The selection of design improvements from analysis of accidental reports involves some very basic problems of generalization. The causal tree resulting from an accident analysis (apart from being a reflection of the analyst's subjective frame of reference) is only a historical record of a particular chain of events, it is not a generally valid functional model of the system. Repetition of the particular course of events found in the record can be avoided by removing any of the causal events or by blocking any of the involved causal paths. Such a solution very likely would be an ineffective ad hoc cure of symptoms since the likelihood that the same course of events would be activated in the future is nil. Generalization across several reports is necessary to identify effective cures. However, even if this is done, the solution is likely to be ineffective because the records are not models of functionality and, therefore, not reliable for prediction of responses to changes. In particular, they *do not reflect internal feedback loops* that may activate compensatory responses by the people involved to the "improvements" of the system. This clearly points to another view of human error, taking into consideration the *close relationship between errors and the process of adaptation* to the features of the environment. We will return to a detailed discussion of this problem in a later section.

To sum up, "human errors" do not represent a well defined category of behavior fragments that can be isolated for scientific investigation. In a modern dynamic environment, where discretionary decision making to a large degree is replacing routine tasks, definition of a correct or normal way of doing things is difficult, and the focus of research should be on the understanding of the way in which features of the work environment shape human behavior and the conditions under which normal psychological mechanisms result in unsuccessful performance. The aim in the present context, therefore, is not to analyze human errors and to create data bases so as to remove errors by proper work system design but, instead, to design work systems that support the actors in *coping with the effects* of their actions when their performance under particular circumstances turn out to be unsuccessful. In this respect, we have found that a better understanding of the relationship between human adaptation to a dynamic environment and human errors is required.

Adaptation and Error

From the discussion of mental strategies in Chapter 3 it follows that the individual actor in most work situations is faced with many degrees of freedom with respect to the composition of normally successful work procedures. The flexibility and speed of professional expertise evolve through a process of adaptation to the peculiarities of the work environment within the envelope defined by these degrees of freedom. During this process, "errors" are intimately connected to an exploration of the boundaries of this envelope (Rasmussen, 1990a,b). Adaptation implies that degrees of freedom are removed and performance optimized according to the individual actor's subjective *process criteria*. Unfortunately, the perception of the qualities of the *work process* itself is immediate and unconditional and the benefit from local adaptation to subjective performance criteria is readily perceived in the situation. In contrast, the effect of activities on the *ultimate product* of work (and on side effects) of local adaptive trials can be considerably delayed, obscure, and frequently conditional with respect to multiple other factors. Shortcuts and tricks-of-the-trade will frequently evolve and be very efficient under normal conditions, while they will be judged serious human errors when they, under special circumstances, lead to severe accidents. The Clapham Junction railway accident (HMSO, 1989) presents several clear examples of how a safe work procedure for signal system modifications, including multiple precautions against human errors, gradually degenerates due to adaptation to a locally more efficient work practice. In this case, the criteria for the design of the normative division of work and of the prescribed work procedure was redundancy and reliability, whereas the criteria actually governing performance happened to be effective sharing of work load and saving time (cf. the discussion of criteria shaping cooperative patterns in Chapter 4).

Adaptation of Individual Performance. Purposive human adaptation manifests itself in error mechanisms at all levels of the cognitive control of performance. For *knowledge-based problem solving* during unusual conditions, an opportunity to test diagnostic hypotheses is important for adaptation and for the development of expertise. It is typically expected that qualified personnel (such as process operators) during unfamiliar work situations can and will carefully test their diagnostic hypotheses conceptually by thought experiments before acting; particularly, if the effects of their acts are likely to be irreversible. This is, however, often an unrealistic assumption, since it may be tempting to test a hypothesis by action in order to avoid the strain and unreliability related to unsupported reasoning in a complex causal net. No explicit stop rule exists to decide when to terminate conceptual analysis and to start acting. This means that the definition of error, as seen from the situation of a decision maker, is very arbitrary. Acts that are quite rational and important during the search for information and test of hypothesis may appear to be unacceptable mistakes in hindsight, without access to many details of a turbulent situation. The post hoc identification of a decision error on the part of, for example, an operator, pilot, or train driver is typically the result of the analyses of accident investigation commissions who have ample time and support in hypothesis testing in the form of tools for mathematical analysis and dynamic simulations.

A similar situation is found when analyzing cases of medical accidents. Diagnosis and treatment of patients can be compared to piloting patients through troubled water (Paget, 1988). The state of health of patients depends on numerous obscure conditions and treatment can have side effects that are difficult, if not impossible, to predict. Therefore, a diagnosis has the character of a hypothesis and treatment is therefore a test of that hypothesis. Frequently, the knowledge needed for a proper diagnosis is obtained too late and the stop rule guiding termination of data collection, analysis, and onset of action is only available after the fact. The termination of conceptual hypothesis evaluation is a kind of intuitive trade-off between indecisiveness and risk of failure.

There will be ample opportunities for modification of procedures even for *rule-based control* of activities during familiar situations for which normative work procedures exist. The development of know-how and rules-of-thumb depends on adaptation governed by an empirical correlation of salient cues to successful acts. During normal work, actors are immersed in the context for long periods, they know the flow of activity and the relevant action alternatives by heart. Then they do not have to consult the complete set of defining attributes before acting. Instead, to minimize effort, they will seek no more information than is necessary to discriminate among the perceived alternatives for action in the particular situation. This implies that the choice is underspecified outside the normal context and, therefore, when situations change (e.g., due to disturbances or faults in equipment), reliance on the usual cues very likely will lead to error.

Even when an effective sequence of actions has been instituted in a *manual skill*, increasingly complex and integrated patterns of movements will evolve through adaptation to the temporal and spatial features of the task environment. When optimization criteria are speed and smoothness, then the limits of acceptable adaptation will only be known from the errors experienced when it once in a while is crossed. In this case errors are the natural consequence of a speed–accuracy trade-off. Thus, some errors have the function to maintain a skill at its proper level. They cannot be considered a separable category of events in a causal chain because they are integral parts of a feedback loop. The monitoring task during anesthesia is another example. A review by Rizzi (1990) shows that slack monitoring is claimed to be an important contributor to the complication rate during anesthesia. The individual patient's need for monitoring varies within very wide limits. It will not be possible to extend the normally used monitoring effort to cover all contingencies, and a tail of the distribution will end up as being "monitoring errors" after the fact. For activities normally performed under severe time pressure, there is no other way to control the proper monitoring effort than to watch the average error rate. Work load pressure will necessarily cause the staff to readjust their monitoring effort until the limit of acceptable performance is found.

The evolution of increasingly long and complex patterns of movements at high levels of skill that run off without conscious control has another effect. Attention is directed towards review of past experience or planning of future needs during such lengthy automated patterns, and performance is sensitive to interference (i.e., capture from very familiar cues).

In conclusion, human adaptation guided by efforts to optimize some subjective criterion (such as mental work load, joy of learning, and time spent) leads to a kind of boundary seeking behavior. The limits of acceptable optimization of performance are defined by the error rate found acceptable in the particular context.

Adaptation in Cooperative Performance. Human errors are normally identified with reference to the behavior of the individual actors within a work system. Usually, however, several people will cooperate and the performance of the individual is strongly influenced by their patterns of cooperation. As discussed in Chapter 4, the division of work and the adoption of roles by individuals depend very much on adaptation to local work conditions according to subjective criteria. This results in a dynamic, situation dependent cooperative structure that strongly influences the performance of the individuals.

The basic structure of *division of work* and adoption of roles depends on the functional requirements of the work domain, such as its relational structure and degree of internal coupling, its topography, the work load involved, the accessibility of information, and the necessary information traffic. Shifting the boundaries between roles taken up by the individual actors will be used to resolve resource–demand conflicts and to match performance to individual preferences. The subjective criteria governing this adaptation will be very situation dependent and directly related to the particular work process, such as perception of differences in work load among colleagues, the amount of communication necessary among agents for coordination, and subjective preferences for certain activities. This adaptation of the role allocation and the coordination of work to local requirements during normal conditions can lead to severe consequences under more unhappy circumstances. In the Clapham Junction case (HMSO,1989), for instance, safety checks following modifications of signal system wiring were planned to be independently performed by three different persons, the technician, their supervisor, and the system engineer. Work force constraints and tight work schedules, however, led to a more "efficient" division of work. The supervisor took part in the actual, physical work, and the independent checks by him as well as by the engineer were abandoned. In addition, the technician integrated the check (i.e., a "wire-count") into the modification task itself although it was intended to be his final separate check. In short, adaptation to a more effective division of work under time pressure causes the redundancy required for high reliability and prescribed in the formal instructions to deteriorate.

In conclusion, for the analysis of human errors in a cooperative context, it is important to understand the criteria under which the dynamic relocation of work roles takes place. Formal constraints and instructions give no assurance for stable allocation under dynamic work pressure. Errors on the part of the individual actor can sometimes be explained by a locally very "rational" adaptation of the cooperative structure to meet, e.g., high workload pressure.

When cooperative work is considered to be the distributed control of a loosely coupled work environment, the exchange of *information necessary for concerted action* can be identified from the chosen division of work and the control requirements of the work domain (see Chapter 4). However, the question remains: How much of this information is explicitly communicated? In a dynamic work situation, much information is communicated through body language and by other nonverbal means. In a shared, dynamic context, the need for explicit communication becomes low, and shorthand comments are loaded with information. The verbal messages exchanged among members of a control room or flight deck crew, for instance, are very rudimentary, because the intention and the activities of fellow operators are perceived by fringe conscious perception of the focus of their attention and the objects they work on. The introduction of integrated computer based work stations on flight decks has uncovered the importance of this informal communication, because it is no longer possible to read off the intention and actions of a fellow pilot who uses a small computer screen.

Communication of values and intention is not only required for mutual coordination of activities, it is a basic precondition for the error correction feature that seems to be inherent in social organizations. When individuals are making frequent errors due to the adaptive nature of behavior, one could fear that errors would propagate willingly in a social system and frequently add up to a major incident. Fortunately, this does not appear to be the case. Individual actors are correcting faults in the messages received and they know how to correctly interpret ambiguous messages and instructions from their implicit knowledge about policies and their perception of other people's intentions and goals. The ability to directly perceive other people's intentions appears to be a facility having a high survival value through biological evolution.

It is, therefore, clear that communication of intentions becomes very important for resolving ambiguities and correcting mistakes, not to speak of trusting information and advice. When decision makers who are mutually familiar are communicating to solve a problem, subtle differences in the way a question is asked is of definite importance to the answer. For example, an X-ray film analyst has given the example of being given a particular film, and knowing pretty well what question will be asked by the diagnostician. The important information for him/her to answer, however, is *how* the question is phrased. This kind of context dependent focus of a question is normally communicated by nonverbal means, but has a drastic effect on the reliability of communication.

To sum up, the communication of intentions among members of a cooperative team is very important for reliable performance. Within a well-established team, this communication often depends on very subtle, informal, and nonverbal cues that are easily disturbed. Again, such disturbances are likely to lead to performance that is judged to include errors on part of the individual actors. In the present context an important point to consider is the danger of disturbing this informal, nonverbal communication of intentional information when face-to-face contact is replaced by communication through integrated information systems.

DESIGN OF SAFE AND RELIABLE ADAPTIVE SYSTEMS

We can sum up some important points by now turning to the design of reliable and safe work systems. For the operation of modern, integrated, large-scale systems, controlling the effects of "human error" is becoming increasingly important. At the same time, many aspects of human error are closely related to the necessary ability of the staff in dynamic work systems to adapt effectively to its changing needs. This adaptation, by nature, leads to boundary seeking behavior and, consequently, human errors cannot be eliminated. Instead, the solution will be to design a coupling interface between the staff and the work content that supports *control of the effects of errors*. The design problem is to make it possible for actors to cope with the work system at the periphery of normal, acceptable performance; that is, to design error-tolerant systems.

The classic approach to this problem is to study the effect of the various human error types that are relevant for a given activity and to arrange the work environment for proper visibility and reversibility of their effect (i.e., by a failure mode and effect analysis). However, going from analysis of human behavior and the related likely error modes toward requirements for the design of the work interface is only realistic for tasks depending on tightly constrained procedures (e.g., for control of technical equipment). In most other cases, many degrees of freedom are left for adaptive modifications of procedures, and the design considerations must be focused on the visibility of the boundaries around acceptable performance irrespective of the work procedures. That is, the boundary defining acceptable behavior must be defined by analysis of the work system and behavior of the involved staff must be supported in a way that the boundaries are respected or are made error tolerant irrespective of the content of the current task.

Migration of Behavior

It is now clear that design of work interfaces to cope with the effects of human error is more likely to be effective, if they are based on an analysis of the work system to define the boundary of acceptable performance together with an analysis of work activities at a more general level than the particular task procedures. That is, at a level that captures the generative mechanisms of performance and their potential threat with regard to violation of boundaries to loss of closed-loop control, irrespective of particular acts. This transition is similar to the transition from Newtonian particle physics to the laws of thermodynamics.

In any work system, it is possible to identify a work space within which the human actors can navigate freely. This was mentioned briefly in Chapter 1 (see Fig. 1.3). This space is illustrated in Figure 6.3 as a space delimited on one side by the limit

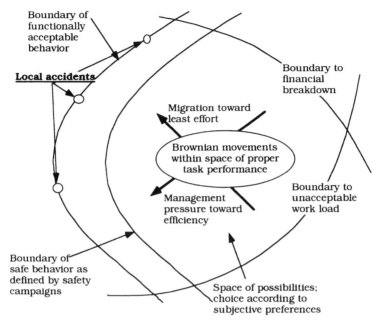

FIGURE 6.3. Activity can be characterized by "Brownian movements" within the work space, subject to work load and effectiveness gradients.

to behavior that is functionally unacceptable. Beyond that limit, the control of the productive processes of the work system is lost, work is unsuccessful or accidents happen. On the other side, the space is delimited by a boundary beyond which behavior is inefficient in a competitive environment and another boundary beyond which the work load is too high. The choice among several possible work strategies for navigation within the envelope specified by these boundaries depends on subjective criteria related to process features such as time spent, work load, pleasure, and the excitement of exploring new territory. Activity, therefore, will show great variability due to local, situational features which, in turn, lead to frequent modifications of (and shifts among) strategies. Activity will be characterized by local, situation induced variations calling to mind the "Brownian movements" of the molecules of a gas. Such variability will give ample opportunity for the actors themselves to identify "an effort gradient" while management will undoubtedly always make a "cost gradient" very clear to the staff.

Therefore, the result will very likely be a systematic migration toward the boundary of acceptable performance and, when crossing an irreversible boundary, work will no longer be successful due to a "human error." If the boundary is reversible, the actors can recover if/when the effects are noticed. If, however, they are irreversible, an accident, large or small, occurs. This systemic migration toward the boundary is observed when efforts to improve safety are compensated for by adaptive changes of behavior. When radar was introduced in commercial shipping, the result was more effective transport under foggy conditions rather than improved safety. Antilocking brakes were introduced to increase driving safety while the empirical evidence (e.g., under controlled circumstances by a taxi company in Munich) indicated no significant decrease in accident frequency but a significant increase in driving speed and speed changes (Aschenbrenner, Biehl, and Wurm, 1986). In psychological traffic research, this tendency has been explained by "risk homeostasis," which is an adaptation seeking to maintain a stable level of perceived risk (Wilde, 1976, 1988). This finding can be an artifact caused by too narrow a focus on modeling behavior from accident and error analysis. When seen from the wider conception of adaptation presented here, performance is likely to be maintained close to the boundary to loss of control in a kind of "homeostasis" controlled by a perception of the dynamic control characteristics of the interaction and not by an abstract variable such as "risk." Thus touching the boundary to loss of control is necessary (e.g., for dynamic "speed-accuracy" trade-offs). As Wagenaar (1989) found, people are *running* risks, not *taking* risks; that is, action is not controlled by conscious or subconscious risk perception.

The significance of contact with the boundary is indicated by the response to an introduction of separate bicycle paths in some streets in Copenhagen. For some streets, this change caused a decrease in car accidents but an increase of car–bicycle and bicycle–pedestrian accidents in those street crossings where the traffic had to merge (Ekner, 1989). In addition, if the effort to move back the boundary from normal behavior results in a more abrupt boundary, for example, because higher speed is possible under marginal conditions (antilocking brakes, radar), then the resulting level of safety may be impaired due to corrupt recovery characteristics.

From this discussion we conclude that efforts to improve the reliability and safety of the coupling between the staff and the work environment should be focused on an identification of the boundary to loss of control from an analysis of the work system and of the criteria that drive the continuous adaptive modification of behavior. Efforts for improvement must be directed toward the control of performance in interaction with the boundary and not on the removal of errors. For a further discussion of this approach, it is practical to distinguish between activities directly in contact with the productive work processes and those at higher planning and coordination levels.

Control of Productive Work Processes

Control of the behavior at the periphery of acceptable performance for people directly in control of the productive work processes is important for two different reasons. On the one hand, they are exposed to some hazards from work accidents when losing control of an activity, and general work safety depends on proper control at the boundary. On the other hand, activities of people involved in the physical work processes can trigger accidental causes of events that propagate through the system and ultimately cause major accidents. Thus efforts to improve performance at the boundary to loss of control can be based on different strategies.

1. One approach is to *increase* the sensitivity of actors for the boundary to loss-of-control by means of motivation and instruction campaigns that create a

gradient close to the boundary compensating for the work load–cost gradient. Safety campaigns may increase sensitivity to the onset of loss-of-control and, thereby, serve to increase the margin to loss-of-control. This improvement, by its nature, will only be temporarily effective because its influence will tend to fade away; it only works as long as pressure that acts against the *functional pressure* of the work environment is maintained. Therefore, such a motivation based struggle for a good "safety culture" will never end.

However, this approach is often the only one available for work domains having little internal structure and a large variety of loosely coupled work processes (such as constructions sites and manual work shops) at the middle-left side of the continuum of Figure 2.9. Due to a lack of stable structure and the large variety of work processes involved, the identification of the boundaries often cannot realistically be based on analyses of the particular work processes. These boundaries must be based empirically on statistical, epidemiological analysis of past failures and their causes in terms of inappropriate work practice. The theoretical foundation of this empirical approach has recently been formulated by Reason (1990) and very high levels of work safety can be attained by meticulous safety management efforts. The basis of the approach is to identify the categories of errors and work conditions from previous cases that are most likely to bring performance to the limits of acceptable performance and, then, to train actors to recognize them and to avoid unreliable work practice. Thus empirical evidence is used to control the occurrence of unreliable and unsafe acts (see the left-hand side of Fig. 6.6). Based on empirical evidence as we know it, the safety management efforts and, consequently, the resulting level of general work safety, are normally checked by the observed rate of accidents, such as lost time injuries and fatalities.

2. A second approach is to introduce *indicators, prewarnings* to indicate operation too close to the boundary to loss of control with the accompanying admonition to move back performance from this boundary. However, if the indicators are not related to the dynamic control of performance or to the active criteria of adaptation, the typical reaction could be a perception of artificiality and, therefore, the indicators may be less effective than desired.

3. A third approach is to *make the boundaries touchable and reversible*. A more direct way of control is to make it possible for actors *to sense the boundary directly* when approached and to give them opportunities for learning to recover. The trick in the design of reliable, adaptive systems can be to give the actors an opportunity to identify boundary characteristics and to learn coping strategies rather than constrain their behavior through a set of rules for safe conduct. To achieve this, it appears essential that the actors maintain "contact" with hazards in such a way that they will be familiar with the boundary to loss of control and will learn to recover.

Perception of the boundary characteristics at the level of *sensorimotor control* of the interaction in manual activities implies the need to maintain the spatial-temporal perception–action loop intact. For instance, in car driving, perception of minute changes in steering–braking characteristics give effective, early warning of changing road surface conditions and direct indications of the limits of control—but only if the direct perception is *not impaired* by servosystems.

For proper *control of action sequences*, access to a complete set of situation-defining attributes can serve to prevent cue–action responses to underspecifying signs and signals (those that do not completely define the situation). Thus activities should be controlled from a true mapping of the internal functional constraints defining the responses of the work environment and not through the use of convenient surface features. It is particularly important to consider this solution when dealing with work systems where contact to the work system is mediated through instrument panels and computer displays.

In other words, this design strategy depends on the creation of a truthful, ecological work interface that supports direct perception of the functional states of the work systems and the boundaries of acceptable system states that are also likely to be the peripheries of reversible responses to actions.

Work Planning and Management

The effectiveness in work depends on many different activities such as staffing, scheduling, supply of resources, and maintenance of equipment, which are taken care of by several functions at the management level. The consequence of errors in activities at this level often pose hazards to other persons than the decision makers—and, most likely, at another point in time. Yet, activities at the management level are without direct contact with the sources of hazard in the productive work processes. Work depends on reports, documents, and computer support systems. Furthermore, planning is typically focused on longer time horizons and on projects involving many different work activities. Therefore, adaptation at the management levels to criteria, such as work effectiveness and cost, can influence the safety level of the work staff considerably.

While protection against simple work accidents at the work process level can be controlled at the local level, as discussed earlier, the protection against major accidents, such as explosions, capsizing of ferries, or hotel fires, has to be based on multiple lines of defense to ensure that a single human error or technical fault will not be consequential. In loosely coupled systems, such defenses will be based to a large degree on warning systems, safety equipment, and administrative measures (redundancy in task allocation and work procedures). In tightly coupled, industrial process plants, defenses are normally based on automatic control and safety systems, physical barriers, and so on. In both cases, safety depends on the maintenance of safety measures that are functionally redundant from a productive point of view. Violation of such safety measures may happen when functional redundancy is removed by "improvements" in work practice to achieve higher effectiveness under normal conditions (e.g., by reducing the maintenance of safety equipment under periods with economic problems). In other words, reductions–relaxations are made in work practices that are not directly at the boundary to loss of control in the local context. In other words, precursors to major accidents are not small-scale incidents. Therefore, violations of defenses against major accidents are related to singular activities within the boundaries to local loss of control, as illustrated in Figure 6.4. In addition, a violation of a redundantly configured defense against major accidents

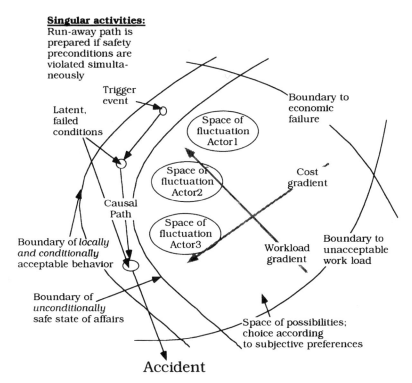

Singular activities:
Run-away path is prepared if safety preconditions are violated simultaneously

FIGURE 6.4. In a complex organization, several actors are migrating more or less independently within the space of acceptable performance. In systems designed according to the defense-in-depth principle, major accidents are caused by simultaneous violation of singular points within the boundaries defining acceptable performance, as seen locally.

is not locally detectable and special precautions are necessary to make the effect of violations directly perceptible to the actors and to work planners and managers.

As for local work safety, there is a natural migration toward accidents in the activities of work planning and management. In a complex organization, actors belong to several departments and adapt more or less independently within the space of acceptable performance. Reports from the analysis of recent large-scale accidents clearly show that accidents have not been caused by stochastic coincidences of failures of all defenses. Instead defenses have systematically eroded due to deficient maintenance during a period of excessive competitive pressure, so that accidents are the results of operating outside the design envelope due to adaptation in management decision making to these kinds of pressures.

When safety defenses depend on the activities of different groups and departments of an organization, the margin to loss of control of safe operation facing one decision maker is contingent on the activities of other planners. It will be difficult to "make visible" the boundary at the higher management levels unless an information system is designed that interrelates decision support throughout the organization.

How this can be done depends on the location of the work domain in the map of Figure 2.7.

RISK MANAGEMENT STRATEGIES

Different work systems based on different technologies and activities pose quite different hazards, thus different modes of safety control and risk management have evolved. To a large degree the efforts spent by society depend on the integrated losses across accidents within categories of accidents. Three main categories of accidents can be identified as shown in Figure 6.5.

The three categories in Figure 6.5 are characterized by the number of accidents contributing to the overall loss during a certain period and by the pace of technological change compared to the mean-time-between-accidents within the different categories. A clear distinction can be made between the three categories with respect to the strategy chosen for hazard control (Rasmussen, 1994b).

1. General work safety involving frequent, but small scale accidents: The average level of safety is controlled from epidemiological studies of past accidents. That is, a closed-loop, feedback control strategy is used.

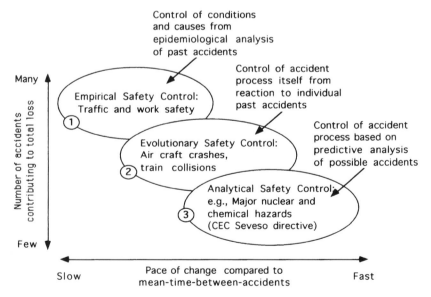

FIGURE 6.5. This figure illustrates the basic features of different hazards that led to different risk management strategies.

2. Protection against medium size, infrequent accidents: In this category, safer systems evolve from design improvements in response to analysis of the individual, latest major accident.

3. Protection against very rare and unacceptable accidents: Risk management in this category is based on multiple defenses against release of energy or hazardous substances. Defenses are based on predictive risk analysis not primarily on empirical evidence from past accidents. That is, safety depends on an open-loop planning strategy.

RISK MANAGEMENT IN LOOSELY COUPLED WORK SYSTEMS

In loosely coupled work systems in the left-hand side of the domain spectrum, the hazard related to loss of control of activities originates in a great variety of work processes, and the singular activities that endanger the defenses within the total space of activities shown in Figure 6.4 will not be stable and cannot be explicitly defined. Current approaches to develop a strategy for risk management at the work planning and supervisory level are therefore based on campaigns to make managers conscious of the error types, which are empirically found to lead to unsafe work planning practice. They are also informed about rules of safe conduct derived from past experience. Thus all in all, risk management is focused on "safety on the average" by empirical strategies (see the left-hand side of Fig. 6.6). This mode of hazard management in work planning is an extension of the empirical approach to work safety at the operational level that was discussed above. Reason (1990) recently described a consistent theory for empirical identification of a safe margin to loss of control, and of the threats against this margin in terms of "resident pathogens" in management practices, derived from analyses of past incidents.

This approach will encounter some basic problems when an organization is subject to excessive competitive pressure. Normally, analyses of accidents do not identify only one single "root cause," but a considerable number of errors, violations, and latent conditions interacting in a very unique way. This number of contributing causes can be increased ad libitum for any accident by searching further back in the causal tree. Yet, many of the causes and conditions found from accident analyses may not contribute to future accidents. Attempts to control safety by campaigns seeking to avoid the empirically identified causes and conditions in the future will very likely be confronted with the "false alarm" fallacy. When working under pressure in order to be cost conscious, one simply cannot be so careful as to always avoid all "resident pathogens" that were identified empirically from prior cases. The large repertoire of potential "less-than-adequate" conditions that can be identified empirically is demonstrated by the large number and complexity of the causal trees that Johnson has included in the Management and Oversight Risk Tree (MORT) analysis tool developed from his life-long experience in the US National Safety Council (Johnson, 1980). In addition, as discussed in Chapter 4, defenses based on administrative measures aimed at controlling the division of work, supervisory monitoring, and safe

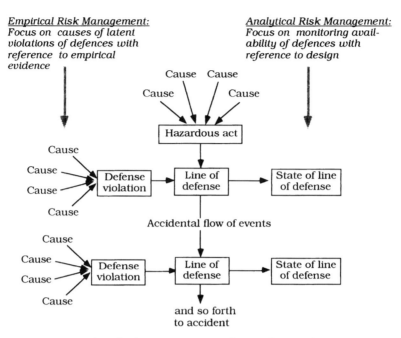

FIGURE 6.6. Two modes of risk management modes are illustrated: Empirical control is focused on the quality of defense related work with reference to general failure types. Analytically based control derives a focus of the quality control from monitoring the effect of work on availability of defenses with reference to design specifications.

work procedures are likely to be overruled by adaptation to the other, active criteria governing cooperative work.

Empirical safety control has a natural limit that can be illustrated by an example. Several farmers were electrocuted when working on agricultural watering systems (Casey, 1993). When moving the system from one location to another, apparently some workers occasionally raised the 38-ft. long thin-walled aluminium pipes to a vertical position and touched high voltage lines. No one expected farmers to raise the long pipes to a vertical position for transporting them to the next field. Only when actually visiting the location and interviewing people did it become clear that farmers usually raised the pipes to a vertical position to release rabbits hiding in the pipes— lethal consequences resulted when done below high voltage transmission lines. This hazard could not be found a priori from empirical evidence and the example illustrates the limits of empirical risk control.

Recently, a trend towards a "no accident is acceptable" strategy (Visser, 1991) has been found for work safety. In this approach, effective protection against violations of defenses against loss of control of the productive processes must be planned selectively for explicitly defined hazards. Control of safety is based on a functional analysis in the design of appropriate defenses against the physical processes that can release the hazard. Thus potential accidents have to be "designed" before the defenses can be designed. With reference to our example: First, the hazard is identified—electrocution under high voltage lines. Next, the accident is designed by seeking in the work context the components (in terms of equipment, action sequences, and so on) that can construct the accidental course of events—in this example, pipes longer than the distance to the transmission lines. Defenses in the present example would involve the use of shorter pipes in fields below transmission lines and not teaching people general human error categories.

The point is that effective safety control beyond a certain limit must be based on focused analyses of the hazards and work contexts actually involved and not on general analyses of past experience. Potential violations of defenses as a consequence of improper work planning can then be detected by monitoring the plans with reference to the specific design basis and not to the general pool of past evidence (see Fig. 6.6). In such cases, the singular activities endangering defenses illustrated in Figure 6.4 can be analytically identified from the functional analysis and the related design basis; they can be made visible to work planners.

LOW RISK OPERATION OF HIGH HAZARD SYSTEMS

The analytical risk management strategy is actually applied for high hazard systems, such as chemical process plants and nuclear power plants at the right-hand edge of Figure 2.9. For such systems, the effect of the possible accidents is so large that the related, accepted mean time between accidents is very long compared to the lifetime of a plant.

Due to the large effects of these potential accidents, systems are designed according to a "defense-in-depth" philosophy. A release of the potential hazard is prevented even if the effects of several technical faults or human errors are present simultaneously. Protection is based on several automatic lines of defenses: (1) redundant equipment is introduced; if one piece of equipment fails, a spare or stand-by is ready to take over automatically; (2) if control of energy or mass accumulations fails in spite of this precaution, it can be detected by monitoring critical parameters, such as increasing temperature or pressure and the process can be shut down by automatic emergency actions; (3) also, if this barrier fails, energy or mass can be retained by containment; or (4) diverted by barriers, and so on. Only a coincidence of errors and faults violating all lines of defenses will release a full-scale accident and, therefore, hazard control is directed at keeping the barriers intact. This strategy is only realistic for systems in which the hazard originates in a few, well-known and well-bounded physical process, such as tightly coupled technical systems. Then the margin to loss of control and its dependence on the operational states of the various defenses can be calculated by a predictive risk analysis.

In a system designed according to the defense-in-depth policy, safety preconditions related to different parts of a protective system, including passive standby functions and functional redundancy, can be affected individually by diverse work activities (maintenance, calibration, and testing) with individually acceptable influences on the overall level of safety. If, however, they are violated simultaneously by different actors, an accident can be the result.

Therefore, low risk operations management of a high hazard system in a competitive environment requires that the theoretical performance boundaries and, in particular the operational state of defenses, are individually made active and responsive to violations to prevent work planners from "running risks" when dealing with nonsafety related issues, such as general resource management and work scheduling.

This involves the introduction of a new information environment for monitoring and work planning at the management level, which can make the general state of defenses with reference to the safety design strategy visible to all agents. A safety index derived from the inverse probability of an accident can be estimated by an on-line, simplified "default" risk analysis and made visible for risk management purposes and for documentation of the positive influences of such a risk management (Fussell, 1987). This can be the necessary incentive for a continuing risk management effort but, in addition, visible margins to safety boundaries can increase operations efficiency by removing the need for excessive margins to invisible boundaries.

Comparison of the empirical risk management strategy normally applied for less structured work systems with the analytical strategy applied for highly structured systems, such as industrial process plants. An analogy can be drawn to the difference between the empirically based diagnostic strategies found in the medical domain and the analytically based variationist strategies of technical diagnosis. See the discussion in Chapter 3 (cf. Reason's use of the medical metaphor of "resident pathogens").

Incentives and Commitment

Recent analyses of major accidents indicate that safety problems remain even if the boundaries of safe operation are made visible to decision makers. The closed loop involved in the control of high hazard systems apparently have some additional basic problems.

As discussed in Chapter 1, effective functioning of a feedback system depends on (1) explicit objectives, (2) a reliable measuring function, and (3) a controller operating within the capability limits. To ensure safe operation, all of these mechanisms must be in place. Apparently, the safety measure advocated above is not enough. In some cases, management was actually informed that operations were taking place beyond the boundary (e.g., in the Zeebrügge case by repeated memos from captains) and yet operation was continued. The problem seems to have deeper roots related to management capability and commitment (see Rasmussen, 1994c).

After a recent ferry fire (Scandinavian Star) and an oil spill at the Shetland Islands, a representative of the Danish marine safety authorities in a TV interview expressed the fear that marine safety would decline because ships were now operated by banks and investors rather than professional shipping companies. Is the required level of safety actually financially acceptable?

Recent accident cases, such as Zebrügge, Clapham Junction, appear to emphasize the question: Are managers willing to spend the effort required for effective risk management? In many cases, when judged after the fact, liabilities and losses could reasonably be anticipated; accidents were foreseeable and obviously preventable. In theory, one would expect the fear of potential liability to serve as a substantial incentive for a company to voluntarily undertake management initiatives to minimize risk. However, despite ample evidence of a liability explosion—especially in the United States—companies continue to experience numerous accidents, indicating that the liability incentive in reality is incomplete or obstructed in several respects. Why do companies fail to act voluntarily to prevent risk and its economic impacts? Because of an economic necessity during periods of high competitive pressure? Because of the nature of "naturalistic decision making?" Due to the legal constraints of management toward shareholders? Is it a problem that CEOs are legally responsible to shareholders to ensure that their decisions are economically sound and, therefore, judgment must be based on a rather short-time horizon?

Are we facing a basic problem of time scales in technological development (such as political election periods, personal career spans of managers, planning horizons for companies and public services) and, finally, the acceptable mean time to accident in the individual company? Is it realistic to expect managers with a personal career planning horizon of a few years and with a legal responsibility to the shareholders to be economically sound in the short run and, perhaps, in the face of an economic crisis, to balance decisions rationally against the risk of a major accident over a horizon of a century? More emphasis on multiple-criteria management including ethical accounting seems to be necessary (Bogetoft and Pruzan, 1991).

Even if the commitment is found, the professional background is likely to be less than adequate at the level of an organization at which managers have access to information about all the activities related to the maintenance of defenses protecting high hazard processes. This is the case partly because the knowledge required is not maintained during normal management activities at higher levels of the organization, partly because of the normal recruiting policy for managers in a dynamic industry where high level managers are chosen among law and business school graduates, and are not promoted from the technically competent staff.

Functional Requirements at the Periphery of Normal Operation

The need for error tolerant systems adds some additional functional requirements to an ecological information system.

- The boundaries of normal closed-loop performance should be made visible to the individual actors and violation of the boundaries should be reversible.
- In systems based on multiple defenses, the boundaries of acceptable performance for an individual actor depend conditionally on the activities of other actors. A prerequisite for making the boundaries visible at the local level, therefore, is an integrated information system based on a model defining the relationship among the states of the individual barriers, based on the design philosophy underlying the defenses.

Chapter 7

The Design Process and Its Guidance

INTRODUCTION

We will now turn to the process of designing information systems for assisting humans at work. In particular, we will discuss a new type of generally applicable design support that incorporates the analytic framework discussed previously. For this aim, a review of the current effort to model the design process will be useful as will a brief discussion of existing—to use a broad term—human factors guidelines and design criteria.

THE DESIGN PROCESS

We will review the results from the analysis of two actual design problems to set the stage for the further discussion of the requirements for appropriate guides in the design of advanced information systems.

The first example (see Fig. 7.1), an *accounting device for a retail shop*, starts as a traditional function → form design task. A new tool is required for measuring what is left on a roll of textile without unrolling and measuring it with a yardstick. The purpose is to facilitate the annual status accounting, which is a considerable problem in a company with thousands of rolls in each retail shop. As the design process unfolds, details in the design implementation initiate ideas that result in the propagation of potential changes throughout the organizational structure of the company. The proposed solution is to weigh the rolls and, at the same time, take a calibration measurement of the weight–length relation, followed by a calculation of remaining length of material. In considering all the different designs and shapes of roll-cores from different production sources, a correction is needed of the weight measured by

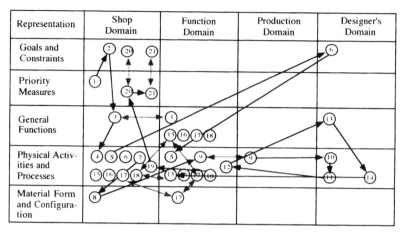

FIGURE 7.1. The trajectory in the problem domains from design of a retail shop accounting device. The numbers refer to a sequence of topical statement in the minutes of design meetings.

subtracting the weight of the core. The identification of core model and retrieval of its weight through the use of a bar-code label is chosen, together with a computer-based calculation. This invites the introduction of the equipment directly in the individual sales transaction. The bar-code solution for roll-core typing provides improved customer service via a simultaneous identification of the material and a print-out of the instructions regarding use of the material on the bill. At the same time, an automatic recording of all sales transactions in a centralized data base opens up for a more optimal stock administration, as well as a more rapid adjustment of the distribution of items to the individual shops according to changes in customer preferences.

This example illustrates how a design task, initially formulated as a classical function → form transformation problem, propagates from ideas initiated by the local, specialized function of a separate piece of equipment. This is accomplished through considerations of a more general integration of its function into the wider activities of its user until, finally, changes in the commercial strategies of the company are considered. A design task formulated in rather specific functional terms evolves into a much more complex discussion of basic company strategies, involving specialists in commercial and organizational issues.

A very similar pattern was found in a manufacturing company concerning the design of an improved *manufacturing process for a door lock* (see Fig. 7.2). A manufacturer of locks wants to modernize the production facilities by introducing computer-controlled machinery. The resulting increased production accuracy gives the possibility of an update of the lock design with more "bits" in the lock code. This, in turn, opens the potential for more complex lock systems having key codes grouped for different personnel categories within an organization. This can influence the personnel administration and the insurance conditions of a customer. In addition,

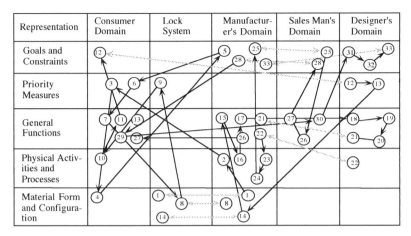

FIGURE 7.2. The trajectory from a manufacturing design problem. The numbers refer to a sequence of statements in minutes of design meetings.

computer-based manufacturing can provide a change in the strategy for designing the individual lock system. In the old design, the foreman of the production workshop would choose the set of key codes for a particular lock system from tables on paper forms and simple heuristic rules regarding useful and inadequate combinations. In the new system, the combinatorial space is too large for heuristic design. Production planning of particular systems must now be based on a mathematical model which, however, can be implemented as a planning program in a portable computer. However, this now makes it possible for the lock manufacturer's traveling salesperson to plan a particular lock system and choose the relevant code system in direct interaction with a customer and, thereafter, hand over to the workshop a complete production specification in the form of a program listing for the numerically controlled production equipment.

In this example, the design problem is initiated by the introduction of new production equipment, which reflects back onto an updated product design which, in turn, give a potential for a different application philosophy in the customer's organization having influences on personnel administration and insurance coverage. In addition, the new planning tools, resulting from the more complex planning, open up the possibility for a new role allocation in the manufacturing organization, in which production planning is moved from the workshop floor to the sales department.

In the following sections we will use these examples as background for a discussion of some important design issues.

MODELS OF DESIGN

In his textbook on design, Hubka (1982) argues:

The predominant center of interest in recent design research has been the generation of general procedural models. This may be seen from the many models that have been published.

Procedural models represent design as an orderly process of "transforming the information of a customer's statement of requirements to a full description of the

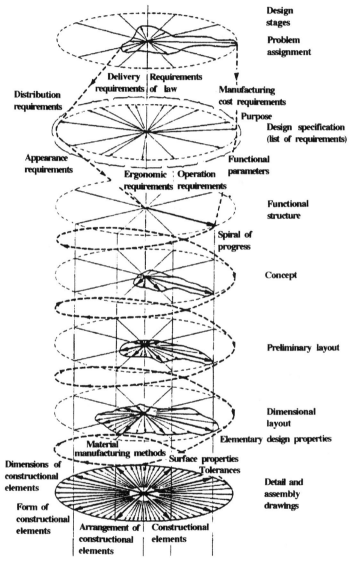

FIGURE 7.3. The classical "sprial model" of the design process. Reproduced from Hubka, 1980, with permission.

proposed technical system;" see the "spiral model" of this in Figure 7.3. Such approaches separate and isolate the function–form transformation from the much wider design contexts utilized in the examples of Figures 7.1 and 7.2.

The rationale for this traditional focus in design research will probably be found in the university teaching environment. Models in terms of rational, normative procedures are well suited for teaching in the established engineering and industrial design schools. Not only are the domains included in the models used in the curriculum of a university course limited but, in addition, normative procedures are effective for the introduction of novices to the rational and analytical tools of the profession. When normative procedures have "initialized" the newcomers, they will subsequently be able to adapt to their particular work contexts and develop their own individual heuristics and, in time, discover the tricks of the trade.

However, the examples show clearly that the decisions taken during an actual design task cannot be represented by a prescriptive and well-ordered sequential progression from problem formulation to solution. Instead this and similar cases show that a more realistic approach will be based on separate yet compatible representations of the knowledge domains involved and of the decision strategies and heuristics used for navigating in these domains. The examples clearly illustrate that engineering design is characterized by iterations among several different domains, which are only known in sufficient detail by different individuals, usually with different professional backgrounds. Since design is a search

for some kind of harmony between two intangibles: a form which we have not yet designed and a context which we cannot properly describe (Alexander, 1964),

a design task will very likely begin with a mutual exploration of context by the members of the group representing different perspectives on the design problem. A similar experience has recently been described by Bucciarelli (1984) from an ethnographic study in a design team:

The task of design is then as much a matter of getting different people to share a common perspective, to agree on the most significant issues, and to shape consensus on what must be done next, as it is a matter of concept formation, evaluation of alternatives, costing and sizing all the things we teach.

In a later study, Bucciarelli (1988) elaborates on this point of view, introducing the concept of "object world" to point to the context within which the individual engages the design, a concept which is similar to the "domains" behind the trajectories in Figures 7.1 and 7.2. He has also identified the need for iteration between different domains. For example, in the design of security equipment for airports, it was necessary to consider the items to look for from two points of view (1) focused on the likely strategies of terrorists and (2) on the detection of objects in airports. Difficulty in detecting "explosives" by X-rays, as formulated in the design specification, disappeared after consideration of the terrorist context and a subsequent reformulation of the requirement to detecting "bombs" because of the necessary wires and switches. This he calls the "naming discourse":

The invention of a name for a part of the design, for a piece of the action, is designing.

One additional example from our analyses will serve to support this view of design as the identification of a solution that matches the contexts of different domains or "object worlds" (see Fig. 7.4).

A telephone company needs a cable joint to be used for assembling undersea optic fiber cables and, together with the cable manufacturer, contact is made with a design team at the mechanical engineering department of a technical university. The functional product specifications are stated—such as the number of light conductors, the predicted cable stress in case the cable is caught by a ship anchor, the deployment equipment and procedure to be used and, finally, the physical–chemical specifications of the sea environment.

Available deployment equipment constrains the acceptable external diameter. On the other hand, the predicted low reliability of laser-welding operations at sea requires room for coils absorbing some spare length of each fiber to enable rewelding fibers individually; this implies the need for an increased internal diameter of the joint. A completely new design of the mechanical configuration of the joint is therefore necessary in order to find a less space-consuming fixture to absorb the specified cable stress. In addition, delicate laser welding operations are required on board the cable deployment ship under rough sea conditions. These requirements lead to a complex

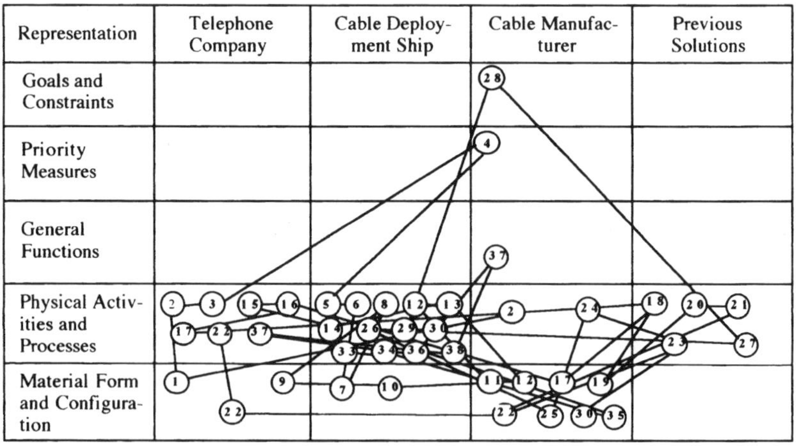

FIGURE 7.4. The trajectory of work during the design of a cable joint. The figure illustrates clearly that the process is not a well-ordered transformation of specifications into a product, but a lengthy iteration among domains which are normally isolated. It is also seen that classic function-form transformation is only a very limited part of the task.

iteration between form–function conditions, where the cable joint is viewed as being (1) a part of a deep-sea telephone line, (2) an object of a rough sea assembly task, (3) the object of the available manufacturing processes, and (4) preferably a modification of prior, proven cable joint solutions.

This example illustrates a rather clear-cut and typical design task in that a change of one material in an assembly that normally can build on a well-established design tradition requires an updating of the physical design. As it turns out, a complete reconsideration is required within several contexts, each with its expert—an engineer from the telephone company, the foreman of the assembly team of the cable deployment ship, and the production planner of the cable manufacturer. Given the design requirements, they have to identify the implications within their respective domains for a solution that does not exist. Therefore, for a considerable period during the design, their role is that of a team of mediators in a domain exploration, supported by a number of design proposals, mock-up models, and experiments with partial solutions.

This process of iteration among the object worlds of different individuals is not treated in the normal design modeling approaches in any great detail. In a way, the problem is circumvented by perceiving the total process as having distinct phases: The design process (feasibility study, preliminary design, and detailed design) and the production–consumption cycle (planning for production, distribution, and consumption) which are considered separately:

> Generally, a new phase is not begun until the preceding one has been completed, although some finer details have to be attended to, while the next phase is in progress. Asimov (1962).

Thus a product design can move through different units of an organization as well as have the interest of different research teams—as exemplified by the separation of interface design from the rest of the functional design and implementation. However, this will not be appropriate in an integrated system design environment.

Consequently, there is a need for the development of tools to support the named mutual exploration of "object worlds" or, in our terms, work domains during the design process. However the development of such an aid is not just a question of compatible data bases and computer systems as much as it is a fundamental problem of structuring knowledge base contents to suit the search queries of the designers in their open-ended "naming discourse."

"Design" During Work, Exploration of Resources, and Information Retrieval

This mutual exploration of work domains has wider implications than for the "design task" in its traditional sense of designing a product for subsequent manufacturing. The decision processes involved in many different situations in a modern, dynamic workplace also require a nontrivial exploration of the work domains of other actors, groups, or organizations. When work is no longer planned and centrally organized in accordance with a stable work practice, but instead depends on continual adjustments, up-dating, high tempo local replanning, there is a considerable element of design and domain exploration involved—at all levels.

Current discussions about a more effective planning of the treatment of patients in hospitals are considering the establishment of integrated health care systems through which the general practitioner can have access to information about the resources and schedules of hospitals in order to plan a patient's hospitalization in cooperation with the patient. This reflects an attempt to solve a domain exploration problem. To plan hospitalization, one has to find a plan that matches the context of the patient's normal "domain" of work, leisure and social relations, and current state of health, a context to which the general practitioner has access during an interview. At the same time, the plan has to match the hospital context; that is, the resources, waiting lists, expertise of doctors, and the time schedules that normally are only known by the hospital staff (see Fig. 3.13 and compare to the design example of Fig. 7.4).

Another example is the effort of the foreman illustrated in Figure 3.5 to effectively control the availability of the resources required for his/her local plans. In his/her negotiations with the foremen up- and downstream from his/her station, he/she is involved in the design of plans that match the resources–constraints of the domains of the different involved actors.

Finally, emergency management that serves to mitigate the effects of major accidents, such as an earthquake or a major industrial accident, involves a significant element of domain exploration. Planning activities such as fire fighting, evacuation, environmental protection, hospitalization, traffic control, public information services, and so on involves many different institutions and services with their particular, often very complex, domains. For example, in case of a major fire in a chemical plant manufacturing hazardous substances, the fire chief needs to consider the plant domain including the equipment, processes, and substances found in the plant; he/she has to consider the threat to the domain comprising the environment—information on population, building layout, access roads, and meteorological data and has to consider the technical resources at the location having to do with the water supply system, the sewage systems, and their sensitivity to hazardous substances. This is a very tricky problem during stressed conditions and several attempts have been made to design information systems composed of data bases on hazardous substances as well as services providing access to expert advice. The problem, however, has generally been that the commander who has the context but not the expertise in the other domains cannot phrase a query precisely enough and, therefore, will very likely receive either too much or too little information in response.

Thus it can be seen that the exploration of several domains or "object worlds" is not only a problem in a classic design task but is becoming increasingly important for many decision makers in modern, dynamic work environments. Basically, this exploration problem can be formulated as an information retrieval problem involving (1) the identification of relevant features in terms of their means–ends relations within the *work domains* relevant for a given *task situation* of the decision maker, and (2) the identification, based on an analysis of their decision strategies, of the query language most likely to be used in seeking information.

This means that the design of information systems to support domain exploration is not only important for design in the traditional sense, but also for planning in many other work situations. As a basis for identifying the requirements for such an information system, a prototype system was developed to support information retrieval in public libraries. The cooperation of a librarian and a library user who normally negotiate in a mutual exploration of the personal domain of the user, as well as the domain of the available book stock, is a good example because it is a well-defined problem context and has been subject to a careful analysis. This system and its empirical basis are described in detail in Chapter 9, which also gives numerous references to the relevant system analyses.

Design Strategies

It is clear from the examples given that the design process is neither a well-ordered progression from a problem formulation to the implementation of a solution, nor is it a conscious, rationally planned process.

In fact, an experienced system designer is deeply embedded in their own context and very often will treat their current requirement as an update or modification of prior designs. The system designer will have many preconceived ideas and solutions at several levels in the goal–function–form hierarchy. In other words, they will not approach design as an orderly top-down synthesis, but instead will consider the process as a sideways modification of prior solutions to similar problems. They will implicitly or explicitly have to conform with constraints from many different sources. These include the ultimate user needs, the company policy and product style, the financial policies of the company and its preferred part suppliers, the hot issues of their own profession and their subjective preferences, personal style, and creative image. All of this influences their design choices and cannot be ignored from a rational design point of view. Many degrees of freedom for choice are present within the space of acceptable product solutions—especially in this computer age.

The *creative* phases of expert designers' work depend very much on intuition and experience. Hillier and Leaman (1973), discussing architectural design, call this intuitive basis of design the "prestructure." They argue that the rational process models of design are ineffective because:

> The syncretic generation of outline solutions in the early stages of design is made to appear illegitimate and undesirable on the grounds that any "rational" procedure to design must seek to generate the solutions as far as possible from analysis and synthesis of problem information and constraints.. . . . The designer's prestructures are not at all an undesirable epiphenomenon, but the very basis of design. Without prestructures of a fairly comprehensible order, it is not possible to identify the existence of a problem, let alone solve it.

The view that intuition and subconscious processes are important for creativity and discovery has long roots. The mathematician and philosopher Poincare is the source of several classic anecdotes describing the importance of subconscious processes in discovery. He argues that scientific discovery depends on new combinations of known ideas. However, since an infinite number of possible combinations exists, discovery depends basically on selection:

> Discovery, as I have said, is selection. But this is perhaps not quite the right word. It suggests a purchaser who has been shown a large number of samples, and examines them one after the other in order to make his selection. In our case the samples would be so numerous that the whole life would not give sufficient time to examine them. Things do not happen this way. Unfruitful combinations do not so much as present themselves to the mind of the discoverer (Poincare, 1904).

However,

> . . . sudden inspiration never happen except after some days of voluntary effort which has appeared absolutely fruitless and whence nothing good seems to have come, where the way taken seems totally astray. These efforts then have not been as sterile as one thinks. They have set going the unconscious machines and without them it would not have moved and would have produced nothing.

This rather closely resembles our perception of the creative phases of design as culled from discussions with expert designers. Later, another French mathematician, Hadamard (1945), elaborates on Poincare's discussion and describes four phases of creative discovery: (1) *preparation*, which involves a conscious exploration of available ideas and concepts, (2) *incubation*, "no work of the mind is consciously perceived," but something is happening which, at a later instant, results in (3) *illumination* when ideas emerge, and (4) need to be *made precise and evaluated*.

If we generalize from this discussion to the information system design process, we can see that the phases of "*precision and evaluation*" are the parts of the design process that are best suited for a normative procedure formulation while the more creative (and equally important) portions including the "*preparation phase*" and the involved "object world" explorations require special treatment. The most effective way to help designers seems to be (1) to support in some way their intuitive explorations of the space of solutions and (2) to supply tools for a systematic evaluation of the results of their more or less intuitive design efforts. The mental processes underlying the creative phase can be compared to browsing in a shop or library. One has a problem or need that cannot be made explicit. Support of this process can only be realized if a recognizable structure can be defined for the space to be explored.

One ingredient in our framework is eminently applicable here. For the creative phase of design, a knowledge-base structure is needed that will aid the transfer of search heuristics among different "object worlds." This is precisely what is intended with the part–whole/means–ends map discussed in Chapter 2. If an identical structure can be imposed on the different object worlds, initial learning, exploration, and "naming" will be facilitated. If the knowledge bases of the various domains are organized according to a single uniform structure, the use of an integrated information system will be possible.

A means–ends structure is already implicit in several approaches to structuring design models in the form of more limited function–form considerations. Alexander (1964), states that "every form can be described in two ways: from the point of view of what it is, and from the point of view of what it does." Hubka's (1980) normative modeling of the design procedure is based on an "abstraction and classification of technical systems," which is a direct transfer of the biological function–form distinction in terms of Phylum and Class for general functional properties, Genus for internal function structure, and Family for anatomical structure. However, it appears that this conceptual development has been restricted by its origin in a biological terminology that does not include intentions and purpose—both of which are indispensable for a design framework.

To sum up, design has elements of a creative, rather haphazard nature (this is not meant in any negative sense). Sometimes, a design is a mere updating (redesign) of a previous result; it can also be a creative invention on the basis of an individual insight, a technology-driven development (laser, multimedia), or an aesthetic self-realization (architecture). Therefore, the underlying design rationale for this book is—rather than striving to impose a well-ordered normative design process on designers, the creative phases should be supported by imparting a structure to the involved object worlds. Efforts to formalize them should be spent on after-the-fact design rationalizations and evaluations (see Chapter 8).

DESIGN GUIDES

To repeat, two important conclusions can be culled from the previous sections and, incidentally, are shared by experienced designers of real world systems (e.g., Whiteside, Bennett, and Holtzblatt, 1988; Woods and Eastman, 1989). These conclusions can be summarized as follows:

1. Design is an inherently variable and opportunistic process. Thus, design solutions depend on a host of unpredictable factors that are unique to the particular problem (cf. Gould, 1988). In each of the cases reported above, discussions of relatively local design issues resulted in widespread repercussions that could not possibly have been anticipated. The open-ended and ill-defined nature of the problem involves an exploratory process that is highly variable and, therefore, unpredictable (= unamenable to proceduralization). Any effort at providing guidance to influence this process must at least take these facts into account. Failure to do so will likely result in a lack of significant impact on the design process and, as a result, on the quality of future information systems.

2. Experienced designers tend to adopt an intuitive, recognition-primed mode of decision making (i.e., they have been there before). While this may at first seem contrary to the first conclusion, it means that, in any given choice situation, they will tend to make decisions on the basis of a simple selection among the alternatives they see and recognize in the context. Therefore, in the course of their education and

professional career, if they have not developed an understanding of or intuition for—again the broad term—human factors issues, then it is unlikely that they will ever—if not provoked—naturally consider such aspects in a thorough way.

In the following sections, we will firstly briefly evaluate how well traditional design guidelines and associated design criteria meet these requirements and then we will present an alternate approach building on the framework discussed earlier.

Limitations of Human Factors Guidelines

As defined here, traditional guidelines are prescriptive recommendations for identifying and incorporating human factors requirements into a product–system in a way that improves the overall quality of the design. Guidelines can range from the very specific to the very general (i.e., from hard data concerning specific ergonomic features to more general design criteria to checklists, structured questionnaires, and flow diagrams of the design process). Usually these various types of aids are general purpose and, therefore, context-free in content—issued as they are by governmental and defense agencies or by user groups within a large encompassing domain (e.g., industrial process control). For typical examples, see Boff and Lincoln (1988), various US MIL-standards and other documents on human engineering, the EWICS' effort on "Guidelines for the Design of Man–Machine Interfaces" (1981) and, finally, HCI guidelines such as Smith et al. (1983). Let us consider how these kinds of aids have been evaluated in the field.

Traditional design guidelines apparently do very little to foster intuition and creativity and/or a re(usable) knowledge base. On the contrary, they represent an attempt at providing advice about human factors issues to people who have little or no human factors expertise. As a result, experienced designers will tend to ignore guidelines and base their design decisions on the factors that they are used to considering (Meister and Farr, 1967). This suggests that human factors will be put off until someone realizes that there is a problem, but by that time it may be too late to remedy the situation in a cost-effective manner.

Perhaps guidelines exhibit their greatest weakness in connection with the conception of portions of design as an inherently variable and opportunistic process. As mentioned earlier, the specific design solutions that are developed for any one application depend on a host of unpredictable factors that are unique to that particular problem. Since guidelines are intended to be generally applicable across a wide number and variety of design problems, they cannot possibly capture the rich sensitivity to context that is required for effective design. The point is well described by Gould (1988):

> Guidelines cannot deal with choices that are highly dependent on context, as many choices in interface design are. Human performance adapts strongly to details of the task environment. There are simply so many details, and this adaptation is so little understood, that guidelines cannot hope to anticipate all of this (Gould, 1988, p. 782).

In brief, design is heavily context-dependent, whereas guidelines attempt to be context-free. An excellent example of this point has been illustrated by Grudin (1989), who has pointed out the limitations of one of the most frequently mentioned design criteria, namely, that interfaces should be designed to be *consistent*. Through the use of several detailed examples, Grudin demonstrates convincingly that

> interface consistency is a largely unworkable concept; the more closely one looks, the less substance one finds (Grudin, 1989, p. 1164).

The basic reason for this occurence is that there are other considerations that may be more important and that should, therefore, override consistency as a design criterion. Most of these other factors are contextual, depending on the characteristics of the user and the work context. Thus, it is the users' work contexts that should be the primary constraint on design, not interface consistency. We would go one step further in hypothesizing that the difficulties identified by Grudin with consistency generalize to most, if not all, design criteria. Exceptions are always possible and design trade-offs need to be made with a deep and clear understanding of the available resources, as well as the constraints imposed by the work context. Other work leads to the same conclusion: Woods and Eastman (1989) found that

> traditional human factors and human–computer interaction guidelines and handbooks were of extremely limited use during actual design projects (Woods and Eastman, 1989, p. 29).

This is not to say that traditional guidelines are useless; only that they are extremely limited as a sole source of support for the design phases focused on in these cases.

Another problem has to do with the trade-off between specificity and generality. Results from surveys indicate that, in certain situations, some designers would rather have specific rules that they can follow in a rote manner than general principles that must be instantiated and translated before they can be directly applied (Smith, 1988). This is probably particularly true for designers who do not have a background in human factors and, therefore, the expertise necessary to make the jump from general principles to design decisions in an effective manner. The obvious difficulty with making guidelines too specific is that generality is lost. The reverse problem has to do with the question of whether our understanding is deep enough to even justify specific recommendations.

As a result of this inherent and necessary tension between the specific and the general, it is generally acknowledged that guidelines must be interpreted within the context defined by the design problem in question, and that the use of guidelines needs to be mediated by good judgment (e.g., Gould, 1988; Smith, 1988). In short, guidelines cannot take the place of experience but they can take advantage of it if properly formed. As Meister and Farr (1967) state,

> The designer . . . tends to respond rather immediately to the design situation on the basis of experientially derived stereotypes. He comes to design armed with an inventory of most favored configurations which he applies rather routinely until and unless he runs into a problem (Meister and Farr, 1967, p. 77).

Therefore, a more productive form of guidance will have to instill and utilize a sense of intuition in designers so that framework issues regarding user resources, limitations, and inclinations are naturally considered in the design process. This possibility will now be discussed. ·

An Alternative: Maps of the Design Territory

As an alternate to traditional guidelines, we propose an approach based on the framework presented in earlier chapters that is intended to help designers attain a more structured and, hopefully, a more generalizable view of their (and others') user–work interface designs. As part of their continuing acquisition of experience and expertise, we wish to assist designers in gaining a better appreciation for the importance of understanding and, thereby, controlling the relations between (1) user interfaces and (2) the work system properties and user characteristics.

In previous sections, it was argued that displaying the invariant structure and the affordances of a work domain is a more effective way of supporting actors during discretionary tasks in a dynamic work environment than presenting normative procedures. In other words, the implication was that *maps support navigation in a work space more effectively than do route instructions.* By recursion, this argument can be applied to the support of design itself and, consequently, we suggest the use of the concept of *maps of the design territory* within which designers will be able to navigate as an alternate to following traditional route instructions or guidelines.

In addition, we propose the use of so-called WTU-based prototypical representatives of utilized interface designs and/or design candidates, which can be characterized by their relations to *W*ork domain, *T*ask situation, and *U*ser category. These can be located and presented with reference to their location in one of the maps of the design territory. When used within the context of the framework discussed earlier, the maps together with the set of samples can hopefully help designers gain an awareness and intuition about their design territory in somewhat the same way as skilled travelers develop a prototypical map of world capitals and their characteristic localities. Our own initial collection of display candidates will be presented later (see Chapter 13). It is our intention to assist designers in generalizing from previous analyses–experience when designing new ecological information systems and in transferring successful solutions from other domains and task situations. In this respect, the hypothesis underlying the present approach is that by having a set of conceptual maps, a designer will obtain an improved awareness of important similarities and differences between information system designs relevant for different WTU combinations and, at the same time, assimilate a working understanding of the influences of all the dimensions of the framework—both system and human related.

A Historical Perspective

In the process of reviewing the literature to find the origins of diagrammatic representations for use in interface designs, it turned out that the proposed approach to design guidance is not at all new. In a historical overview of the role of nonverbal thought in technology, Ferguson (1977) finds that "origins of the explosive expansion in the West" lie in a tradition of distributing heavily illustrated machine books, such as those of Ramelli (1588) and Leupold (1724–1739), and other engineers' notebooks. The tradition started with Leonardo, who published numerous drawings of machines and machine elements—often in an original "exploded view" format in order to improve on traditional pictorial presentations. Early in the eighteenth century, Polheim constructed three-dimensional models of a "mechanical alphabet" to assist designers. To cite Ferguson (op. cit.):

> Just as the writer of words must know the letters, said Polheim, so must the designer of machines know the elements that are available to him. To Polheim, the 'five powers'—lever, wedge, screw, pulley and winch—were the vowels, while the rest of the elements were consonants.

Hindle (1981) found several attempts to map the world of available machine elements in the late nineteenth century. Bigelow (1840) published a classification of "Elements of Machinery" and Fulton (see Sutcliffe, 1909) wrote:

> The mechanic should sit down among levers, screws, wedges, wheels, etc. like a poet among the letters of the alphabet, considering them as the exhibition of his thoughts, in which a new arrangement transmits a new Idea to the world.

Hindle points out that many of the well-known inventors of machinery used their spatial representational abilities acquired in other contexts—as map makers, architects, artists, and so on. Fulton and Morse were active artists before they became successful technological innovators. Hindle comments:

> It became clearer how artists, used to filling their minds with the images they struggled to express on canvas, might understand the process of filling their minds with mechanisms in order to design machines.

Thus our approach seems to be a repetition of history, but at another level of abstraction.

MAPPING THE DESIGN TERRITORY

The following sections present an outline of a design guide based on a set of maps characterizing the territory through which a designer has to navigate. The basic idea is that design is a largely intuitive process that is shaped by the context of the designer's previous experience, ideas from other designers' successful results, the product line and practice of the company, and so on, all of which will generate a background of experience and expectations peculiar for the individual designer. A guide then can only be a framework within which the designers themselves can structure and locate a pool of personal impressions. The discussion in subsequent sections is illustrated by figures and sketches that we have found useful for discussion with designers, computer system developers, and professionals from work places

Map 1

	Domain Characteristics		
Task Characteristics	Natural Environment. Assembly of loosely connected elements and objects.	Environment includes man-made tools and artifacts, generally low hazard, reversible, trial and error acceptable.	Work Environment is highly structured system; Tightly coupled, high hazard, potentially irreversible, trial and error unacceptable.
	Environment structured by the actors' intentions; rules and practices Environment structured by laws of nature		
Detection, activation			
Data collection			
Situation analysis, diagnosis			
Goal evaluation, priority setting			
Activity planning			
Execution			
Monitoring, verification of plans and actions			

FIGURE 7.5. Map 1 spanned by the work domain characteristics and the decision task.

during field studies and system design. Thus, what we present here is neither a detailed, operational guide nor a set of explicit design tools, but only a navigational outline for each designer to operate within. We will discuss the content of maps serving to structure the following issues: (1) the spectrum of work domains, (2) the organization of knowledge bases of information systems, (3) the navigation in such knowledge bases, (4) knowledge representation in design, and (5) the composition of display formats.

Map 1: Work Domain Characteristics

A map of the territory of work domains has been developed in the previous chapters and is shown in Figure 7.5. It is used to illustrate the similarities and differences between various work activities (see Fig. 7.6 for examples). The map is characterized by the basic properties of different work domains along a horizontal continuum. The decision elements that are in focus for support system design are displayed along the vertical axis.

At the extreme left of the map *loosely coupled* systems are found that comprise a selectable set of separate tools used by an autonomous user who defines the problem himself/herself and composes or selects the necessary work environment according to subjective preferences. The illustrated prototype WTU for a domain in the left-

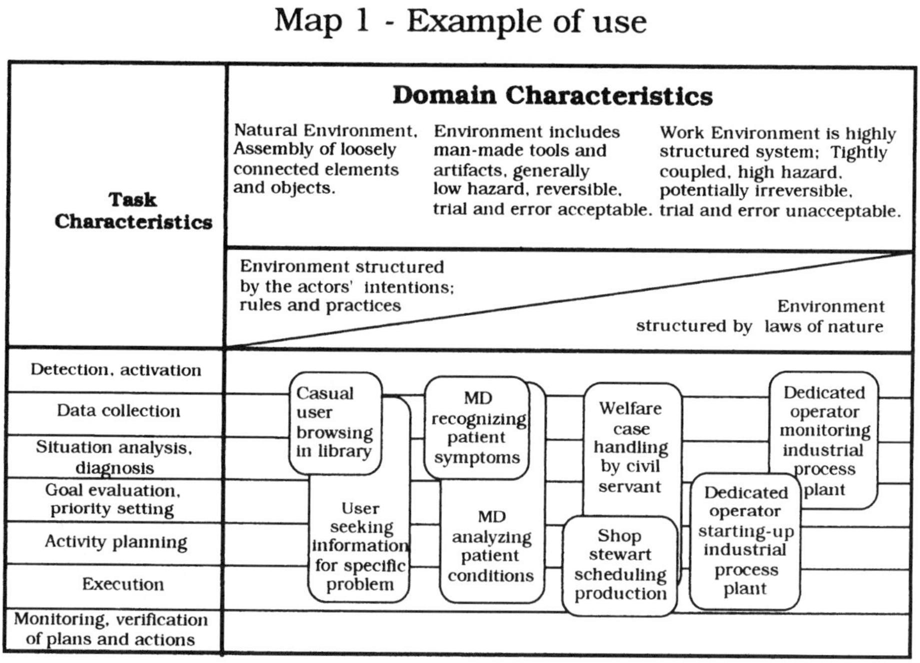

Map 1 - Example of use

Task Characteristics	Domain Characteristics		
	Natural Environment, Assembly of loosely connected elements and objects.	Environment includes man-made tools and artifacts, generally low hazard, reversible, trial and error acceptable.	Work Environment is highly structured system; Tightly coupled, high hazard, potentially irreversible, trial and error unacceptable.

Environment structured by the actors' intentions; rules and practices

Environment structured by laws of nature

Detection, activation						
Data collection	Casual user browsing in library	MD recognizing patient symptoms	Welfare case handling by civil servant		Dedicated operator monitoring industrial process plant	
Situation analysis, diagnosis						
Goal evaluation, priority setting	User seeking information for specific problem	MD analyzing patient conditions	Shop stewart scheduling production	Dedicated operator starting-up industrial process plant		
Activity planning						
Execution						
Monitoring, verification of plans and actions						

FIGURE 7.6. Use of map 1 for an attempt to identify prototypical WTU-sets, i.e., work-domain/task-situation/user combinations which can be used for a kind of modular design of work stations for diversified, discretionary task and job configurations.

hand third of the diagram is a library information system for the general public. To the extreme right are *highly structured and tightly coupled* technical systems, such as industrial process plants, with dedicated operators who are paced by the system. Between these two extremes more or less *constrained autonomous users* are found in environments dictated by various combined weights of user intentions, company or institutional goals and policies, legislation, and/or other forms for regulation. The illustrated WTU example is a generic case-handling system. For each of the examples, typical *work situations* are listed, which will involve the vertical dimension representing the *prototypical decision tasks* that must be identified and considered with respect to interface design. A third orthogonal dimension coming out of the page, which represents a user's competence, is not shown. This dimension is included in the WTU samples in Chapter 13.

Map 2: Organization of the Knowledge Base

It was argued in Chapter 2 on the analysis and representation of the work domain that the means–ends network is a useful representation of the functional inventory of a work domain, and thus defines the content of the global knowledge base of a support system from which the information needed in a particular situation can be retrieved

Map 2

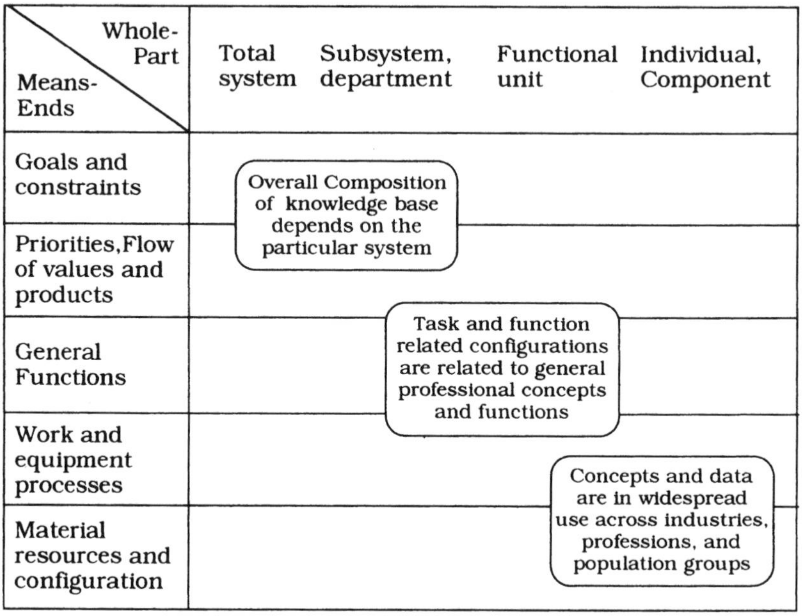

Means-Ends \ Whole-Part	Total system	Subsystem, department	Functional unit	Individual, Component
Goals and constraints	Overall Composition of knowledge base depends on the particular system			
Priorities, Flow of values and products				
General Functions		Task and function related configurations are related to general professional concepts and functions		
Work and equipment processes			Concepts and data are in widespread use across industries, professions, and population groups	
Material resources and configuration				

FIGURE 7.7. Mapping the general organization of the knowledge base of a work support system.

by a user. The basic input to the knowledge base often is elementary data at the lowest levels of the network, such as, measured process data from a process plant, bibliographic data on individual books in a library stock, or accounting data from particular commercial transactions. To support decision makers effectively, information should be available from the knowledge base at the levels of abstraction and aggregation at which the decision are to be made in the means–ends network and in response the a query whether this is posed in a WHAT, WHY, or HOW question. The organization of the knowledge base depends on the particular domain, but a few general features can be identified.

Very often, a correlation is found between the levels of abstraction and of aggregation of the information needed for decision making (e.g., see Fig. 2.1). From this follows some general guidance for choosing aggregation and abstraction levels for representation of information in a knowledge base (see Fig. 7.7). At the lowest levels, data and concepts used are normally in very widespread use, according to general practice, industry standards, and population stereotypes. When data are aggregated for use at higher functional units, they will be transformed according to the concepts and task demands of various staff professions and company practices. At the highest levels of planning and monitoring global system performance, the organization of the knowledge base will depend on the goal and value structure and the overall functional structure of the particular work system; that is, the location of the system along the horizontal dimension of Map 1. The use of Map 2 cannot be illustrated visually, because the use is implicit in the concepts used for representation of the means–ends network (see Chapter 2).

Some more characteristics of the organization of knowledge bases for different work systems will be discussed in the next section, which considers road maps for navigation in the knowledge bases. In addition, we will see in a subsequent section, that the map of Figure 7.7 will also be relevant for composition of individual display formats.

Map 3: Road Map for Navigation

A critical problem for an information system designer is to visualize the structure of the knowledge base to a user by the displays made available for a particular activity. To solve this problem, system designers must formulate for themselves a map of the global organization of the information system as represented by the means–ends network and a consistent road map that can guide the navigation of a decision maker or user. No generally valid map structure can be suggested, but some basic patterns can be presented to illustrate the maps that are relevant for different typical work activities (Figs. 7.8, 7.9, and 10.4).

Map 3: Navigation by means-ends relations

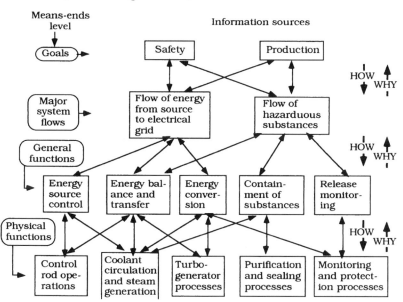

FIGURE 7.8. The means-ends relations of tightly coupled technical systems represents the invariant structure of the information system and will support navigation during operation monitoring and disturbance control.

This global structure of the information system must be easily recognizable and "navigatable" by the user to avoid "getting lost." This problem was discussed in detail by Woods (1984). He formulated the need for a "cognitive momentum" so that users can go effortlessly from operations with one display to another guided by mutual cross-referencing features in the individual displays (e.g., by "hypermedia" techniques, see Chapter 11). In addition to this, we find it important to make the global structure of the knowledge base transparent to the user. The source of regularity used for design of the knowledge base and the information retrieval functions depends very much on the work domain and the task situation to be supported. We will discuss some examples in this chapter.

Monitoring and Disturbance Control of Tightly Coupled Process Systems. For a well-structured technical system in operation, the relevant knowledge base is representing the active part of the means–ends network as coupled for operation. The route taken by an operator through this knowledge base during disturbance control cannot be planned in advance. This route will depend on the particular, maybe rare, cause of disturbance. Therefore, navigation through the numerous displays necessary for monitoring such systems depends on the conception, by the operators, of the structure of the display system which, therefore, should closely match the means–ends network. Professional operators will know the functional structure of the plant by heart. The organization of the numerous displays should be organized with reference to this invariant structure to facilitate easy navigation. Primary, measured data should be integrated and transformed so as to present the state of the functional units directly at the various levels. Additional support can be given by presenting a map of the knowledge base according to the means–ends organization in Figure 7.8, and to indicate to the operators those information sources that present new information (e.g., state-changes, disturbances). For detail, see Goodstein (1985b) on "functional alarming."

Start-Up of an Industrial Process System. This activity is paced by professional operators and is normally constrained by intentions of the plant designer as represented in specific start-up instructions. Start-up of a process system involves the aggregation of elementary parts into process units. When connected and started, the operational states of more units are aligned and the units are connected to higher level functional units, and so on, until the entire plant is operating. The structure of the road map through the knowledge base to be followed by the operators is a convergent tree (see Fig. 7.9). This tree will be implicitly known by the operators, but interface designers should identify it carefully so as to be able to design a consistent, interrelated set of displays at all functional levels. Identification of the tree often requires "reverse engineering" to extract the tree explicitly from the operating instructions that normally are formulated in terms of a detailed sequence of actions. Consequently, this analysis, depends on detailed technical knowledge. When reference is made to the discussion of displays below, the figure indicates the display formats matching the concepts and representations at the different levels of the knowledge base.

Map 3: Navigation by task constraints

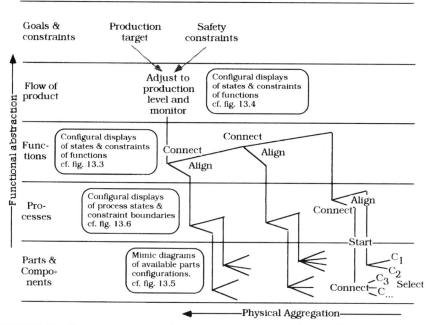

FIGURE 7.9. The tree structure of the task to start a process plant. It illustrates the related navigation in the knowledge base. It also indicates the concepts and representation used in the display formats at the various levels, ranging from mimic diagrams of the configuration of physical parts to increasingly abstract and integrated functional displays.

Assembling Products in Manufacturing. This is another activity that depends on a convergent tree structure for navigation in the knowledge base (see the diagrams of Figs. 3.2 and 3.3). The tree structure through the knowledge base in this case reflects the paths of the parts through the production system, as planned by the production planner. The information available to the actors at the various work stations should reflect the parts in process and the actors' planning intentions at the adjoining work stations to satisfy the varying horizons of attention of the actors, as shown in Figure 3.5. The following example is similar to this, except that the road map for navigation in the knowledge base is different.

Legal Case Handling, for Example, for Work Accident Compensation. The structure of the knowledge base of legal case handling, for example, the analysis of work accident and the decision about the subsequent support of the victim, is a decision tree representing the various decision points in the case handling together with the alternatives open to choice at the decision points, (i.e., a divergent tree for analysis of a problem and choice of solutions). This tree structure reflects the intentionality behind the legislation, and making it explicitly visible to all involved decision makers will be a useful support of navigation in the knowledge base. In a project aiming at support of a consistent up-dating of the British legislation for social welfare, a formal representation of the legislation by first-order predicate calculus was developed, indicating that an explicit representation of the structure of decision trees for the related case handling is realistic. The content of the knowledge base in this case will include the attributes of the alternative choices offered at each branch: the pool decisions from previous cases with characteristics, the number of cases waiting at each branching point with reference to their priorities and, possibly, for each decision maker some representation of the actual planning intentions of the decision makers at adjoining stations. In this respect, the situation is very similar to that of the manufacturing foreman of Figure 3.5.

In such cases, making the tree structure explicitly available to the individual actors would facilitate not only navigation in the information system, but also the coordination of activities, especially if a way to represent and communicate intentional information can be found.

User-Driven Activities in Libraries. In the examples discussed above, the road map for navigation in the knowledge base depends on an invariant structure found in the systems representing functional relationships or the intentional structure of legislation or manufacturing practice. For the activities of autonomous users in, for example, libraries, no such structure to shape the navigation road map is given a priori. Instead, the invariance must be imposed on the system from an analysis of the users' preferred task structures, and the result should be reflected in the *user-system dialogue*. For the BookHouse system, the structure of the information system is illustrated in Figure 10.4. However, since the system is intended for users with very different backgrounds, coming from various population groups, explicit presentation of this structure in the interface will be less useful. Instead, as described in Chapter 10, the user is given the opportunity to learn the structure in an implicit way through the interaction by means of *a metaphor*, in this case a house metaphor, which is also used to apply direct mutual cross-reference between the individual display formats. As a back-up for this implicit communication of system structure a verbal *help system* is introduced.

Map 4: Knowledge Representations in Design

In discussions of the concept of mental models and its role in design it is often not clear when the mental model in question is, e.g., the user's mental model of the system, the designer's mental model underlying the design, or the designer's conception of the user's model of work requirements. To resolve such questions during discussions among designers, users, and system programmers we have found a map of the many different representations that are relevant during design and use of a system to be very useful. Such a conceptual map is shown in Figure 7.10. and is used in Figures 7.11 and 7.12 to characterize the paths of transformation of models during design of a work support system for different work domains. Several, more detailed submaps are needed within this basic map of representations.

The figure is organized in three vertical columns. The center column shows the

Map 4: Representations used in Design

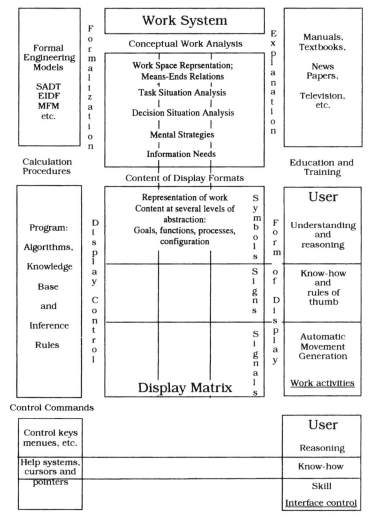

FIGURE 7.10. A map of the knowledge representation to consider for information system design.

representations of the properties of the work system as described within the different dimensions of the work analysis framework; this includes the *means–ends structure* of the work domain (a very important submap; see Chapter 2 and Fig. 3.13), the *task situation in work terms*, the *decision tasks* (see submap representations in Figs. 3.6 and 3.7), and the *cognitive strategies that can* be used. Figure 7.10 indicates how the means–ends network identified by the work analysis will produce information on the physical anatomy and topography, the processes and tools in the various activities, the general functions and, finally, the goals and value structures—all of which will define the related *contents* of the knowledge base of the information system that must be accessible through the interface in the different interface formats serving the relevant activities and coping strategies.

The right-hand column of the figure indicates the representation of the work domain that is available in textbooks, manuals, training material, and in the professional literature. This kind of material will influence the mental representations of system users in some work domains. The figure also indicates potential influences on user representations from general information sources such as radio, TV, newspapers, and other media—this is especially relevant in work domains with casual, perhaps nonprofessional, users. Such representations can often be based on established general population stereotypes, which can also be utilized in the interface display designs. In addition, this column shows the users' mental representations.

The matrix in the center of the figure pinpoints the complex matching problem in interface design where the information *content* has to match a user's discretionary choices of means–ends levels to work at. At the same time, the *form* of the display has to match the needs of users with varying levels of expertise. The figure illustrates that the choice of the *form* of display representation is to some degree independent of its *content* and, therefore, different display representations can be chosen to match the interpretations involved in skill, rule, and knowledge-based operations.

The left-hand column sketches the formalized, computational representations that are necessary for driving the information systems. For a formal, functional analysis different methods have been developed, such as Systems Analysis and Design Techniques (SADT, see, e.g., Ross, 1977), the related Integrated Definition Format (EI-DEF, see, e.g., SoftTech, 1981) or Petri Nets (see, e.g., Boucher and Jafari, 1990). A recent extension of the SADT/EIDEF modeling techniques introducing the more flexible "concept maps," which better match the framework presented in Chapters 2 and 3, has been proposed by McNeese et al. (1990) and McNeese and Zaff (1991). Lind's multilevel flow formalization of the means–ends relations in a work domain is also an important candidate for the design of computer programs to drive display systems (Lind, 1982, 1992, 1993). In addition, the recent development of hypermedia tools should be considered, as discussed in Chapter 11.

Normally, a design will aim at a direct communication between user and work through the center matrix. Occasionally, however some control of parameters or functions of the program system is required and the relevant information must be available at the interface; see the bottom row of Figure 7.10. This communication path is useful in systems serving professional users who will have the system adapt to their own particular needs. Thus, the importance and the functions involved in the computer program–user interaction depend on the degree and mode intended for the adaptation of the system to users' needs and characteristics.

Examples of the Use of Map 4.   To illustrate the use of this map, a couple of examples of typical design paths will be presented.

Map 4 - Examples of use

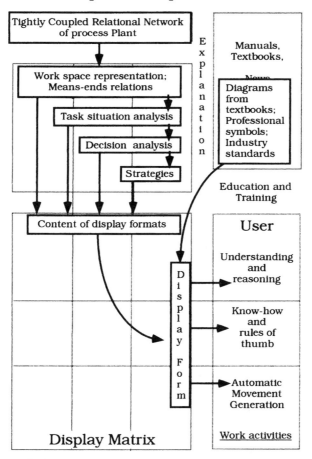

FIGURE 7.11. Paths for the identification of display content and form for tightly coupled technical systems.

An Information System for the Control Room of an Industrial Process Plant. A technical system located at the extreme right side of the spectrum in Figure 7.5 is functionally integrated and tightly coupled at all levels of the means–ends network (see Fig. 2.10). That is, the functionality at all levels is determined by the tightly coupled physical configuration of the equipment together with the laws of nature governing the basic physical processes (see Fig. 2.10). In addition, the operational states at all levels are tightly constrained by the automatic control system or by procedural specifications, the functions of which represent the propagation of the intentionality top-down from the high level production and safety goals as interpreted by the designer. The operation of these systems is stable over long periods during which operators develop a large repertoire of know-how and skill. However, such periods are sometimes interrupted by potentially hazardous disturbances that require competent knowledge-based responses.

THE CONTENT OF INTERFACES. The version of Map 2 on Figure 7.11 indicates that, for this kind of system, the design of the functions and interfaces of the information system depends largely on a careful analysis of the *system* characteristics, the *tasks*, and the work *strategies* that will serve the system. The content of the interface displays must represent the resources available for all the relevant tasks, to guide the operator's planning. That is, the mapping between means–ends levels of the basic design must be represented at the interface. Also, to support control of the system functions, displays must be available showing the relationship between the actual functional state, the targets states, and the boundaries to be respected (e.g., the capability limits of equipment). Finally, to support planning during unforeseen disturbances, information must be available in support of all strategies that can be useful for the operators. To sum up, the content of the displays required for the various task situations is identified by an analysis of the system properties and the design intentions behind it.

When compared with the other domains, there will be considerably less concern about user–operator personal needs and aspirations than with ensuring their competent supervisory control. Note, therefore, the education and training link to the operator. Although, as stated, the intentionality of these actors is dictated by the system, problems can occur when an operator actually has to take on the role of a designer (e.g., in situations requiring an ad hoc, goal directed change of the physical configuration of the system). Thus, some of the intentional structure can temporarily become operator dependent. At any rate, the display contents will depend entirely on the analysis of the plant system.

THE FORM OF INTERFACE DISPLAYS. It is frequently proposed that the *form* of effective interfaces should be based on an analysis of the users' mental models to ensure a proper match. This is *not* the case for many large-scale technical work domains. The basic problem in such systems is to ensure that the operating staff, during their daily work, acquire mental models that enable them to cope with infrequent, unforeseen conditions. For this purpose, the interface must be a faithful mapping of the functional relations and the intentional constraints governing system behavior. The form of representation should be chosen so as to force the operators to develop mental models during normal work that are effective during infrequent disturbances; that is, the form should be chosen according to the ecological display principles discussed in Chapter 5. The best sources of inspiration for the selection of proper representations for this purpose are probably found in textbooks and manuals representing long traditions in teaching, training, and design discussions.

An Information Retrieval System for a Public Library. The design of a system to support the retrieval of books in a public library that will serve the needs of casual users is taken to be an example in the left-hand side of the spectrum in Map 1. In this

Map 4 - Example of use

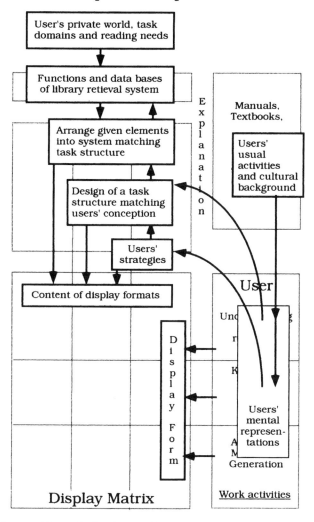

FIGURE 7.12. The paths for design of a system serving an autonomous user.

domain, no laws of nature control the structure of the domain and the intentionality depends entirely on the user's needs (Fig. 2.10). Therefore, system designers have many degrees of freedom in designing the functionality of a retrieval system and will naturally seek to match the users' conceptions of the domain and the task. That is, the functionality of the system should be chosen so as to match their preferred strategies for searching through a book stock as identified from studies of their retrieval behavior in libraries. The concepts used to represent the book stock should reflect the users' formulations of their needs in reading. Therefore, in contrast to the previous domain, the transformation of representations propagate *from* the user *toward* the system (see Fig. 7.12).

With respect to the content of the displays, it is therefore necessary to design a set of displays that can support the various search strategies useful to the user, and to match their content to the users' conception of the book stock with reference to their needs. As shown in Figure 7.12, the functionality of the system and, consequently, the *content* of the display, will be determined from analysis of the users conception of their needs and the strategies they *can* use for retrieval.

Similarly, the *form* of the display formats should match the users' mental models from other similar work environments, that is, the form must be chosen to match familiar metaphors. The Map 2 design path for this kind of domain is shown in Figure 7.12, which illustrates how the transformations, in contrast to the technical process system, are directed *from* the conceptions of the user *toward* system representations. These points will be discussed in more detail in Chapter 11.

General Information System for a Casual User. In the library example, the task and user could be well specified. This is not the case in information systems for casual users, such as those developed for home TV terminals as museum guides and travel planning aids. Therefore, a structure must be imposed on the interface to help users to navigate, to find, and to manipulate information, as well as to remember where to find previously consulted items. In general, a metaphor is useful to make it possible for any user to transfer intuition, habits, and skills from other situations. Well-known examples are the desk-top metaphor, as well as current graphical user interfaces (GUIs). A more ambitious approach is the *multiple metaphor–multiple agent* approach in the Japanese Friend'21 personalized information environment (Friend'21, 1989). To avoid the problem of making system functionality transparent to an unknown user, "agents" provide various services, while multiple metaphors are used to diminish semantic distance. Agents and metaphors can be switched dynamically to suit user and situation demands. A TV-metaphor presents information in a TV-news context both pictorially and with a robot providing comments and explanations. A newspaper metaphor permits retrieval of information via a "turning pages" facility, while an album metaphor assists the user in filing, sorting, and retrieving pictures. Figures 13.12 and 13.13 illustrate this approach.

These examples are borderline cases and all work domains inbetween will involve more or less autonomous actors with considerable degrees of freedom within the constrains from legislation, regulation, and company policies, as well as from the available work resources. The paths to take for design of such systems are clearly a complex mixture of those discussed for the boundary cases and, therefore, explicit consideration during design of the various forms of representation of Map 2 has proven to be helpful.

Map 5: Display Composition

Finally, a map to structure the visual composition of the individual displays will be discussed. Various modes of interpretation of information are utilized by users depending on their level of expertise, their cognitive style, and on the particular task situation. This interpretation will change with work domains, professions, and cultures and a conceptual map can assist in guiding the design of graphic interfaces, icons, and metaphors, as well as the use of multimedia. The aim of an ecological display is to present a rich information environment that leaves the choices of level of interpretation and cognitive control to the discretion of the user. For generalization, it will be advantageous to identify some regularities in the relationships between the composition of complex displays and the structure of the work requirements. Some preliminary observations will now be discussed.

Ecological design of interfaces is a research area in rapid development and the approach to representation depends strongly on the context of a design. Therefore, we do not aim to present a detailed discussion, for this the reader is referred to the current literature (a recent collection is found in Flach et al. 1994). Only some general aspects of the problem will be discussed in the following sections.

Decomposition-Chunking. In most activities, the work sequence will involve shifts in levels of abstraction and in aggregation of perceptual–conceptual units together with concurrent changes in levels of intentional–activity related complexity. In natural environments, all observable information is available at all times (see the discussion in Chapter 5 and Fig. 5.3). Thus, the level of abstraction at which the environment is perceived can be varied at will by the actor. The degree of perceptual chunking is discretionary and is closely related to the level at which the actor expresses an intention to act. This, in turn, is determined by the complexity of the automated patterns of movements available to the actor for the particular activity.

A good example that will help to make these ideas more concrete is that of musical skill. Skilled sensori-motor performance with an instrument is characterized by integrated patterns of movements resulting in very high capacity and speed in performance. As the musician's level of proficiency increases, movements are aggregated together into higher order chunks. Whereas the novice must control performance at the level of individual actions, skilled musicians can work at the level of complex sequences of actions. The key requirement for attaining this type of skill seems to lie in the mapping between the musical notation and the associated actions. Thus, experienced musicians are able to form higher order visual chunks of notes and then directly map these onto a concurrent chunking of movements. The musical notation is a form of ecological representation that has evolved through centuries. Rousseau (1762) presented a more "rational" representation for the French Academy based on a logic notation in terms of a number code. It was well received by the academy and a committee was founded to review the system. Apparently, numerical systems for musical notation were subject to a wider discussion at that time. However, according to Rousseau, when the composer Rameau was presented with the numerical notation he argued:

> Your signs are excellent, with respect to representation of tone and interval--; but they are very poor because they require an activity of thought which cannot keep pace with the performance. The location of signs in our usual notation imprint on the eye without

support of this kind. If two notes, one very high and another very low, are connected with a sequence of intermediate notes, I immediately by first glance perceive the gradual rise from one to another. By your system, however, I necessarily have to spell my way through from number to number, one glance will not do it.

This was a nice argument for the ecological display design that actually convinced Rousseau.

Two conclusions can be drawn from this musical example. One conclusion is that very efficient "direct manipulation" skills can evolve, even if the information interface is separate from the manipulation interface. This is important for the design of computer-based interfaces. Another conclusion is that the information presentation should support the integration (i.e., chunking) of visual features into higher level cues that correspond to the aggregation of elementary acts into more complex routines, which occur with an evolving manual skill. In other words, the structure of the information display (within a single display and across displays) should be isomorphic to the part–whole structure of task activities. Within one separate display, this can be accomplished by revealing higher level information as aggregations of lower level information (e.g., through appropriate perceptual organization principles). In this way, multiple levels are visible at the same time and the user is free to guide their attention to their current level of interest, depending on their level of expertise and the current domain demands. Developing such a hierarchical visual structure should facilitate the acquisition of skill by encouraging the chunking process. At the same time, flexibility is maintained by not constraining people to attend to a specific level. For this purpose, the design of an integrated, symbolic display would benefit from a correlation between abstraction and aggregation such as that often found in a work planning sequence. Its manifestation is a trajectory in the problem space map going diagonally from the upper left to the lower right corner, as discussed for the organization of a knowledge base (Fig. 7.7). Choice of the visual representation of the concepts used to organize the knowledge base should be guided by the representations normally used by the professions or groups who are the source of the knowledge concepts and, therefore, a map similar to Figure 7.7 is useful for the visual coding within a composite, ecological display format (see Map 5 in Fig. 7.13). This map is very similar to Map 2 in that it simply illustrates the visual representations that are used for the knowledge categories of Figure 7.7.

As shown in this map, low level display elements would map closely to acts directed toward tools or physical components. Since such items will no doubt be found in a wide context, the form should match population or professional stereotypes. At a somewhat higher configural level, activity is related to functional relations for which stereotype representations (in some cases standards) for drawings and manuals can be found for the given profession. At the global level, activity is related to the constraint structure of the global task situation. Also, Flach and Vicente (1989) advocate this use of the means–ends network as a framework for parsing the functional structure of work domains together with Gibson's concept of nested invariants as a framework for parsing the display composition in a way that will naturally map onto the means–ends relations of a work space.

Map 5: Visual composition of displays

Whole-Part / Means-Ends	Total system	Subsystem, department	Functional unit	Individual, Component
Goals and constraints	Representation of high level functional and value structure depends on the particular system			
Priorities, Flow of values and products				
General Functions		Task and function related visualization derived from professional concepts and practices		
Work and equipment processes			Visualization of parts and tools are derived from standards and stereotypes of industries, professions, and population groups	
Material resources and configuration				

FIGURE 7.13. Map of the relationship between the source of visual codes used at an interface and the level of abstraction–decomposition of the knowledge items to be represented.

A good example of a composite display design illustrating the use of Map 5 is a display format designed to support monitoring of the global operation of an industrial process plant (see Fig. 7.14). The display is used for monitoring the operation of a nuclear power plant and is based on Lindsay and Staffon (1988). The overall pattern reflects the flow of energy through the plant from the reactor core through a heat exchanger and a steam generator to the turbine driving the electrical generator in a unique combination with the overall functional structure of the particular plant. The display builds on a sequence of configural representations of the functional subsystems of the plant based on engineering practice and theoretical (thermodynamic) conventions. In addition, elementary, iconic representations are used, which are derived from general industrial practice and standards. Thus the means–ends/part–whole coverage is considerable within the confines of the single display. We will discuss the configural and iconic symbols used for the display on pages 195–197.

A related example for a less constrained work environment is found in Figure 11.8 which uses a "work room" metaphor for helping the user carry out an analytical search strategy in a library. The set of search dimensions (tools) to be considered in this particular task are found in the work room and the configural elements depicting

Map 5 - Example of use

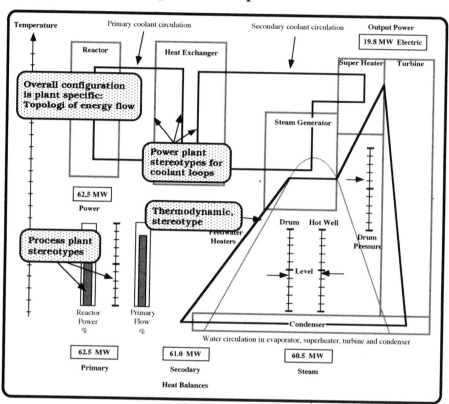

FIGURE 7.14. Organization of a composite symbolic display for an industrial plant. (Adapted from Lindsay and Staffon, 1988, with permission.)

these "tools" are derived from general population stereotypes (for a detailed discussion, the reader is referred to Chapter 11). This example also illustrates that the representation of the knowledge base and, accordingly, the use of Map 5 is distributed over more displays.

Visual Coding. The basic principle for design of ecological interfaces is that the display formats should support direct perception of the states of the work environment with reference to the goals and the relevant alternative actions, in order to support performance by cue–action relations during familiar task situations. In addition, the representation should be transparent with respect to the behavior shaping constraints of the work environment, that is, the internal sources of regularity acting as behavior shaping constraints, in order to support knowledge-based analysis and planning. This latter requirement considerably constrains the choice of visual coding, but even then there will be ample freedom to take perceptual qualities into account.

For knowledge-based reasoning, the visual form serves as a *symbolic* representation—an externalized mental model—of the relational network representing the work content at the appropriate means–ends level(s). The form of the displays should not be chosen to match the mental models of actors found from analyses of an *existing* work setting. Instead they should induce the mental models that are effective for the tasks at hand in the *new* work ecology. If they exist, established graphical designs used in textbooks and for teaching are probably the best sources for these representations.

In addition to the graphical representations that relate symbolically to the contents of the work domain, special consideration needs to be given to those elements of the displays that basically serve as *signs* or *cues* for action at the rule-based level of performance. Actually, symbolic graphical representations of system state can provide such action cues based on defining attributes, and thus serve to prevent errors due to the use of underspecifying cues for actions (see Chapter 6).

For supporting skill-based operation, the visual composition should be based on visual patterns that can be easily separated from the background and perceptively aggregated–decomposed in a way that matches the formation of patterns of movements at different levels of manual skill (see the discussion of decomposition and chunking above).

The actual coding to be chosen will depend very much on the domain in question, on existing traditions for figural and diagrammatic representation found within the related professions, and on the creativity and background of the designer in the context of company practice. However, practically all of these will build on a few basic ways of mapping information onto displays. Pepler and Wohl (1964) describe four such types. With some modifications, these comprise:

Scalar coding, that is, the basic coding is *geometric* (i.e., magnitude to length, angle, or area). The underlying data is *quantitative* and can be discontinuous (discrete) or continuous in nature. Reference information identifying coordinates, scales, and units is alphanumeric. There are many variations on this theme. The time dimension can be explicit or implicit.

Matrix coding, that is, the basic coding is positional (i.e., data to cell in a two-dimensional row-column arraylike arrangement). Cell coding can consist of any of the other types.

Diagram coding, that is, the basic coding is *representational* and/or relational. Diagrams comprise connected elements coded as lines, symbols, text, and so on, to reflect structure or function.

Symbolic coding, that is, everything from narrative-like formats employing words and numbers to pictorial, analogical, or abstract entities (*icons or buttons or metaphors*) serving as symbols or signs relating to domain or task.

Composite coding includes formats that combine one or more of the above.

The general applicability of the basic types should be obvious. Scalar displays are an indispensable tool where the data is (can be made) quantitative; however, certain variations will be domain dependent. Likewise matrix-type arrays are widely employed for depicting spatially organized information (e.g., spreadsheets); moreover they form the basis for menu arrangements, groupings of command buttons, and the like. Where relevant, diagrams play an indispensable role in supporting mental models of structure and function, while all sorts of symbolic representations are of course commonplace. Of special interest are metaphoric representations that can be considered as a kind of "cover story" that couples the functionality built into a system to a functional structure relevant to another domain. Well-known examples are the desk-top metaphor for office systems and prevalent personal computer graphic-user-interfaces (see Clanton, 1983; Mihram and Mihram, 1974; Mihram, 1972; and Hollan, 1989 for further details). However, it should be realized that a potential danger with metaphors is that they can cause interferences in situations where their interpretation is misleading.

Visual Coding: Some Examples. As mentioned previously, it is not our intention to present guides for display design. This will depend strongly on the characteristics of the domain and the background of the users. For this, readers are referred to the current literature on professional representations and on perceptual characteristics (e.g., see Flach et al. 1994). In the following sections we only present some examples of the borderline cases of Map 1 in Figure 7.5 to illustrate the basic ideas.

Monitoring Operation of an Industrial Process Plant. As explained earlier, for highly structured technical systems the work system itself is the exclusive source of the behavior-shaping constraints to be represented on the displays. A visual support of, for example, monitoring and diagnosis during operation of a process plant, can be given by an ecological display including configural representations serving direct perception of the relationships between the actual functional state (the relationship among measured variables) and the intended target and limit states in a compatible fashion. For technical systems, for which the sources of regularity of behavior and limits of acceptable operation depend on physical laws, the representation of invariants and states will normally be in terms of diagrammatic depiction of structure and scalar representations of the relationships among quantitative variables. To enhance perception, the intentional structure—as reflected in these target operating states and boundaries to unacceptable operation—should be represented by overlays on the representation of the actual functional structure and states. In addition, these quantitative relationships should be represented by virtual objects that change their shape and/or move in a symbolic topography that matches the characteristics of direct human perception. Since the trajectory of exploratory searches in the abstraction–decomposition work space will primarily be along the top-left/bottom-right diagonal, the ecological displays should support this by ensuring that span of attention and shifts in abstraction are correlated.

MONITORING OPERATION OF A POWER PLANT. As an example of the visual coding of a complex configural display we will consider the example given in Figure 7.14 in some detail. As we will see below, this display is based on presentation of primary data, but even then, the configural pattern can be perceived at the level of flow of energy, of state of the coolant circuits, of the physical implications of temperature readings, and so on, at the discretion of the observer. At the same time, the display takes advantage of a standard format (i.e., a temperature–entropy plot) that is used by engineers to graphically display the thermodynamic properties of heat engine cycles. Another important feature, as mentioned previously, is that this kind of graphic display can be interpreted at several cognitive levels. Since it is transparent (i.e., it represents internal functional relationships that are to be controlled) it will support knowledge-based, analytic reasoning and planning. Furthermore, because it reflects this relationship in a perceptual pattern, cues that will be correlated to familiar action sequences are likely to be defining patterns rather than convenient signs. Therefore, it can effectively prevent "under-specified" action mistakes. The properties of the basic coding concepts are described in the following paragraphs. A general discussion of displays for process control is found in Goodstein and Rasmussen (1988) and in Rasmussen and Goodstein (1988).

CONFIGURAL REPRESENTATION OF A COOLING CIRCUIT. Figure 7.15 illustrates the use of graphic patterns to represent the mutual relationship within a set of primary data. The figure represents the states of the cooling circuit shown in the upper left corner of Figure 7.14. This display is based on the measured, primary sensor data in such a way that the shape of the emerging graphical patterns supports perception of higher level functional features. Graphical patterns show the circulation paths of coolant and water. The background boxes represent the heat producing reactor core and the primary heat exchanger. The rectangular overlays represent the cooling circuits (arrows indicate the coolant flow direction). The position of the corners of the rectangles indicate temperatures with reference to the scale: The reactor heats the coolant from temperature T1 to T2. The coolant temperature drops a little in the pipe to the heat exchanger, from T2 to T3, and further from T3 to T4, when delivering energy to the heat exchanger. The cold coolant further looses a little temperature T4 to T1 on its way back to the reactor. A skilled operator will be able to perceive the operational state of the entire circuit by one glance, while they can also, if needed, consult the individual primary data.

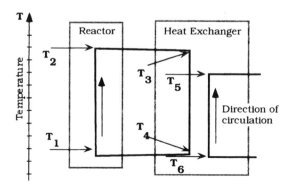

FIGURE 7.15. Configural representation of the cooling circuits of Figure 7.14.

CONFIGURAL REPRESENTATION OF A THERMODYNAMIC RANKINE CYCLE. The central function in a power plant is the energy conversion from boiling water to electricity by means of a steam generator, a turbine, and a generator. The design and optimization of the operation is based on a thermodynamic representation in terms of a Rankine cycle which represent the pressure–temperature–entropy relations in the energy conversion system. Beltracchi has promoted the use of displays based on Rankine cycle diagrams for the control of power plants in several papers see, e.g., Beltracchi (1984, 1987, and 1989).

The Rankine cycle concept is based on the physical laws describing the properties of a two-phase, water-steam mixture, shown in the left-hand side of Figure 7.16.

The right-hand side of this figure shows an overlay of a Rankine cycle thermodynamic process and reflects the laws of nature governing the pressure–temperature behavior of the steam generator–turbine–condenser circuit. The display supports the crucial task of matching the temperature–pressure conditions of the steam to the requirements of the turbine and to the safety limits of the steam generator (i.e., to the intentional parameters). It illustrates the combination of configural representation of the actual operating state of a complex system with the intentional target and limit states.

Figure 7.17 illustrates how the dynamic propagation of a disturbance through the system can be directly perceived by an operator. The upper figure illustrates the change in the coolant flow–temperature pattern when the primary coolant flow is suddenly increased. The power production can now be transported by the coolant

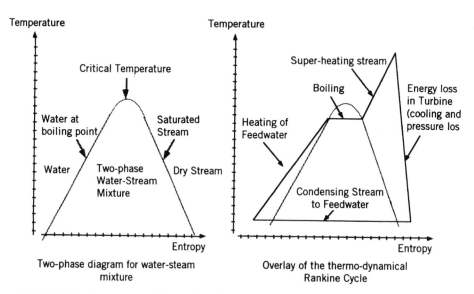

Two-phase diagram for water-steam mixture

Overlay of the thermo-dynamical Rankine Cycle

FIGURE 7.16. The source of the configural representation of the steam generator in Figure 7.14: The two-phase diagram of a water–steam mixture with an overlay of the thermodynamic cycle of a "Rankine machine."

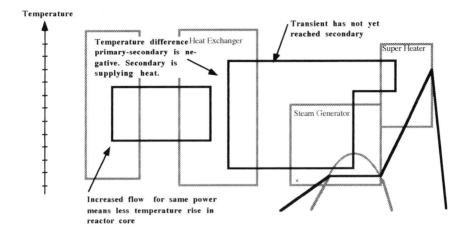

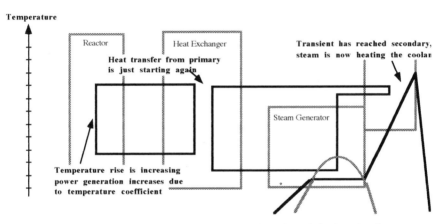

FIGURE 7.17. The figures illustrate the potential of configural displays for direct perception of the propagation of a disturbance through a thermodynamic system.

with a lower temperature rise and the temperature profile of the primary coolant circuit collapses, compared to Figure 7.14, which shows the normal state. The secondary circuit is not yet affected and its higher temperature forces power backwards through the heat exchanger. This can be seen directly from the change in sign of the temperature difference between the secondary and primary side of the exchanger. The lower part of the figure presents the state a little later. The transient has now reached the secondary circuit while the primary circuit is restoring due to an increase in power production caused by the decreased temperature and a negative power–temperature coefficient.

Information System for Information Retrieval in Libraries. We have discussed the visual coding of displays for control of technical systems because this is a typical example of the use of configural patterns to map the invariant properties of a work system faithfully. At the same time, it represents the relationships among quantitative variables. The aim has been to make it possible for the reader to contrast this approach to the visual coding of the displays that are useful at the opposite edge of the map of Figure 7.1; that is, an information system for information retrieval in libraries, as described in detail in Chapter 11. In this chapter we only present some basic features for immediate comparison.

While the display coding in the previous example is tightly constrained by the need to represent the internal structure faithfully, display coding for systems such as the library system leaves the designer with considerable freedom to create a structure and to choose an analogical coding familiar to the user (a metaphor). For the BookHouse described in Chapter 11, the overall visual coding is based on a store house metaphor, while the choice of the elementary elements have been based on the creativity of the system designers and artists involved. The resulting figural elements have then been tested experimentally toward the response of representative user groups and accomodated accordingly.

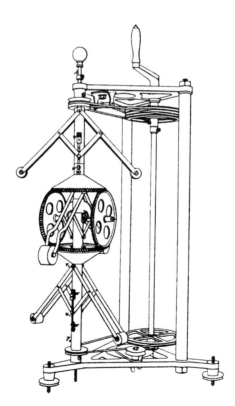

FIGURE 7.18. Boltzmann's visualization of Maxwell's electromagnetic equations. (Reproduced from Miller, 1989, with permission by Birkhauser.)

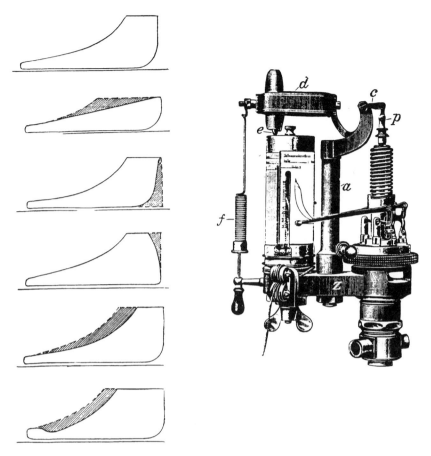

FIGURE 7.19. An old ecological display design: A graphic display of the volume–pressure relationship in the steam engine cylinders has been a common tool for operating steam engines efficiently for nearly a century (reproduced from Häntzschel-Clairmont, 1912).

Visual Representations in Research

It is argued in previous sections that the symbolic representations found in textbooks can be an important source of ideas for configural representations, and we have made a tentative effort to identify good sources. More work needs to be done in this direction.

Some interesting research has been reported on the visual representation of *scientific concepts* focused on imagery with reference to perceptual structures in the physical world (for a recent review of imagery in science, see Miller, 1989). The planetary model of the atom and Boltzmann's mechanical model of electromagnetic coupling (see Fig. 7.18) are examples. Discussion of the use of diagrammatic representations in science and technology is sparse, and yet many diagrammatic representations have been used as "externalized mental models" for visualizing the

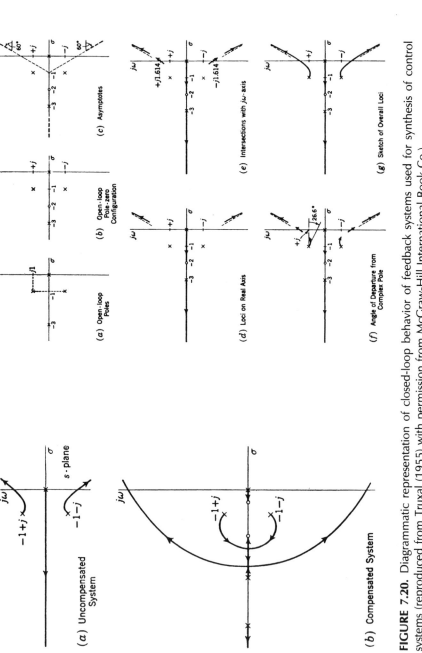

FIGURE 7.20. Diagrammatic representation of closed-loop behavior of feedback systems used for synthesis of control systems (reproduced from Truxal (1955) with permission from McGraw-Hill International Book Co.).

effects of manipulations. That is, a mathematical, relational representation is reinterpreted in terms of causally related objects responding to "events," such as design decisions and changes of scientific hypotheses.

A classical example is the volume–pressure diagram for adjustment and maintenance of steam engines—used for over a century (Fig. 7.19). With this "integrated display," a large repertoire of heuristic rules evolved for diagnosing deficiencies and identifying maintenance needs from the shape of the curve. Another example well known to control engineers is the pole-zero root locus representation of the mathematical representation of the dynamic properties of a feedback loop. For this representation, rules for changing system properties in an intended direction have also evolved, which are directly related to the visual shape of the representation (see Fig. 7.20).

Optimal operation of industrial processes and the limits of acceptable operation will normally be linked to mathematical analyses of the relationships among quantitative variables that can be measured. The limits to efficiency and conditions of optimal operation of steam engines were derived by the science of thermodynamics. Therefore, abstract diagrams representing theoretical relationships are developed when mathematical, relational representations enter the scene. Diagrammatic representations to guide functional reasoning were suggested by Babbage (1826), who pointed out the necessity for an abstract, functional representation in addition to the visual illustration of the machinery itself. Also Gibbs (1873) explicitly stressed the need for diagrammatic representations of phase relationships in thermodynamics. A careful review of the history of scientific, diagrammatic representations with respect to the relationships between subject matter characteristics and configural form and their role in causal reasoning could be fruitful to guide the design of ecological interfaces, in the same way as catalogs of machine elements were used to guide designers of mechanical machinery during the period of industrialization.

Summary on Design Maps

We have now dealt in some detail with an alternate approach to aiding the designers of modern information systems, which is based on a set of maps of the design territory that can be tailored to a designer's individual context, needs, and experience. To conclude this discussion, a brief (preliminary) discussion of the possible use of these maps can be helpful. Figure 10.1 depicts a set of simplified versions of maps for a library system together with an instantiation of the navigation parameter. All of these are linked by some kind of hypermedia mechanism to provide free access and no route is prescribed. In general, the intention is that these maps be used for planning, categorizing, searching, and generalizing design (and also evaluation) information. If a domain or application is relatively new, then one particular sequence of use may be preferable in order to establish the details that are missing and carry out the necessary analyses to fill the gaps. In other situations, information can be extended–adapted to suit new requirements. It is up to each designer to decide what is best for them.

Chapter 8

Evaluation of Design Concepts and Products

NEEDS FOR EVALUATION

The underlying design rationale for this book (as stated in Chapter 7) is that instead of striving to impose a well-ordered normative design process on designers, efforts to *formalize* should be spent on parallel and/or after-the-fact *evaluations*. Seen in this light, evaluation will now be dealt with in some detail. In general, one can be interested in carrying out an evaluation for several reasons:

- To assess whether a system conforms to some standard–goal.
- To compare alternative approaches.
- To identify–assess eventual improvements–changes in a system.

Unfortunately, very little is published in the scientific literature on the evaluation of major information systems, even though it has been found that they often are not accepted by their users, and that they have not been worth the funds spent (e.g., see Rouse, 1981, 1982). However, the need for systematic evaluation methods is increasing rapidly. There are several reasons for this:

- The rapid pace of technological development that makes a smooth empirical and incremental development of systems difficult.
- New technology that upsets the traditions and the common framework that encourage mutual understanding between designers and users–consumers during periods of technological stability.
- A greater interest in the compatibility between the demands the "system" places on the users and their ability (and inclination) to utilize the system in the intended fashion.

- An increased concern with large centralized systems having potential by high hazards (explosions, releases, loss of property, money, information or, at the worst, life).

Issues relating to safety, security, and economy, as well as associated cost–benefit trade-offs in human–machine system design, sometimes create the need for "hard data" to assist designers in choosing among alternatives and in assuring potential customers, as well as concerned regulators, that a new design concept will match their requirements. This in turn creates the need for empirical evaluation techniques based on simulation of system–user interaction scenarios.

Evaluations can be comparative if several systems are to be checked for a differential result or they can be absolute in testing whether a single system will be able to, or does in fact, achieve a given level of performance. Likewise, evaluations can be subjective, if they are based mainly on user and/or expert judgment, or objective, if attempts are made to measure (quantitatively) the degree to which objective criteria are met (time to do, error rate, etc.).

Evaluation in itself is a rather general term. Therefore, two further descriptors are often used to clarify the goal of the underlying activity.

- *Verification* is an assessment of the degree to which the results meet the requirements of the design specification. Verification is supposed to answer the questions: Is the design right? Does the product meet the design intentions?
- *Validation* is an assessment of the degree to which the design achieves the original system objectives. Validation is thus supposed to answer the questions: Does the product meet the needs of the end user? Is it the right design?

In a recent review of the evaluation problems in high hazard systems such as nuclear power, Tanabe (1991) emphasized the need to explicitly evaluate the potential side effects of system functions during abnormal operational conditions. That is, in the validation of system objectives, explicit considerations are necessary for certification. Thus:

- *Certification* is a particular type of validation with a focus on the constraints around the original system objectives. This explicit focus is particularly important when advice systems are introduced, which are based on heuristic rules, as in expert systems. While it is practically possible to validate the systems within the design basis, it is very difficult to certify that the response of heuristic rules to unpredicted situations outside the design basis will not have unacceptable side effects. Thus, certification of software and hardware will very probably become a major concern for regulatory bodies.

EVALUATION METHODS

The evaluation of the quality of a new system can be done *analytically* and/or *empirically*. The role of these two approaches has previously been discussed by Rouse (1984) and Rouse, Frey, and Rouse (1984) (see Fig. 8.1).

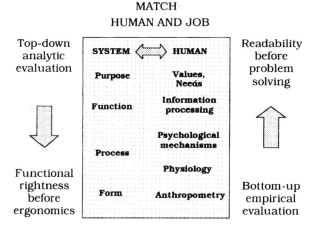

FIGURE 8.1. The complementary nature of analytical and empirical evaluation.

They consider several levels of human–system interaction requiring separate evaluation in order to check the human ↔ work match and it can be seen that these are compatible with a representation of the two components in terms of several levels of abstraction (Chapters 1 and 2). The two approaches are seen to be complementary—analytical is top-down; empirical evaluation is bottom-up experimental. Figure 8.2 is an alternate representation, which begins to link evaluation to the various dimensions of the framework for work analysis covered previously. Various boundaries are shown that can be labeled with specific types of evaluation goals–situations conducive to either analysis or experiment. Considering the latter, the innermost boundary corresponds to the evaluation of actor-related issues in an environment that corresponds most closely to the traditions of experimental psychology. The remaining boundaries successively "move" the context further from the actor to encompass more and more of the total work content in some kind of increasingly complete simulation. We will return to this later.

Rouse distinguishes among the following aspects of the human–work interaction:

- *Compatibility* with human sensory and homomorphic characteristics; that is, can the displays be read and the controls operated?
- *Understandability*: can the user understand the text and graphics of the interface? Does the user understand the provided functions?

These aspects may not need an empirical evaluation if appropriate experience, human factors handbook data and/or checklists were available during design. However, new media may need checking. The following aspect will require the most attention:

- *Effectiveness*: Do the functions provided meet the task requirements of the end user? *Can* the system be used?

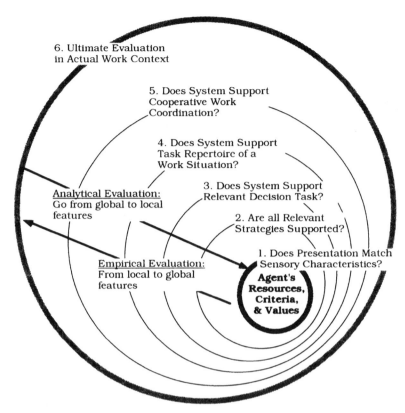

FIGURE 8.2. The relationship between analytical and empirical evaluation and the work analysis framework.

To this, we can add:

- *Acceptability*: Does the design approach and the assumed work organization match the performance criteria and preferences of the users? *Will* the system be used?

Analytical Evaluation

Given that a design normally will reflect more of a "sideways" modification–improvement of an existing system than a top-down process with good documentation of conditions, assumptions, reasons for, and so on, a systematic analytical evaluation of the final result with reference to the dimensions of the framework can be necessary. In this exercise, an analyst will not concern themselves with the ergonomic aspects of displays, letter sizes, digital versus analog information presentations, and the like, before they are satisfied with the functionality of the proposed system at all the relevant levels of consideration. Thus they will plan an orderly evaluation according to the scheme shown in Figure 8.2 and proceed systematically inwards from global-to-local features. They may then seek help from the empiricist in attempting to verify that the form of the displayed information for the users has the proper compatibility and understandability to support their preferred strategies and tactics for coping with the system in an effective and for them acceptable way. Thus issues related to the *contents* of the information will be evaluated analytically, while issues related to its *form* involve context, user experience, and preferences and, therefore, need to be looked at more from an empirical point of view.

An analytic approach is demanding in resources. As discussed in Chapter 3, the use of prototypical work settings and activities as the basis for design and evaluation is a possible approach. If/when design becomes more of a problem-driven activity, the prospects for a concurrent analytic evaluation will be improved.

Empirical Evaluation: First Considerations

As indicated in Figure 8.2, the work framework represents the various behavior-shaping factors in the form of a nested set of constraint envelopes centered around the individual actor. This suggests the definition of a matching set of nested boundary conditions for experimental scenarios that can serve an empirical evaluation in its various shades. The result of this can be seen in Figure 8.3. As a side comment to this figure, it should be noted that complex experiments for evaluating advanced information systems can be wasted if the evaluation is carried out too early at too "high" a boundary.

For empirical evaluations, it is necessary to establish an experimental work situation creating a defined constraint envelope around the subject and to study whether the subjects' responses to this envelope leads to the mode of behavior that was assumed as the design basis. For a comprehensive experimental evaluation of a system, it is necessary to define a suitable sequence consisting of a set of boundary conditions with increasing distances from the actor in order to be able to evaluate more encompassing features of the system. As suggested, this experimental evaluation should be compatible with the analytical evaluation (which, hopefully, is made concurrently with the design) in order to manage the content–form trade-offs.

If we concentrate our discussion on effectiveness and satisfaction issues, then the following possibilities exist (see Figs. 8.2 and 8.3):

Effectiveness. Empirical evaluations of effectiveness can be made at various envelopes, which increase step-wise in their "distance" from the actor.

- Effectiveness at the *mental strategy level*: Are the effective work strategies supported? In this case, classical, well-controlled experiments can be run. Task situations that call upon the relevant strategies are designed and the subjects instructed with respect to the goal of the task and the use of the strategies. For such experiments, tasks can very well be "artificial" without relation to the work situation for which the system is intended. The question is: Given that the user adopts a particular strategy, does the interface enforce the relevant mental model

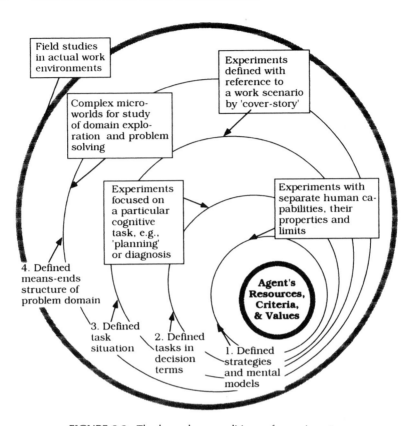

FIGURE 8.3. The boundary conditions of experiments.

and is the information necessary for the strategy present in an understandable form?

- Effectiveness at the *cognitive task level*: Does the system provide an information environment that support the actor's decision making in the task? Are situation analysis, goal evaluation, and planning supported for familiar, as well as less familiar situations? For this, more complex task situations must be simulated, which also serve to evaluate the goal formulations of the subject and allow for a faithful interaction between the action alternatives presented and the situation analysis.

- Effectiveness at the *work situation level*: Does the system adequately support the actual work situation? Is its capacity adequate? The question here is whether the system supports the entire task repertoire: Are the tools adequate?

Individual Acceptability: Will the system be used? For this level, the evaluation must be based on actual work scenarios generated from an actual work analysis. The aim is not a task simulation but a *work place* simulation. The tools will be similar to the evaluation of effectiveness at the work situation level, but the complexity and scope of the simulation will be greater.

Social Acceptability: Does the system concept support the cooperative coordination of activities? Will the system be accepted by the social work organization? This is the ultimate level of evaluation complexity and must include not only a complex work place simulation but also facilities for simulating different interperson communication modes.

Evaluation Measures

Empirical evaluation is a broad topic and the following relevant questions need to be answered before such an evaluation is started: What is to be evaluated; a product, a concept, or a partial solution? In this connection, the following questions can be relevant:

- What constitutes an unambiguous definition of goals and objectives that can be transferred to the evaluation level? What is the evaluation supposed to establish?
- What are the (categories of) situations to be evaluated?
- How will performance be defined? How will it be measured. What will be the link between evaluation *objectives* and *measurable performance variables*?
- What are the effects of the intermediate variables (training, experience, task, environment, etc.)?

Empirical evaluation involves a comparison of measures of performance. Which measures to choose depends on the utilized approach to the experimental design. Very often, hard data are required by the "customers" of an evaluation and these can include measures of effectiveness, time spent to solution, error frequency, and so on. These are particularly relevant when experiments are designed around "practically isolated relationships" and involve an averaging quantitative performance measure across experimental sessions. However, very often the evaluation of complex information systems will have to be based on "causal models" that is, on scenarios constituted by sequences of events, acts, and decisions. Evaluation here depends on a minute analysis and understanding of the individual experimental trajectories and the results will be found in the form of comparisons of the categories of behavior, mental strategies, psychological mechanisms, and so on, as identified by the analysis and the hypothesis of the design. In these cases, detailed analyses of the individual trajectories and generalizations across samples are more important than quantitative data.

Validity of Empirical Evaluation

From the previous discussion, it is clear that the empirical approach to evaluation has its limitations. It seems to be best suited for separate tasks or functions for which a reasonable level of operational skill can be developed and appropriate performance criteria found. For more complicated situations, such as decision support in disturbance management, the empirical approach is most difficult to carry out convincingly and realistically because of all the uncontrolled (uncontrollable) variables. This leads to the danger that if the results of an evaluation are not in accordance with the intuition–expectations of the experimenter, then conditions will be readjusted until they match (confirmation bias). No explicit stop rule exists for the termination of these adjustments of the experimental conditions. When the experimenter strives to explain negative outcomes, they may very likely discover situational factors that need tighter control. This is a natural consequence of causal explanations (see Chapters 1 and 5) and does not imply a manipulative experimenter.

To sum up so far, it seems obvious that some optimum combination of the analytical and empirical approaches to evaluation is required although a specific recommendation is difficult to give. A natural approach will be to perform the analytical evaluation more or less concurrently with the design and, in the case that prior experience or analyses of work in existing systems do not support the necessary design decisions, to test the new concepts empirically during design. This was done for the BookHouse (see Chapter 12). In general, this can be called a form for hybrid evaluation.

Hybrid Evaluation

Now it seems clear that the analytical and the empirical approaches to evaluation are compatible and, in most cases, can both be used in different phases of a design–evaluation process. However, in some cases, it can be advantageous or necessary to integrate the two approaches for the same evaluation problem.

The need to evaluate the properties of a new information system with respect to performance during rare, critical events that punctuate the routine operations of some systems such as high hazard process plants is an example. While reliable performance under such conditions is the raison d'etre for advanced operational support systems in such installations, it is difficult to recreate the situation experimentally where highly skilled personnel are caught unaware by an unfamiliar, high risk situation. Under such conditions, an indirect test can be advantageous; that is, a hybrid evaluation based on an analytical identification of the preconditions for successful performance followed by an experimental test of the match of the relevant functional elements of the system to such preconditions.

For a professional staff, a display will activate over-learned cue–action responses during the normal, familiar work situations. In fact, this is the definition of their expertise. In contrast, when rare events interrupt the normal routine, the display must support knowledge-based diagnosis and planning. A direct empirical test of a new display for knowledge-based support during rare events by a simulated task situation is difficult. Experiments with users trained in the old work environment are unreliable due to interferences from their "imbedded" cue–action repertoire, while performance testing with unbiased novices is unreliable because of their lack of competence. The best approach appears to be to design experiments to test the preconditions for knowledge-based performance in the specific task situation and include a test of the match with a proper mental model and a test of the adequacy of the support given by the system for different relevant strategies for diagnosis, goal evaluation, and resource planning. In effect, this involves the replacement of one direct performance test at the task constraint envelope of Figure 8.3 by several tests at constraint envelopes closer to the actor. All of this must be developed from an analytical decomposition of the user–system interaction followed by a set of separate psychological experiments with carefully selected tasks and subjects to test each of the performance elements. This transition requires control of the constraint propagation among the levels of representation of the actual work condition (i.e., a reliable model of task performance).

A hybrid evaluation test along such lines has been used by Vicente (1991) for testing ecological interfaces for diagnosis. In his experiments, a direct diagnostic evaluation is replaced by a test of a match of an interface to the mental models of subjects—all substance matter experts—by means of a memory test (based on DeGroot's experiments with chess experts). This consisted of a test of whether these experts, on the basis of a short presentation, could accommodate the data presented by a display in their mental model and regenerate the individual data on request— doing it better than novices. Later experiments with this memory recall test (Moray et al., 1991) showed that great care should be taken to reliably identify the cognitive mechanisms that are to be considered for the analytical decomposition and in the selection of the separate experiments. For instance, even when a display matches the mental model of experts, they may not be able to recall data if they do not have a repertoire of familiar and labeled system state patterns that they can use to initialize their mental model for regenerative memory recall.

Evaluation and Design: A Dynamic Process

If a match between requirements–hypotheses and results during analytical or empirical assessments is not obviously possible, then changes will be required—either in the system or in the actual work environment. This, in turn leads to the realization that evaluation actually should be a dynamic process; that is, a continuing design refinement throughout the design process itself, in the transition from design to operations and during the subsequent operational period. In this way, a formalized evaluation of a design can serve as a link between design and operations to support a continuing review process of actual operations versus assumptions and conditions.

This link between design and operation is particularly important for systems that possess a potential for accidents for which the probability must be kept very low. Such systems are normally designed according to a defense-in-depth safety strategy based on protective systems, stand-by equipment, and barriers so that a coincidence of violations of all protective measures and barriers must occur in order to have an accident. Thus, such an accident will be a very rare event and, as described above, a hybrid evaluation of system properties and operator responses with respect to accidents must be used. The analytical evaluation can be constituted by the predictive risk analysis that is an established part of the design of many high hazard systems.

Such a predictive analysis is based on a number of assumptions about the functional properties of the system, its maintenance, and operations management. The empirical part of the evaluation depends on a comparison of the actual operation and management of the system with the assumptions made in the analytical evaluation. This hybrid evaluation must continue through the entire operational life in order to avoid any migration toward the accidents discussed in Chapter 6, because the time horizon with respect to the mean time between accidents for some high hazard systems is of the order of magnitude of 10^6 or 10^7 years, that is five to six orders of magnitude longer than the operations planning horizon.

The fact that all the large scale accidents that occurred in recent decades have been caused by degradation of the design basis, indicates that a more systematic, continuous hybrid evaluation of high hazard systems must be instituted to reach and maintain the intended level of low risk operation.

Boundary Conditions of Evaluation Experiments

As we have begun to see, both simple and complex experiments can fill many roles during the design and evaluation of information systems to meet the earlier mentioned requirements for compatibility, understandability, effectivity, and acceptance. This is particularly the case when new versions can not be based to any great extent on incremental changes of prior designs. The design of a new information system depends to a certain extent on the transfer of reliable models, concepts, and solutions from a current (or a similar) application to the new context. However, as discussed earlier, given an existing system as the "starting point," the identification of behavior-shaping constraints and subjective preferences, some of which may no longer be active (= vocalizable) due to the formation of habits and established practice can be difficult. Therefore, a continuing experimental evaluation—either of hypotheses and models emerging from field studies or of system features and user preferences arising during system design—will often be necessary. In addition, experiments can be required to test hypotheses about the effectiveness of new tools, that is, about users' preferences for certain mental strategies when new interfaces and tools change the demand–resource match, and so on. Following the conceptual design, experiments based on prototypes of varying complexity can be used to validate the design before expensive system manufacture is undertaken. Thus the experimental type and design can be matched to the scope and nature of the problem.

It is outside the scope of this book to suggest particular experimental designs [Fig. 8.4 illustrates some general approaches that have been employed (see Rouse, 1984); see also the section in Chapter 3 dealing with mental strategies]. However, the use of the work framework will be utilized in subsequent sections to support system designers and evaluators in specifying appropriate experiments and/or accessing the experimental results available in the literature. In addition, examples from the various categories will be given.

Such experiments can take on many different forms—from small scale experiments to test local design features or particular human attributes to complex microworld setups for evaluating hypotheses about user adaptation and preferences for

METHOD OF MEASUREMENT	EVALUATION ISSUE		
	COMPATIBILITY	UNDERSTAND-ABILITY	EFFECTIVENESS
Static paper evaluation	Useful & efficient	Somewhat useful but inefficient	Not useful
Dynamic paper evaluation	Useful & efficient	Somewhat useful but inefficient	Not useful
Data-driven part-task simulation	Useful but inefficient	Useful and efficient	Marginally useful but efficient
Model-driven part-task simulation	Useful but inefficient	Useful and efficient	Useful but inefficient
Full-scope simulation	Useful but very inefficient	Useful but inefficient	Somewhat useful but inefficient
In-use evaluation	Useful but extremely inefficient	Useful but very inefficient	Useful but somewhat inefficient

FIGURE 8.4. Alternative evaluation techniques (see Rouse, 1984).

certain strategies in a complex environment. A consistent framework is necessary to enable *generalization* and the *transfer of findings** among actual work analyses and different experimental designs. Usually a section of a work environment is separated and transferred to an experimental setup. Proper experimental control requires an explicit definition of the boundary of the particular cut and the decoupling that separates it from the influences of the rest of the environment. The boundary along which the cut is made varies considerably, depending on the aim of the experiment. Therefore, a suitable framework for identifying the experimental boundary conditions must be a subset of the framework for the analysis of work in general. As discussed earlier, Figures 8.2 and 8.3 illustrate that boundaries defined by the dimensions of the framework (Fig. 8.2) can be made compatible with the constraint envelopes used for defining experiments (Fig. 8.3). Figure 8.5 uses the work analysis diagram of Figure 1.9 to illustrate in more detail the constraint envelopes with reference to the various dimensions of the analysis. Among other things, it indicates that a closer look at the role allocation, as well as management philosophy implied in a given experiment and their relations to the actual work scenario, becomes more and more important the further the defined constraint envelope is moved away from the actor.

Figure 8.6 illustrates how the experimental framework of Figure 8.3 can be used for the definition of experimental boundaries. In the example shown, which involves an experiment for a given task situation, only a part of the actual work system is directly involved–simulated, while the rest is included in some way as "stationary' features of the cover story or backdrop utilized. The intention is that the situation for the user regarding the task situation, as well as the inner boundaries—decision tasks, choice of strategies, and so on—is the same as it would be if they were confronted with the corresponding portion of the real work system. If this can be achieved, then there should exist a good possibility for transferring the results of the experiment to the actual system design—and/or to other systems with similar constraint boundaries.

---

*Some material on generalization, as seen from a behavioral science point of view is presented at the end of this chapter.

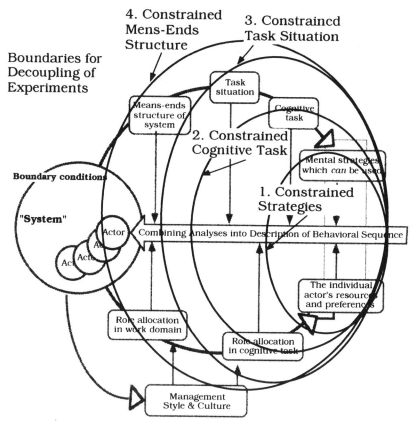

FIGURE 8.5. The various levels of experimental constraints superimposed on the framework for work analysis.

A TYPOLOGY OF EXPERIMENTS

In addition to identifying the boundary conditions with reference to the constraint envelopes shown in Figure 8.3, it is necessary to clarify any eventual links between the *experimental* work domain as seen by a subject and their *private* world of experiences. It is obvious that while an experiment is embedded in the world of the experimenter–evaluator, it interacts at the same time in some way with the private world of the subject. In traditional psychological laboratories, the setup of the experiment and the instructions to the subject are intended to decouple the experimental conditions from this private world. However, in complex experiments (which utilize so-called "microworlds'), "cover-stories" are used to define the contents of the task through their inherent links to the subjects' private world experience.[†] Indeed it can

---

[†] A note in passing: We are not completely happy with the term experimental "subjects." They may be *subject* to the experimenters' manipulations but they are *objects* of attention and analysis.

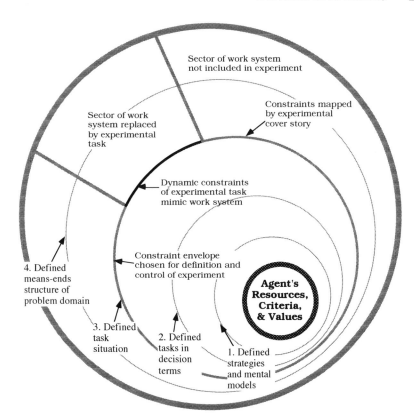

Sector of work system
not included in experiment

Constraints mapped
by experimental
cover story

Sector of work
system replaced
by experimental
task

Dynamic constraints
of experimental task
mimic work system

Constraint envelope
chosen for definition and
control of experiment

**Agent's
Resources,
Criteria,
& Values**

4. Defined
means-ends
structure of
problem domain

3. Defined
task
situation

2. Defined
tasks in
decision
terms

1. Defined
strategies
and mental
models

FIGURE 8.6. Transfer of results from laboratory experiments to the design of actual work systems depends on a well-specified envelope of constraints that is shared by the actual work system and the experimental task in the context of the cover story used to instruct subjects for the experiment. The figure illustrates the use of the framework to make the boundary conditions explicit.

be relevant to take a closer look at the relations between the different types of experiments and the object worlds of *both* the experimenters and their subjects. To make the interaction among these two domains explicit and to characterize the dependencies of different experimental categories on this interaction, the means–ends structure discussed in Chapter 2 can be useful. Thus the three-columned means–ends diagram shown in Figure 8.7 can be seen to comprise 1) the experimenter's domain, 2) the subject's interpretation of this particular experimental domain and, 3) the subject's private world. This figure illustrates that the experimental setup, its processes and semantic functions are common to the worlds of the subject and researcher. However, this is not (necessarily or usually) the case for the related priorities and goals. In the following sections, the boundary conditions relating these worlds will be reviewed for different kinds of experiments. All of this has implications for being able to make generalizations for other contexts.

Another complication in experimental design is that basically different representa-

MEANS-ENDS RELATIONS	PROPERTIES OF EXPERI-MENTER'S DOMAIN	SUBJECT'S DOMAIN	SUBJECT'S PRIVATE DOMAIN
GOALS & OBJECTIVES, CONSTRAINTS	*Experimental Objectives and Constraints*	*Subject's Goals*	*Goals and Constraints of Work and Leisure*
PRIORITY MEASURES, MONETARY VALUES, PERSONNEL, MATERIAL	*Experimental Evaluation Measures*	*Subject's Priorities*	*Subjective Priorities and Performance Criteria*
GENERAL FUNCTIONS AND ACTIVITIES	*Experiment* *Functions involved in Experiment*		*Functions Of Work and Private Life*
PHYSICAL ACTIVITIES AND PROCESSES OF EQUIPMENT	*Physical Processes of Experimenal Equipment*		*Processes known from Work and Private Life*
APPEARANCE, LOCATION & CON-FIGURATION OF MATERIAL OBJECTS	*Configuration of Lab. and Equipment*		*Familiar Objects , Configurations and Topography*

FIGURE 8.7. The means–ends structure of an experiment.

tions can be used to model the work environment as well as the actor. As a proponent of the study of human behavior in complex microworlds, Dörner (1987, 1989) and Dörner et al. (1983) have in a provocative way contrasted traditional experimental psychology with microworld research by characterizing the first as *"dissections"* in a way that is unacceptable for a complex, nonlinear and nonadditive system such as the human, while the latter depends on *"condensations"*; that is, simplifying structures in the same way as a woodcut maintains similarity with the total picture without being muddled by insignificant details (cf. the distinction between decomposition and isolation vs. abstraction and separation illustrated in Fig. 1.2).

However, when viewed from the present perspective, this distinction is not so much tied to the complexity of the experimental setup as it is to the distinction between the use of *relational* or *causal* models (see Chapters 1 and 4). In the present context it is important to reconsider the basic differences between these modes of representation. The relational model depends on an abstraction and a separation of a relationship expressed in terms of (typically mathematical) interdependencies among measured variables. Causal models build on a decomposition and a discretization of the world into objects, and of their behavior into events and acts. Such models are expressed in terms of regular connections of events including sequences of, for example, decision making, action planning, and strategies. In general, their descriptions are verbal reports. Either of these modes of modeling can be chosen for the work domain and, independently, for the human actor in any and all types of experimental scenarios, as will be seen from the discussion in the following sections.

A simulation of the behavior of a work domain must necessarily be based on a representation of its fundamental source of regularity. Consequently, domains that are shaped around well-structured technical systems (aircraft, process plants, etc.), where the basic functional relationships can be isolated from the disturbing complexities of the world, are normally simulated by sets of differential equations describing, for example, dynamic vehicle behavior or mass and energy balances. By contrast, a less structured work domain, such as a hospital or an office, will more readily be describable in terms of verbal statements about decision and transactional sequences.

Similarly, the behavior of human actors in a well-defined system and task, such as aircraft piloting, where movement patterns relate directly to system dynamics, can readily be represented by a relational model (the manual control and optimal control models discussed below), while the control of less structured systems are described more effectively in terms of discrete decision making.

Experiments based on the two categories of models require different techniques. Models in terms of "practically separated relationships" are abstractions that are considered to be decoupled from the influence of the surrounding world. The law of gravity is, for instance, only valid if the interferences from friction and air resistance are neglected. Experimental design tends to control the "separation" of the relationship and to eliminate disturbances from individual and situational differences. The results of psychological experiments are often expressed in terms of averages across a population unless individual differences are studied; hence, the need for statistical significance (see also the concluding section in this chapter on generalization).

Similarly, regular causal relations are abstractions and depend on analyses of singular cases by decomposition, followed by a categorization along and across the individual scenarios in order to identify recurrent categories of events and their causal relations. Generalization is done by defining prototypical samples. As discussed in Chapter 1, categories cannot be defined by exhaustive lists of attributes and counterexamples can therefore always be found. In this mode of representation, small changes of parameters can cause a quantum jump among categories. Experimental verification of such causal models is difficult because the models actually represent relations among types, whereas each simulation run is only a particular token. Therefore, a verification of a causal model requires a test of the category boundaries by a cluster analysis of trajectories resulting from parameter variations and several simulation runs and, thereafter, a comparison of the clusters found with different subject and situation characteristics. The quest for statistical significance is replaced by a search for prototypical significance. See Brehmer et al. (1991).

Similar techniques are used for experiments with isolated relationships to guarantee that the experimental samples will be drawn from the same category of relationships. The basic difference is one of focus. In the relational approach, analysis and generalization are focused on average performance; in the causal approach, the focus is on the analysis and understanding of the particular trajectories as a prerequisite for generalization.

DEFINITION OF EXPERIMENTAL BOUNDARY CONDITIONS

The various categories of evaluative experiments will now be discussed and exemplified to illustrate the use of the framework for specifying the experimental and the boundary conditions, as well as other factors dealt with above. The categories are labeled with reference to the boundaries shown in Figure 8.3.

The design process that unfolded during the development of the BookHouse system described in Chapter 9–12 is a very clear example of design as a creativity–resource trade-off. Many degrees of freedom remained for the designer when we were given the results of previous field studies and the objective to design a computer-based tool for the retrieval of fiction by the casual user of a public library without requiring any special instruction. These had to be resolved in a trade-off between available programming tools and resources (the hypermedia tools were not available at the time of design); the established practices in the (Danish) library institutions (including indexing practices); the uncertainty about user acceptance of new concepts (such as retrieving books by visual browsing in iconic representations), and so on. Thus the design process resulted in a long sequence of iterations among design ideas, prototyping of part solutions, and experimental evaluation of ideas, hypotheses, and solutions. The sequence of experiments performed turned out to be a very good illustration of the need to consider the conditions underlying experimental designs at the various boundaries between the user and the ultimate work context in the library. The sequence of experiments is discussed in detail in Chapter 12, but references are made in the following review of experimental design for the evaluation of new systems.

Boundary 1: Controlled Mental Processes

It is safe to state that the bulk of experiments in this category—with certain exceptions—are not carried out by system evaluators in connection with any specific application. Instead, most investigations fall into the realm of academic experimental psychology whose practitioners—with support from governmental (military, transportation, energy agencies, etc.) and other research funders—carry out a wide variety of studies to extend what we know about basic human capabilities and limitations and to make the results available without much direct concern for the eventual contexts of the end users. In general, these aim at identifying, describing, verifying, and quantifying a (limited) hypothesized psychological phenomenon. The utilized methodology essentially derives from the natural sciences and the results are packaged in a more or less standardized reporting format. Strong emphasis is placed on measuring observable phenomena and testing the statistical significance of the quantitative findings. Experiments in this category are based on carefully chosen and well-controlled experimental tasks that selectively stress the application of particular psychological mechanisms—the "isolated relationship." For such purposes, the tasks can be somewhat artificial—indeed they can be categorized as perception, tracking, dual choice, association tasks, and so on—as illustrated by the various editions of the *Journal of Experimental Psychology*. The formulation of the subject's instruction is at the procedural level and is very explicit. It serves to define the constraint boundary around the experimental situation and isolate it from (1) the general, personal knowledge background, and performance criteria of the subject, and (2) any eventual higher level considerations within the experimental domain itself (see Fig. 8.8). A particular feature is the close relationship that often exists between the processes of the experimental equipment and the experimental task—this is not found in more complex

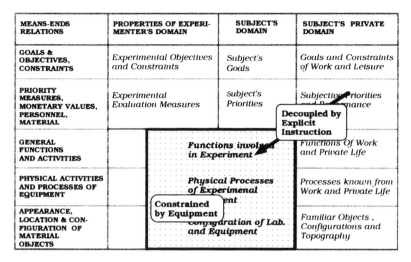

MEANS-ENDS RELATIONS	PROPERTIES OF EXPERI-MENTER'S DOMAIN	SUBJECT'S DOMAIN	SUBJECT'S PRIVATE DOMAIN
GOALS & OBJECTIVES, CONSTRAINTS	*Experimental Objectives and Constraints*	*Subject's Goals*	*Goals and Constraints of Work and Leisure*
PRIORITY MEASURES, MONETARY VALUES, PERSONNEL, MATERIAL	*Experimental Evaluation Measures*	*Subject's Priorities*	*Subjective Priorities and Performance*
GENERAL FUNCTIONS AND ACTIVITIES		*Functions involved in Experiment*	*Functions Of Work and Private Life*
PHYSICAL ACTIVITIES AND PROCESSES OF EQUIPMENT		*Physical Processes of Experimenal ent*	*Processes known from Work and Private Life*
APPEARANCE, LOCATION & CON-FIGURATION OF MATERIAL OBJECTS		*Configuration of Lab. and Equipment*	*Familiar Objects , Configurations and Topography*

Decoupled by Explicit Instruction

Constrained by Equipment

FIGURE 8.8. Instantiation of the means–ends structure of an experiment.

simulation scenarios. Consequently, the use of only one experimental method for each study is commonplace.

Examples from this area include controlled experiments to selectively measure human performance parameters and capacity limits—such as speed, accuracy and signal-to-noise ratios in motor control, capacity of working memory span, learning, and associational performance. Some of the "hard data" resulting from these investigations is included in general purpose human factors handbooks (Pew, 1974; Boff and Lincoln, 1988; and Boff, Kaufman, and Thomas, 1986) summarize results from these kinds of experiments.

Several experiments were conducted within this category for the BookHouse design discussed in Chapter 9. One of the objectives for interface design was to actively support this approach by probing preattentive, unconscious cognitive processes in order to improve the utilization of implicit knowledge more efficiently. Therefore, a number of experiments were made to determine users' associations to pictures from keywords chosen from the users' descriptions of book contents. In addition, picture–word association experiments were conducted to evaluate the consensus between designers' conceptions of associative relationships between pictures and keywords and those of different groups of potential users of the system. A controlled, multiple choice word association experiment was used to evaluate the words associated by different groups of users (children and adults) to the icons intended for interfaces (see Chapter 12).

Boundary 2: Controlled Cognitive Tasks

This category moves one step away from the actor towards the total work situation by focusing on the study of separate decision functions, such as diagnosis, goal evaluation, planning and/or the execution of planned acts.

The studies of *manual control* are examples. Pioneered by McRuer and Krendel (1957) and Crossman and Cook (1962), these studies have typically been focused on practically isolated relationships based on control theoretic concepts. In these investigations, the boundary conditions of the experiments are rather well defined because the behavior shaping constraints are to a large degree given by the simulated properties of a well-structured process.

Another extension going from boundary 1 to 2 results in a focus on studies of *problem solving* and *decision situations*. Here, the cognitive processes are more complex and less constrained by the environment. An illustrative example in this area with implications for generalization and transfer to actual work conditions has to do with experiments in diagnostic reasoning.

The boundary condition of these experimental scenarios is typically defined by the cognitive task given the subjects while the possible diagnostic strategies are in general only implicitly specified by the particular presentation of the problem. Since these experiments are often initiated from problems identified in actual work performance, a "cover story" is used for instructing the subjects. However, this "package" usually refers to the work context without any rigorous analysis of the differences in the "real-life" versus experimental behavior-shaping constraints that shape the criteria for the choice of strategy. If these differences in constraints cannot be explicitly formulated, other means for identifying the differences in mental strategies employed in practice and in the experiments must be used (verbal protocols, discriminant analysis, cluster analysis, etc.).

An example of this problem arises in connection with experiments on diagnostic behavior based on the *social judgment paradigm*. The approach has been based in regression analyses of the influences of presented cues on judgments made by subjects in laboratory tasks. Studies have been made in several professions—stockbrokers, clinical psychologists, and physicians (e.g., see Brehmer, 1981). In these experiments, cues identified as diagnostically relevant by expert judges are presented to subjects in a set of trial cases, generally in the form of verbal descriptions on cards. From the experimental sessions, a statistical model describing diagnostic behavior is identified. In general, linear statistical models from a multiple regression analysis have been found adequate to describe performance. The following conclusions from these experiments are typical: The judgment process tends to be very simple. Even though experts identify up to 10 cues as relevant to a diagnosis, they actually use very few—usually only two or three, and the process tends to be purely additive. Furthermore, the process seems to be inconsistent. Subjects do not use the same rule from case to case, and judgment in a second presentation of a case may differ considerably from what it was the first time. There are wide individual differences even among subjects with years of experience. They differ with respect to the cues used and the weights they apply. Finally, people are not very good at describing how they make judgments (Brehmer, op. cit.).

Seen from the present perspective, the constraint envelope around the subjects for these experiments is defined at the decision task level—boundary 2 (see Fig. 8.3). As stated earlier, professional subjects are normally used and the task is well specified as a diagnostic problem stated in professional terms. However, the experimental setting is far removed from the natural context. Therefore, an important question relates to the extent to which this condition forces the professional subjects to choose a different mental strategy for diagnosis. Another illustration of the importance of defining the experimental boundary conditions can be found in the experiments of Wason and Johnson-Laird (1972). They indicate that a logical inference task, which proves to be difficult for subjects when presented in abstract logical terms, becomes very easy when embedded in a familiar context, such as a shopping account approval task.

Thus, the problem for this category of experiments is normally not connected with the validity of the results but instead with the generalization that is possible. Generalization is frequently made in terms of the behavior of professionals in a nominal task—let us say "diagnosis" under actual work conditions. This generalization is unreliable. After comparing the results of laboratory studies with our analyses of diagnostic tasks in hospitals and repair shops, we can identify some important differences that signal great caution concerning any transfer of the laboratory results to the actual professional work context. This does not imply that the results of the laboratory experiments above are not valid for multiple attribute judgment tasks, but rather that isolated multiple attribute judgments are not always a characteristic of a diagnostic judgment in actual work (see the discussion in Chapter 3). Reliable generalization at this level of constraint specification requires the identification of the relationship between the experimental situation, the possible mental strategies, and the *performance criteria* governing the actual choice; that is, the behavior-shaping features. In the example described, any generalization should probably be stated in terms of the ability for intuitive, multiple attribute judgments rather than for professional on-site diagnostics.

Many laboratory studies have studied diagnosis and fault management (see e.g., Rouse, 1981; Sanderson, 1990). They are often referred to as experiments on "fault-finding" or "trouble-shooting." However, it seems clear that the diagnostic conditions in an actual work situation are different from those in an investigation requiring a subject to locate a fault in a constructed logic network. In these experiments, the subject is presented with a symbolic representation of the system in which the problem is found. Therefore, a particular "diagnostic field" (see Chapter 3) is forced upon the subject. This turns the experiment into an evaluation of the subject's ability to operate in this particular diagnostic field and to use any related diagnostic strategy. This does not render the experiment invalid, but a careful explication is required of the boundary conditions of the experiment and the kind of mental strategy the experimental conditions imposes on the subject. Thus a generalization can not be made with reference to performance in the work domain, which is the source of the cover story, but only with reference to the conditions giving rise to the applied strategy.

An interesting approach to an experimental design of a series of simulation experiments to train diagnostic skills and, at the same time, to train the ability to generalize has been taken by Rouse (1982). In this "mixed fidelity simulation" training, the trainees are subjected to diagnostic tasks moving through levels similar to those of Figure 8.3 and ranging from the simulation of faults in logic networks to simulated task situations.

The structure of the experimental work domain faced by a subject in this category is somewhere between the situations illustrated by Figures 8.8 and 8.9. The functions required by the subjects are to a lesser degree related to the configuration and processes of the experimental setup while the explicit task instruction is to some degree replaced by a "cover story" referring to work context (trouble shooting, scheduling, etc.).

In short, we find that the framework suggested here can serve to resolve much of the standing controversy about the value of laboratory experiments for understanding "real life" work—a controversy caused by designers' frustrations concerning the lack of explicit descriptions of the boundary conditions of experiments in terms that could facilitate the transfer of results to their work contexts.

For the BookHouse design, several experiments were performed within this category to evaluate the applicability of the dimensions of user needs as defined from analysis of user–librarian negotiations for specifying search terms for information retrieval (see the description in Chapter 12).

Boundary 3: Controlled Task Situation

The development of advanced information technology and the need for models of cognitive processes in work performance have brought about a significant interest in more complex experiments focused on actual task situations. Consequently, the simulation of complex task situations controlled at boundary 3 in Figure 8.3 has become popular.

Two approaches are also found in this category. Studies based on a separation of relational networks have been made—in particular for modeling human performance in aircraft and vehicle control in terms of an extension of the manual control paradigm to include complex optimal and adaptive control models (for a review see Sheridan and Ferrell, 1974). Because of a rigorous representation in terms of sets of differential equations (due to the well-known dynamics of aircraft) and of the parameters involved in planning and decision making, the boundary conditions of the experiments are explicitly known and the basis for generalization is well documented.

The transfer of the mathematically based modeling methodology to less constrained environments is a problem that is difficult, if not impossible. Consequently, another line of experiments has emerged based on more qualitative causal task scenarios (for a review of activities in Europe, see Funke, 1988). In this category, we find studies of decision making and diagnostic behavior using simulators of high functional fidelity for aviation, chemical process plants, power plants, and so on. In addition, the optimal control models have recently been extended by adding decision and planning modules based on a causal, object oriented modeling approach (Baron et al. 1986).

In other cases, special simulators have been used. The line of experiments with manual control of Crossman and Cook (1962) has been continued using simulated industrial process systems (Moray, Lootsteen, and Pajak, 1986; Moray and Rotenberg, 1989; Sanderson et al., 1989; Vicente, 1991). Also, Rouse and his group have been doing experiments in problem solving in simulated industrial process tasks (Rouse, 1982; Knaeuper and Rouse, 1985; Morris, Rouse, and Fath, 1985). Other examples are GNP, a generic nuclear power plant simulator (Goodstein, 1985a), as well as a simulator for nuclear reactor feedwater systems (Cacciabue, Codazzi, and Decortis, 1990).

Since such simulations are designed as replicas of actual work scenarios, the expectation is often that the behavior observed in these experiments will mimic the actual behavior of people in the "real situation" provided that the subjects have had some experience with the "real system." However, an important factor to consider is that the subject's attention with this kind of simulation normally is focused on a particular task—such as "diagnosis" or "scheduling"—and, at the same time, on a particular part of the system. This will change the constraint set at the boundaries around the experiment. For example, when the simulation is concerned with some work process, the laboratory situation will differ from the real work situation in that the subject may work alone, rather than as a member of a shift team; the motivational structure may be different; the person may work for a shorter time, and so on. Whether these differences are important must be carefully assessed. The only possible way to generalize from such experiments is to seek an identification of the behavior-shaping features at the boundaries of the experimental situation in order to compare them to other situations and systems.

In experiments striving to constrain the actor by only specifying the task situation, the boundary of the experimental domain is less well defined because a "cover-story" is used to invoke the subject's "private" context. In this way, the specification of the boundary conditions is made with reference to the subject's experience with or general understanding of the work scenario that is being mimicked (see Fig. 8.9). Therefore, it is necessary that the functional structure of the experiment matches the

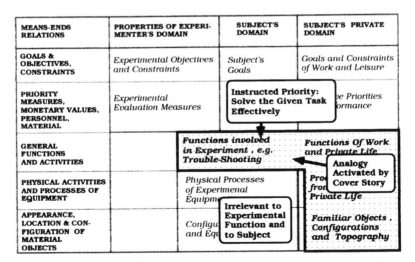

FIGURE 8.9. Instantiation of the means–ends structure for an experimental setting controlled at boundary 3 for studying hypotheses about the relationships between particular task parameters and task performance.

context alluded to by the cover story. This match can only be demonstrated by an explicit identification of the behavior-shaping constraints of the experiment, as well as intimate knowledge about the constraints of the actual substance matter domain. In efforts to cope with the difficulties of mapping the transformations across the boundaries shown in Figure 8.3, Moray and Dessouky (1992) designed a sequence of experiments in a psychological laboratory and in a model workshop of an industrial engineering laboratory.

For the BookHouse evaluation, discussed in Chapter 12, efforts were concentrated on testing retrieval strategies and the indexing policy used in task situations specified by cover stories.

Boundary 4: Complex Work Environments: Microworlds

A more recent category of experiments has been focused on human problem solving behavior in complex simulated work environments in which the entire decision process is activated, including value formation, goal evaluation, and emotional factors.

Prominent examples of this kind of simulation are those designed by Dörner (1989) and Dörner et al. (1983). In one of his experiments, a subject is asked to assume the role of mayor of a small German town called "Lohhausen." In another scenario called "Moro," the subject's task is to serve as advisor to a tribe in Africa. The experiments in this category are aimed at a study of the performance of the subjects in their explorations of an entire problem domain, including their formulations of the present problem, the goal to adopt, and the solution to choose. Because of their very nature, the instructions for these experiments become very open and are given by a "cover story" that relates the experiment to the private world and general experience of the subject (see Fig. 8.10).

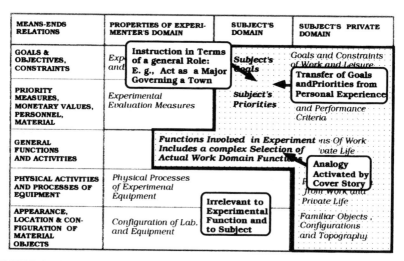

FIGURE 8.10. Instantiation of the means–ends structure for a boundary 4 microworld experiment.

The performance will then be strongly influenced by the particular background of the individual subject. In addition, it is very difficult to control the actual goals and performance criteria adopted by the subjects in a reliable way when either naive subjects (e.g., psychology students) or professionals from the particular work domain serve as subjects. When instruction in a complex experimental task is given by a cover story, personal experience and knowledge about the world are activated with very idiosyncratic effects. Dörner gives some examples of this influence. In the Lohhausen experiment, a subject who was a school teacher focused attention on particular teaching problems in the school system and left the rest of the society on its own; a left wing subject decided to introduce extensive socialistic measures and explained difficulties with their strategy as being caused by reactionary sabotage; a student without any experience in production planning used a simple analogy—her personal experience from rolling their own cigarettes as the basis for their decision making.

Similar experience has been reported by Sanderson (1989), who found that different assumptions about the use of cars and busses in the United States and in England were likely to influence their results when they duplicated experiments on implicit and explicit learning originally done in Oxford by Broadbent, Fitzgerald, and Broadbent (1986). In Sanderson's experiments, it seemed (Sanderson, private communication) that the US subjects did not believe that "citizens" would continue to accumulate at a bus stop until a bus came. Instead they assumed that the longer the time periods between busses, the fewer people would elect to take the bus. The assumption was that everyone had cars. This is probably less likely in the Oxford subject pool, which is composed typically of women with sufficient time on their hands and/or a financial need who desired to take part in psychological experiments at the university.

Several experiments were made for the BookHouse design by simulated retrieval sessions supported by paper-and-pencil simulation of support tools (see Chapter 12 for detailed discussion).

A shift to a more unconstrained environment for evaluating retrieval performance had to be taken to check performance and end users' satisfaction with a complete system. However, this had to be done progressively, first by involving library staff and end users in an evolving prototyping process leading to the introduction of the "final" version as a tool in the ordinary daily library routine. Three different prototypes were developed for the BookHouse data base, each with approximately the same system functionality but with different interfaces: The Command Interface, The Book Machine, and the BookHouse, which reflected differing degrees of similarity with traditional interfaces for data base retrieval.

Boundary 5: Experiments in Actual Work Environment

Experiments at this level depend on techniques similar to field studies and we will, for a general discussion, refer to the methodological hints presented in the previous chapters.

For the BookHouse evaluation several extensive experiments were performed in the actual library context, as described in Chapter 12. Before a text version was released for public testing it was subjected to an exhaustive functional verification in a library to ascertain that the system could meet the functionality requirements. A subsequent evaluation using the general public was then performed to validate whether or not the system concept actually was appropriate for supporting library users in finding good fiction (see the description of the experiments in Chapter 12).

COMPARING SOURCE AND RESEARCH DOMAINS

In evaluative experiments, the processes and anatomy of the simulation replace the physical processes and anatomy of the real work domain. A subject will normally be well aware of this fact, particularly when "in trouble." Frequently, a subject will explain unexpected performance by peculiarities in the simulation or bugs in the computer program. In an actual work environment, there are very subtle many-to-many relationships between the goals, the functional level, and the possible implementations at the physical process level. This makes it very difficult for a subject to judge what is and what is not included in the simulation. Therefore, the subject will have to infer its scope as well as the goals, constraints, and functions that actually are included on the basis of more or less intuitive assumptions about the "source world" as conceived by the designer of the experiment. In this situation, it is important to be able to analyze and describe explicitly how the actual source domain is treated in the research hypothesis and to make explicit how the behavior-shaping constraints are transferred to the research domain. It is only possible to draw conclusions from selective experiments and simulations if we can unambiguously demonstrate that the propagation of constraints from the source domain and the actual work conditions results in a similar envelope around the subject, as does the propagation from the research domain and the experimental conditions (see Fig. 8.11).

Instructions, Metaphors, and Cover Stories

In experiments, subjects are supposed to control a task according to an instruction. The content of this instruction changes with the experimental strategy. In traditional, well-controlled experiments, the instruction can be explicit and unambiguous. In complex experiments with "microworlds," such explicit unambiguous instructions are impractical and "metaphorical" instructions are used. A cover story is a very effective short-hand instruction. Any attempt to formulate an instruction without retreating to a metaphor, but instead to express it in terms that only relate to the features of the simulation display and the actions upon it, turns out to be complicated, lengthy, and incomprehensible. A cover story should be based on a context very familiar to the subject as well as the experimenter, and the relational structure of the simulated work should be faithful to the cover story to avoid unpredictable interferences from the subject's private life.

In general, the problem with complex experiments is not the ambiguity of the metaphorical or cover-story instruction, but the problem of explicating the conditions

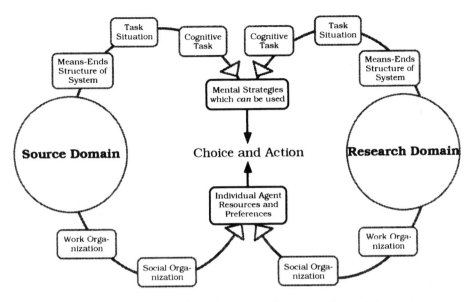

FIGURE 8.11. Propagation of constraints in an experimental design.

for generalization. Generalization cannot refer to the metaphor or the cover story since the experimental results are typically not directly related to process control, trouble shooting, or to mayors controlling towns. Generalizations can only be made with reference to precisely formulated behavior-shaping constraints. In short, the main problem of experiments in complex simulations is to establish an explicit representation of the features of the experimental context; it is not due to the use of a cover story for instructing the subjects. Cover stories work only one way—for instruction. Generalization back to related source domains is another and more difficult question.

EXPERIMENTS FROM A BEHAVIORAL SCIENCE PERSPECTIVE

Generalization from experiments has been a matter of concern within some sections of the behavioral and decision science community. Hammond (1986), Lopes (1986), and Hogarth (1986) have in a set of articles dealt with the problem and now it may be relevant to include a summary of some of their views. Hammond starts by emphasizing the multidisciplinary nature of this field and its underlying host of paradigmatic premises. He attempts to clarify what generalization and operational context are all about and, in that connection, discusses whether the former refers to a generalization of *results* (its empirical sense) or to a generalization of *models* (its mathematical sense). A researcher aiming at the generalization context places great value on a *correspondence* theory of truth (i.e., there is an empirically founded correspondence between an idea and some aspect of the real world). The researcher is primarily interested in the *range of conditions* (in this real world) over which the correspondence holds. This view is treated with disdain by those who place great credence in the *coherence* theory of truth. Their first requirements are that a mathematical model must be available and the math must be correct. Beyond this, as far as generalization is concerned, they are interested in the *range of conditions* included in the model. Hammond mentions the apparent demand in the decision and management sciences today for both—each in some "satisfactory" portion—but he concentrates mostly on the problems encountered in generalizing results derived from experiments (i.e., the correspondence approach).

The time honored method is *intervening* (i.e., determining the correspondence between theory and reality by experiment). Here, by bringing into the laboratory some manageable fraction of the world of interest and by controlling and manipulating (i.e., interfering with) it, the researcher attempts to take sufficient safeguards against their drawing incorrect conclusions about the particular cause–effect relations under study. But, in doing this, the question arises whether the results from the experiments indeed generalize to the conditions of interest (i.e., those not in the experiment). In general, there is a vast difference between these results, so that it is fashionable to make a distinction between the laboratory and the real world, but it is apparently not so fashionable to provide an argument for linking them. In this connection, Hammond decries current practices of attempting to generalize solely over *subjects* (i.e., from subjects in the experiment to subjects not included) but not over *conditions* (i.e., from conditions in the experiment to conditions that were not included). In this connection, he names the representative design approach of Brunswik. In any event, a balanced approach would require that not only the size of the sample (number of subjects) be considered but also the range of variation of significant variables in the circumstances to which generalization is desired. His other major complaint concerns the method-dependent character of behavioral research (i.e., results are more tied to method than to concept; change the method and the results disappear).

Lopes (1986) produces several telling remarks. While generalization implies abstraction and, hence, theory, it does not seem possible that any theory lacking a strong empirical base will be applicable or generalizable to the real world. That is, the theory at its heart must be a theory of something independent of itself. She comments on the currently popular "biases and heuristics" movement that leads to theories of why people deviate from theories, not theories of why people behave as they do. The consistent framing of empirical studies in terms of normative models has produced a science in which human behavior is not so much studied as it is graded and in which the goal of describing behavior is replaced by the goal of explaining why the observed behavior is different from some other behavior. It makes no sense, she says, to use a model of behavior as a benchmark for the behavior it is modeling. If there are disparities, then it is the model that has failed to measure up. She has some warnings concerning empirical research. She says that we must not be content with cover stories and word problems that suggest real world settings without engaging any real world motivations or real world behavior. Instead, we must try to develop tasks and techniques that involve subjects experimentally in ways that mirror their involvement with real world problems. She mentions some studies'of this kind—including one by Scribner (1984) who in a dairy establishment established how relatively uneducated "preloaders" could handle mixed number representations of orders for milk, and so on, by means of spatial strategies. These were later confirmed in laboratory experiments.

The third author, Hogarth (1986), points out that generalizations (at least in the behavioral sciences) decay with time. As he puts it, science is the creation of knowledge that is most usefully coded in terms of *causal statements* that can be elaborated in more or less detail. Generalizations are then expressed in terms of cause–effect relations and decay because (1) it is difficult to identify appropriate causal agents (= working hypotheses) and (2) the observations of simple cause–effect relations are complicated by the myriad of environmental conditions in which they operate. He believes that much can be done to prevent this through careful attention to methods. He discusses models and their importance in the spectrum extending from "data without models" to "models without data." One pungent citation, made by Abelson, "you can really only say that you understand a phenomenon when you can make it go away," requires a model. Hogarth distinguishes between two types of models—symbolic (including mathematical) and replica. In his view, good models will permit comparisons with data at several levels: (1) the level of the assumed underlying process, (2) concerning environmental conditions, and (3) predictions versus observations.

Hogarth has six reasons why we should continue to do experiments (i.e., build replica models):

- A little knowledge is better than none.
- Experiments help avoid metaphysical speculation.
- Experiments can and should be used to illuminate scientific conflicts.
- Experiments can impact on practice.
- Experiments are a form of history.
- Experiments help define new questions.

Chapter 9

Design of a Library System

INTRODUCTION

In this chapter the library information retrieval system described will be discussed with reference to the concepts presented in previous chapters. The framework for work analysis found in Chapter 2–4 has been used to conduct field studies in libraries and the principles for ecological system design found in Chapter 5 was used for design of a new library system, guided by the use of the maps presented in Chapter 7. The actual evaluation of the library system will be discussed with reference to the evaluation principles introduced in Chapter 8. This chapter is organized according to the order of presentation in the theoretical part of the book. We hope to illustrate that this is not intended to be a normative design guide but rather a framework within which a designer can iterate freely to what is found useful during the design process. Actually, the presentation of the theoretical framework has evolved as a rationalization of the models and principles that were found to be useful during work analysis and system design in various domains, including the library.

The library domain is located in the left-hand side of the territory shown in Figures 4.2, 7.5, and 7.6 with autonomous users retrieving information from a large inhomogeneous collection of documents. This work domain suffers from the lack of a coherent internal structure originating in physical laws and explicit system objectives as found in the right-hand side of the spectrum. The library is a complex work domain, due to the number and variety of documents, and the span and change through time of user interests and backgrounds, together with the rapid growth of the information volume.

The work space of a library includes the librarians and their clients; the library users. Library users are largely autonomous even if they are somewhat constrained by the books made available under the library laws and policies for services rendered and by the mediation goals of librarians when asking for their retrieval assistance. This chapter describes the design of an information retrieval system called the "BookHouse." This retrieval system is intended to serve the library users in their own search as well as the librarians when assisting in the search.

In this way, the information system described is intended for the particular work situation in a library involving the identification and retrieval of documents that will serve a given user's need for information. The system was designed with two complementary objectives.

1. The primary objective was to have an experimental evaluation of the effectiveness of the approach in a theoretical framework for cognitive work analysis through the full-scale implementation of a library support system in a typical (Danish) library environment. This objective was met successfully. As of this writing, the prototype system has now been developed into a commercial product as a library program to support end users and librarians in libraries in both information retrieval and indexing tasks.

2. An additional, but equally important, objective was to use the library system to explore the problems of information retrieval found at many work stations in general, not only in libraries, but also in many other work places (e.g., in companies, manufacturing, and hospitals), where information retrieval has become a daily activity (Pejtersen and Rasmussen, 1986, 1989, 1990).

Limitations of Present Library Information Retrieval Systems

It is well known and acknowledged by users and designers that users of systems providing on-line access to computerized public library catalogs and to commercial bibliographical data bases are faced with serious limitations and retrieval problems. See Atherton-Cochrane (1981), Hancock-Beaulieu (1989), Hildreth (1982, 1989), Markey (1990).

One category of problems is related to the content of information retrieval systems. The only source of regularity in the presentation of documents found by users of libraries is the reflection of traditional scientific paradigms made explicit in formal, rigid classification schemes. Thus, the classification neither reflects the changing conditions of work, discussions, and paradigms nor the users' personal value criteria or affiliations. Policies for identification of retrieval concepts are chosen from the point of view of the supplier of the data bases and are founded in long traditions for use of universal systems for classifying and indexing information. Users may follow these schemes, but they are not effective, because they do not reflect perspectives outside selective categories of document content reflected in these schemes.

Another category of problems is related to the vocabulary used in bibliographic records, which is often incompatible with the language brought to the system by the users. The majority of today's commercial systems favor data base records consisting mainly of bibliographical information, short abstracts of content, and a field for controlled keywords. In many local library data bases abstracts and keywords are not available, retrieval is based on broad classification categories. The user may extend or narrow this information by asking for other formats, which will then provide more or less detailed information within the same restricted types of categories. Information about the individual item is too restricted and does not cover the users' demands for information about book attributes [too little content information to judge relevance and too much irrelevant (bibliographical) information]. Users fail to match their own need formulation with the system's subject vocabulary, since concepts and terminology are taken from the controlled vocabulary of a thesaurus and codes from library classification schemes.

A third category of problems are related to the user–system interaction that often support only one single mode of search, that is, the analytical combination of subject terms by means of Boolean operators. Boolean operators force users into the difficult transformation of their pragmatically conceived needs into a sophisticated formal logical notation. Some more advanced systems allow browsing randomly in the data base, but in any case users often experience navigational confusion during the search process (Canter et al. 1985), since clear choices among several routes are not available.

The interface displays cause another category of problems since they usually reflect the system facilities, the options provided, and the content of data base records in formats that are not familiar to the users. The interface is normally aimed at a text based command search mode allowing unlimited access to data base information given that the user can remember the relevant search commands. Sometimes simple text menus are used that allow access to limited parts of the data base but are still on an analytical basis. Interfaces of information retrieval systems will usually not provide the user with a uniform context that helps to understand the language and terminology of complicated search commands, the formal, logical Boolean retrieval methodology, and the thesaurus and classification applied to access the content of records. Interfaces in commercial data bases are considered as "universal" front ends reflecting the different existing data bases. They are not configured as an integrated and uniform set of displays dedicated to different domains and tasks and matching users' search strategies and cognitive capabilities. Lacking understanding and use of system facilities lead to missing opportunities to complete searches successfully.

Information retrieval will be further complicated in integrated work stations by the number of different data base systems linked by communication protocols, which will suffer from the same problems described above, but are increased by the source of diversity. A source of diversity is the use of data bases of very different origin— various institutions, services, and companies, and the availability and format of the information derived from the use of different classification and indexing schemes having little or no basis in the end users' problem space or information needs. In addition, a source of diversity has to do with differing search languages, conventions, and operational procedures, which are not based on any uniform standard. Most often each data base has a unique interface format.

The amount of information in a commercial or local library data base is very large and complex: The number of different, potentially relevant documents to be taken into account is often millions of records. The variety of information attributes is very large since data base systems usually contain many diff rent domains and topics. Such multidisciplinary attributes are abstract and are coupled within and across the individual documents in a network of similar and dissimilar concepts and topics.

All of this makes it very difficult for the nonexpert user to predict the performance of the system. Even experienced users possessing considerable domain and system knowledge can find it to be tedious and resource demanding to plan a search efficiently. A great deal of inference making is needed on the part of the user to find and use search terms that are compatible with the utilized classification and thesaurus structures employed to organize the retrieval attributes in the particular data base. Considerable effort is needed on the part of the user to translate from their conception of need to search commands and formal Boolean combinations of terms. The large numbers of menus and search commands to be learned place heavy demands on users' memory capacity. Extensive training and use of manuals are needed to map the interface to the user's capabilities.

Finally, as will be seen in the description of cognitive decision tasks during retrieval, this description of users' problems with retrieval systems and the constraints they put on a "natural" interaction between the domain information and the user's world has not yet taken into account other major obstacles: The inherent problem of the users' own formulation of their information need and the identification–specification of the information required to solve the user's information need. (Belkin, 1978; Belkin et al., 1987; Belkin, 1990; Ingwersen, 1992; Croft, 1986; Taylor, 1968).

Design of a retrieval system has usually been concerned with the redesign of existing systems by the development of new front ends, by extended data access, or by introduction of "intelligent" intermediary support. Some experiments combine several of these approaches. Much experimental work has been a kind of bottom-up "repair the damage one step at a time" design philosophy driven by problems found in existing applications. Often little progess is made because design is based on records and observations of users' interacting with existing systems. The results of user actions through the intermediary or the computer in connection with advanced technology are not immediately mappable into system responses at any useful level, since these will mainly reflect the complexity of existing retrieval systems.

In conclusion, information retrieval systems will only really be useful and find widespread acceptance when these problems are solved. An analysis is necessary of the domain and task characteristics to provide the basis for a categorization and representation of document contents matching the end users' intentions and needs, together with a design of the user interface configured as an integrated and uniform set of displays matching the users' conception of the task and their capabilities.

The following sections describe a "top down" approach leading to radical design changes by relating all aspects of information seeking in data bases to the framework described in the preceding chapters. Although the description systematically follows this framework, many iterations among field studies, laboratory experiments, and prototype designs were needed. As mentioned earlier, the framework actually evolved through our efforts to make the concepts and approaches that proved effective during the complex design iterations systematic and explicit.

Why Fiction Retrieval?

First, the design of a system for fiction retrieval was chosen as the basis for methodological development because some initial field studies of the information retrieval activity in fiction in public libraries revealed a lack of appropriate tools. It uncovered a poor and uneven exploitation of the library resources spent on fiction which, in general, constitutes about 50% of a library stock.

Second, since no tools or commercial data bases were available in this area, it was an ideal test bed for exploration of the effectiveness of a top-down cognitive systems engineering design in solving the many problems with existing retrieval systems. Furthermore, only few constraints on design solutions were present from the existing tools.

Third, and of particular importance in this context, the problems faced in design of a fiction retrieval system are similar to the problems that are likely in design of new information systems for other modern work domains. Advanced communication and data base technology is now being introduced in many complex work domains as information resource management tools. The objective here is more efficient and cost effective use of information as a company resource; for instance, by reuse of information produced in projects on product development (Burke and Horton, 1986). These systems will often be introduced in domains, where no experience yet exists in the design or the use of integrated information systems for storage and retrieval through front ends to company data bases. The design of such company wide information systems must satisfy the goals and strategies of a cooperation as well as the domain dependent intentions behind the exchange of information within the staff. This exemplifies a need for indexing, representation, and retrieval of highly complex, subjective, and qualitative information dealing with value judgments and beliefs.

The objective of many special studies during the design of the BookHouse system has been to find ways of achieving consensus among different groups of system users about a consistent methodology for the storage and retrieval of highly value oriented information. Sometimes this was accomplished with conflicting organizational and individual, subjective values. In fact, the majority of resources went into the development of a coherent structure and policy for adequate management of information hitherto regarded as too subjective and complex to be structured.

FIELD STUDIES IN LIBRARIES

The field study of work in libraries was focused on the retrieval task and the users' query formulations to identify a representation of the stock of documents that match the users' needs. Since search in libraries often takes place in cooperation with intermediaries, local organizational issues and library policies will impact the search process and will have to be taken into account in system design. The goals and intentionality of the library can then interfere with the goals and intentionality originating in the users domain of work or interest. A means–ends analysis of both the library work domain and of the users' domain will be useful. In the present case, a means–ends analysis can only be made from the users' problem formulation and query statements in the library retrieval situations. For retrieval in a public library domain analysis beyond this representation of users' work domains will be difficult, but for in-house retrieval systems dedicated to the service of the staff in a particular organization, a more thorough means/ends analysis will be useful and possible.

Figure 9.1 shows the objects of field studies and their simplified relationships in a library retrieval situation. Field data were collected in a number of different public libraries (branches, central and special libraries, hospitals, etc.) and amounted to 500 recorded user–librarian negotiations. Negotiations were followed by unstructured and structured interviews of the users and librarians after each negotiation. Recordings were reviewed in cooperation with librarians and considerable effort was spent on their participation and consensus in the explanation of recorded activities and the subsequent analysis of the data (Pejtersen, 1974; Pejtersen and Cramer, 1976).

Goals and Objectives in the Library

In the following, fiction retrieval in the library work domain will be described mainly in terms of goals and purposes, as well as the level of work processes connected with the general function of information retrieval (as shown in Fig. 9.2). This is useful to illustrate how the institutional goals propagate down to the level of specific work processes and tools. This controls the staff's information retrieval activity and, therefore, has implications for designing a support system for information retrieval.

According to the Danish library law, the objective of public libraries is to: "encourage enlightenment, educational, and cultural activity." A subsequent directive

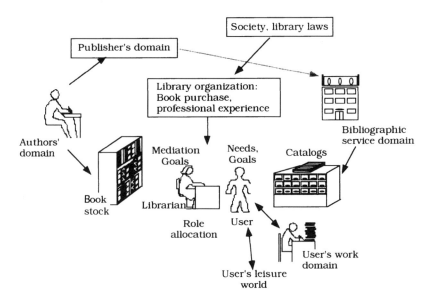

FIGURE 9.1. Information flow in libary domain. The figure illustrates the topics in focus of a field study.

elaborates on this objective and demands that "materials should make possible the independent formation of opinion, general and professional orientation and knowledge acquisition, personal development, artistic and human experience." Traditionally, library laws have expressed the same fundamental view since the beginning of this century. History shows various manifestations of this policy.

For example, society may pursue such objectives as the promotion of the population's subjective and personal benefits from cultural activities, derived from basic needs for recreation and for cognitive and affective development. Another objective is to maintain the existing society with its economic, political, and cultural basis through education. In a democratic society, the freedom of speech has the highest priority in the value system of society. This value system consists of norms produced within various social groups, institutions, professions, and public debates. Consequently, the quality of fictive literature is evaluated from many different value perspectives, depending on affiliations with scientific paradigms, group norms, political parties, literary schools, and criticism. Consequently, in the design of a bibliographic data base, it is necessary to develop a domain structure that reflects the value system of various groups and institutions in society. Some of the socially accepted values may be topical, when society, groups, schools, or political parties promote certain historical and social situations in terms of specific topics and attitudes regarding, for example, feminist literature, and so on. Or they may be aesthetic or cognitive in nature and refer to "extra topical" features such as beauty of language, composition, originality, and artistic sophistication (Pejtersen, 1986c).

MEANS ENDS RELATIONS	LIBRARY DOMAIN PROPERTIES TO REPRESENT
Purposes and Constraints	Objectives: Cultural Mediation; Public Education; Assembly Place for Cultural Activities, Public Information Center Constraints: Budget Limits; No Political, Moral Censorship; Only High Quality Information, Union Agreement; General Work Regulations
Priority Measures Abstract Functions	Flow of funds: sources and applications: Customer payment, fiscal law; Book purchase, Salaries. Quality measures: literary and cultural values; Use of products: volume with reference to quality and population categories; Distribution across fact literature fiction, and other materials
General Functions	User service; Book and art-item selection and purchase; Exhibition programs; Information Storage, Retrieval and Mediation; Administration: Books, employees; finance;
Physical Processes and Activities	Reading books; talking to customers, finding, getting, storing books, use of computer tools and card files; etc.
Physical Resources and their Configuration	Librarians, reference librarians, cataloguers, clerks. Scientific and fiction book stock, card catalogs, handbooks, lexicographic tools, dictionaries, local and remote data bases, Rooms for lending, reading and exhibitions. Shelving and storage facilities. Meeting and reference rooms. Administrative tools and facilities

FIGURE 9.2. Means–ends relations in the library domain; to be represented in system design.

In means–ends terminology, this influences the level of priority measures and abstract functions related to the purchase of materials. Criteria for acquisition and selection of books are stated in the Government directive: "As to fiction, readers should have access to the best among available books, only quality, not religious, political or moral viewpoints must be decisive." No specific methods, except the dictated constraints in terms of prohibited political criteria, or definitions of the quoted terms of objectives are given, so in the library domain there is methodological freedom and professional responsibility to pursue these objectives. Therefore, the role of the librarian as an intermediary between the library and its users has become an important feature for a retrieval system which has to support the attitudes, skills, and methods developed in the library profession, as well as the ways in which the staff makes these goals operational.

Field studies, public debates, and historical sources provide a colorful picture of attitudes and approaches (Pejtersen, 1981a,b). Ever since the rise of public libraries in the nineteenth century, fiction has been the source of debate concerning the tasks of the library profession. The cause of this debate was the fact that the foundation of public libraries was advocated first and foremost as the promotion of the enlightenment and education of people who otherwise would not be regular bookreaders. According to some librarians, the concept of libraries as democratic institutions implies a limitless right to obtain books of all kinds and to stress the respect for the reader's right to make their own judgments. Other librarians contributed to the objectives of libraries by promoting knowledge about facts and history, by promoting social and historical education by means of the sciences and imaginative literature, by making people worthy members of a democratic society, and supporting society's task to "elevate" the population. To others, the final goal was to make all classes knowledgeable to the same extent and to dissolve class distinctions in the field of knowledge and information to achieve a "knowledge wellfare society." For cultural information in libraries current attention is on the competition of books with other (mass) media that is popular with the user population in times where budget cuts are prevalent. The emphasis is on making it possible for books to survive as a popular media along with the activities and cultural experiences offered by the new multimedia technology. Priority measures exist as a trade-off between the demands from user groups for "popular" material and new media, the available budget resources for acquisition of material, and the professional and institutional goals followed by public debates in the media.

Other trends in today's information society advocates the library as a professional center supporting effective access to all information sources relevant to users' individual needs, which can be retrieved in a global network of information, irrespective of the quality of individual items, the value of which is determined "on the spot" by users' needs. This switch removes attention *from* an institutionalized quality control of the match between end users' subjective quality criteria and the objectives of work in the library domain *toward* the development of new functional aids in retrieval systems providing improved support for users' subjective intentions and values related to decision making and task performance. The latest decade has, therefore, produced a number of studies on (1) how to define the *value* of information and develop an information value theory; (2) how to assess information value and develop a value measure based on utility theory, fuzzy sets, semantic nets, linear discriminant analysis, and so on, and (3) how to implement such theories in an actual system design—in particular in professional domains requiring qualitative support in experts' decision making (Morehead et al., 1984).

As this brief review demonstrates, goals and values are rather diffuse and difficult to determine unanimously and in generally acceptable terms, compared with the relatively well-established consensus of values in previous times. The fiction domain is particularly sensitive to this question since authors communicate subjective values through aesthetic means—often in a complex and highly emotional spirit, which is subject to multiple interpretations and impressions on the part of the readers and librarians as well. With the many new information media created by new technologies, it becomes even more difficult to obtain a consensus about the relevant information sources and tools to be implemented in new information systems to support the necessary application of quality criteria in the library's mediation process. A first step in that direction is discussed in Chapter 10 on the design of the BookHouse data base.

Tools in the Library Domain

The lowest level of the means–ends analysis for identification of tools for information retrieval served primarily to identify an amazing lack of tools and the shortcomings of those few that were available. Many attempts have been made during the library history to improve the use of the stock and resources actually spent on fiction in libraries, but with no great success due to a general acknowledgment of the great complexity of this domain (Pejtersen, 1980b). An analysis of the level of physical work activities in the means–ends representation clearly showed the difference between fiction and fact literature with respect to the availability of tools that provide subject matter access to the book stock.

Today, the retrieval of fiction in libraries is still based on an alphabetical classification by authors for shelving arrangements supported by bibliographical data bases with author–title access including a number of additional bibliographical data—and sometimes supplemented by a short annotation about the story of the book. This offers the advantage of uniqueness in the physical location, identification, and retrieval of an individual document, but it is of little help for users with needs related to document contents and/or other book aspects. Requests for specific subjects in fiction like "exciting books about everyday life of children on farms in Guatemala" or "critical books about physical demands in modern sports" are increasing in libraries. This increase is due to the cross-disciplinary education in schools and to the influence of TV and other mass media, where multipurpose users do not perceive any fine grained distinction between the fact and fictive analysis of topics. Field studies showed that browsing shelves was the most frequent approach to retrieval. Since fiction is organized alphabetically according to the author's last name, those authors with names beginning with the first letters of the alphabet are selected first, and those books with exciting and appealing covers are borrowed most frequently. Neither a high recall of potentially relevant documents nor a high precision in relation to users' needs exist in sets of documents retrieved by browsing. This is also not an effective retrieval method with subsequent effective utilization of available information items. New user services will be needed to circumvent the present situation, where the quality of the service in fiction is dependent on the individual librarian's book knowledge, reading taste, and memory capacity, or on the success gained in browsing shelves, using secondary information sources like colors and pictures on books, or genre shelve arrangements.

Information sources available to support retrieval include newspaper reviews of newly published books, short reviews by librarians published by a company for library services as separate paper sheets to support quality judgments in book purchase and acquisition; manual genre catalogs with short annotations; books and lexica on literary history; publishers' and bookstores' commercials and catalogs; the descriptions on bookcovers of content, and, recently, on-line access to genre catalogs in bibliographical data bases. Traditionally, all this information was processed by librarians in preparing for book acquisitions. It was also used as a personal knowledge base for support during the information retrieval task. This was a feasible method with small book stocks, small numbers of new publications, and homogeneous user groups and queries. However, with the increasing amount of book editions, upcoming access to large data bases, and a diversity of user groups with growing demands for subject access to fiction within a diversity of topics, this approach is no longer effective.

In libraries, regular meetings take place for selection of new publications. At such meetings, publications are discussed with reference to library policy and user needs based on formal information sources with descriptions of new material in a format that is matched to the decision making situation at that particular meeting. The content and format of material that support the acquisition of new items has been developed through decades of negotiations among librarians, public debates about library policies, and concurrent changes that have taken place in materials from the bibliographical service center, due to shifting demands from libraries. This material was later judged to be too unstructured for reuse of information in creation of a data base for the BookHouse system, but it became one of the major sources for indexers' effective skimming of document content within the structure of users' needs.

Similarly, observation and recording of these meetings provided information about the content of communication among librarians, concerning material that needs to be available in the retrieval system. Written material and verbal communication in professional negotiations, and criticism of the characteristics and values of documents, will reflect in practice not only their domain taste, competence, and knowledge, which was gained through education and experience, but also will reflect the goals they pursue, their professional identity, their knowledge about users' demands, and their predicted user requests for materials.

Task Situation: Information Retrieval

When the analysis is focused on one work function—the retrieval of books for the general user of fiction in a public library—the delimitation of the prototypical work situation is fairly straightforward in terms of a rather selective activation of the means–ends representation of the library domain (i.e., the level of general functions.

The basic operation is comprised of iterations of (1) generating a user need based on current goals, background activities, and so on, (2) choosing an information source, manual sources, or data bases; (3) carrying out an information search; and (4) evaluating–judging the relevance and compatibility of retrieved documents with needs. After a match is achieved, the next task is to identify the location of the information, collect information on shelves or initiate a mail order or reservation, and so on. This negotiation can be an isolated act (a teacher visiting a library on the way home from school) or, in an expanded work situation, it can be part of an on-going, perhaps daily activity, where diverse information has to be acquired rapidly and frequently from different sources as an important element in a dynamic problem solving operation. See Fig. 9.3.

The situation to be dealt with here is related more to the first alternative (i.e., we will be dealing with a fiction retrieval system in a public library for casual–novice users). However, the second situation has much in common with the topics discussed in the introduction to this chapter (e.g., problems of information retrieval in integrated work stations, which the present approach is equally well suited to address).

Cultural Mediation Task.   In the *cultural mediation functions*, intertwined with the information retrieval task as described above, the library serves as a mediator between

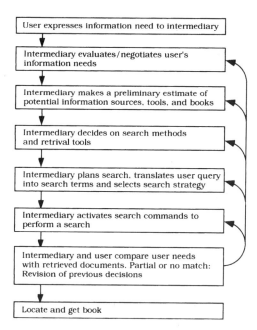

FIGURE 9.3. A typical retrieval sequence in a public library.

the author–publisher and the user. In the *information retrieval function*, the librarian and the user interact. In both these tasks, the librarian supports the user's exploration of the author–book domain to identify items that can serve to satisfy reading needs. The means and tools mentioned previously are not efficient as operational aids in a specific retrieval situation. Their information content is more directed towards the cultural mediation task as they furnish librarians with a critical analysis of documents involving contextual criteria for value judgments and promotion.

Activity Analysis: Retrieval Task in Decision Making Terms

For each of the tasks identified in domain terms, it is necessary to identify the cognitive activities involved in information processing terms including, for example, situation analysis, goal evaluation, and priority judgment or planning, which are essential elements in the information retrieval activity.

A typical information retrieval task will involve some or all of the following decision elements in various sequences depending on a number of factors including the user, type of information need, retrieval tools, topical domain, and the librarian's experience. Iterations will typically take place.

> *Situation analysis.* This phase includes identification of the characteristics of the user; diagnosis of the user's information problem, their task, goal, preferences, and capabilities. Analyses of the library environment and the capabilities of the available information systems in relation to the user's needs are also included. Two main difficulties are found: (1) the problem of identifying the user's often subconscious and intuitively formulated need, and (2) the problem of formulating relevant *strategies* for searching for documents that are not indexed and represented according to needs but according to formal, bibliographical data and crude genre terms. Due to psychological factors, there is often a gap between users' requests and their real needs. One of the core issues in the user's problem solving activities during an information retrieval task is the formulation of the information need in explicit terms. This problem has been well known for decades. Taylor (1968) defined four levels of an information need, of which the "visceral need" and "conscious need" are predecessors of the "formalized" and "compromised" need. This problem was studied later by Belkin (1978), Ingwersen (1986), and Pejtersen (1982). It is known that important information is likely to be excluded from verbalization during rapid retrieval of information from long- to short-term memory because this information retrieval process is also based on preconscious, tacit recognition processes (Ericsson and Simon, 1984). Limited access to mental processes is a core issue in the difficulties experienced by users during the identification and explanation of an information need and the related activity of verbal formulation of a search query. Need identification, as well as the explicit reasoning and formulation of this need, depend on the success of searching one's knowledge from long-term memory in the initial stages of retrieval. Now we can easily run into problems since the main part of the request–book correlation is usually rather intuitive

and many of the reading habits of the users are based on ill-defined attitudes and associations which define the fundamental frame of reference underlying the requests as well as the criteria of satisfaction.

Exploration of the library domain for information is a task that will take place at regular intervals during a search. Typically, retrieval will involve stages in which there is neither exact need formulation nor any unique answers to users' queries or needs. Situation analysis and exploration of the library domain and the user domain are intertwined decision tasks performed to achieve a formulation of the user's goal and the means needed to achieve the goal in terms of requirements to different levels of document attributes.

Planning of search strategies, search tactics and procedures, tools to use and goals to pursue, identification of potential priorities of search paths and outcomes, preparation and translation of query formulation into data base relevant language, formats, and search commands.

Execution of search, control of retrieval activities, of search tactics, and of strategies.

Comparison of match. Verification of the match between the keywords selected for the execution of a search and the keywords in the retrieved document. Examining the results of the search further includes comparison of match between retrieved documents and the user's need to assess the relevance of items based on a summary of content. The actual *validation* of the usefulness of retrieved items cannot take place until the user has read the document.

Relevance feedback. If comparison of match shows the search did not result in a sufficient degree of match, possible revision of search terms, new activity planning, execution of search, and so on.

Identify location of information when a match is achieved.

Collect information on shelves, if not available, use mail order, network reservation, and so on.

These decisions are listed in a logical order ranging from the initial recognition and formulation of a need to the final location of a retrieved item. In practice, these decisions will take place in any combination of sequences and many iterations will be necessary to achieve a proper result depending on the situation.

As an example, in Fig. 9.4, the decision task in information retrieval is represented within the framework of the decision ladder, which is described in a previous section (Chapter 3 and Fig. 3.6).

The analysis leg illustrates the situation analysis of a user's needs and the contents of the available books when searching for a match. The analysis of book content is shown dotted to indicate that it may not take place on the immediate occasion (e.g., by consulting annotations of book contents provided by library services in card catalogs and data bases or by consulting the book stock), that is, in cases when the librarian–user is experienced or is a specialist confronted with a typical user request. Knowledge acquisition and deeper reviews of information content will typically be part of the professional activity at another point in time—for instance, in connection

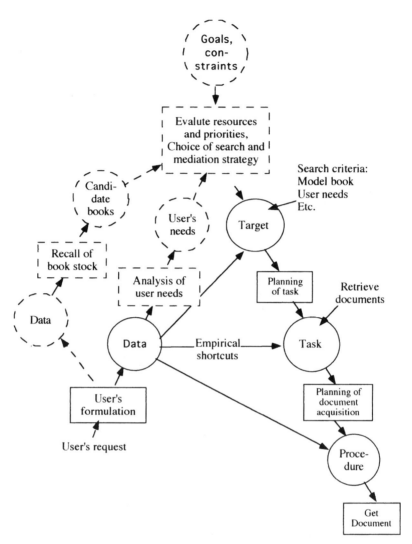

FIGURE 9.4. The decision ladder. The decision task in information retrieval is represented within the framework of the decison ladder. The analysis of book content is shown dashed to indicate that it may not take place on occasion.

with book purchases, a task that has the spin-off effect of updating the librarian's personal knowledge base. The figure shows that the experienced user–librarian will also tend to bypass the upper part of the normative decision sequence. The downward planning leg is concerned with acquiring the physical document or book from the book stock, another library, an electronic medium, and so on. This is a natural end to the retrieval task but, at the same time, it constitutes an independent task that can be managed independently of the retrieval operation, and thus a separate means–ends analysis will be required, if an appropriate support system is to be furnished.

Users' Conception of the Book Domain Structure. For the design of current information retrieval systems, the focus was primarily addressed to the means–ends representation of the relationships among a few relevant levels in the library domain. Figure 9.2 gives a general picture of these relationships, but in this chapter the focus will be on the identification of the *intentionality* in the user driven system as reflected mainly in user queries during actual information retrieval negotiations in libraries. The field studies in libraries demonstrated how users tend to characterize book contents from a number of different angles. This led to the conclusion that books should be classified in a multidimensional way (Pejtersen, 1980a). Five main dimensions were identified in users' requests.

A user's request might concern only one of these dimensions, but fiction inquiries often contained a combination of book features from different dimensions. Users formulated needs within the five dimensions of book characteristics: (1) author's intention, (2) literary school, style, and values, (3) frame, (4) subject matter, and (5) accessibility. These were specified further by users into several categories within each dimension. Only a few users expressed their needs within reference to all the dimensions and categories identified in the data collection.

The highest ("why") level in the user's expression of needs refers to their goals for reading documents in terms of various kinds of emotional experiences and/or education and cognition–information. These needs of course depend on the user's current task and/or product-oriented intentions with reading books. The majority of users' formulations about author intention belongs to the functional purpose of "emotional experience," which presumably can be explained by the fact that this is the most common perception of the function of fictive literature as opposed to fact literature. Scientific literature is typically the communication of factual information. In these cases, an author may include value judgments, but these are usually related in some way or other to objective facts and are not introduced for their own sake. Authors of fiction face no such constraints. Fiction may also communicate factual information, but the informative function usually plays a secondary role.

The typical functions of fiction are (1) to evoke certain states of mind in the reader—joy, compassion, curiosity, anger, grief, excitement, and so on (this is the aesthetic function of fiction that involves the user's emotions during the reading process) and (2) to communicate the author's value judgments on events, circumstances, phenomena, properties, and so on. This is the cognitive, educational function of fiction, which concerns the long-term influences on the user's "attitude towards life" and their ability to act and change interior or exterior circumstances. Thus the purpose and value of fictive literature is relative to the degree to which the information in a text gives the reader an emotional, aesthetic, cognitive, or educational experience. If fictive (or fact literature for that matter) does not decrease the reader's uncertainty, it has no cognitive, educational, or informative function. However, fiction may still have an aesthetic value derived from the reader's response to linguistic beauty and emotions aroused during the reading process.

These goal related "why" motivations are correlated with various types of content, such as events, plot, subject matter, different social, geographical and time frames, the so-called "what level," and depend on the user's current task and situation. This was frequently referred to in most negotiations.

A further decisive factor for a successful reading experience is the accessibility of books with regard to the level of communication employed by the author in relation to the user's reading abilities, the "how" level. This typically includes difficulties in language and/or substance matter content or in literary form relative to the social use of texts. It further includes the physical appearance of documents, such as color and cover or front page illustrations, as well as the name, age, and other characteristics of the main characters in the story. In the recorded user-intermediary negotiations, users evaluated book contents on the basis of information from the covers and used these as pieces of "bibliographical" information in their search for a particular item.

Since all five aspects are of importance to the user—at different levels—an information retrieval system must comprise access to all the relevant dimensions. If we are to create a system that is compatible with users' requests and needs, their ill-defined reading experience and their document characterizations must be made explicit and integrated in the definition of "aboutness" employed in the scheme. As a consequence, the classification scheme should define the boundaries and attributes of a reasonable and handy number of aspects within several of the dimensions that characterize the documents in relation to users' needs (Pejtersen, 1979b).

Figure 9.5 illustrates that, as viewed here, information retrieval essentially is an

Document Content	**User Needs**
Why: Author Intention	Goals: Why ?
Information; Education; Emotional Experience	User's Ultimate Task and Goal
Why: Professional paradigm	Value criteria: Why?
Style Literary or Professional Quality; or School	Value Criteria Related to Reading Process and/or Product
What: General Frame of Content	Content frame:What?
Cultural Environment, Historical Period, Professional Context	General Topical Interest of Historical, Geographical or Social Setting.
What: Specific, Factual Content	Subject matter: What?
Episodic, Course of Events; Factual Descriptions	Topical Interest in Specific Content
How: Level of readability	Accessibility:How?
Physical Characteristics of Document; Form, Size, Color, Typography, Source, Year of Print	User' s Reading Ability, perceptual and cognitive capability

FIGURE 9.5. Document content and user needs. The figure illustrates that information retrieval essentially is an activity that attempts to achieve a "match" between two multi-level entities, one representing the user–reader with their needs/goals/values and the other a document collection described at various corresponding levels of abstraction.

activity that attempts to achieve a "match" between two multilevel entities, one representing the user–reader with their needs/goals/values and the other a document collection describable at various corresponding levels. A suitable aid should therefore aim at eliciting and structuring knowledge on user needs and document attributes on the basis of these descriptive levels.

User Queries in Cognitive Decisions. The domain concepts involved in cognitive decision tasks of need analysis and comparison of match revealed the levels of users' domain knowledge involved in fiction retrieval. Decisions provoked different types of queries and the language used at the various levels differed. A more detailed analysis of the users' vocabulary at each level was important to try to match the users' natural language queries originating in a problem situation in the domain–work context with a limited vocabulary to be used in the data base. When users' queries and language had been identified, the task was then to analyze the relationship between the concepts in users' query language and the actual content of the document collection in question. This task was followed by an attempt to predict the queries that are likely to be made, when a data base with a new content and a new vocabulary are available. Some issues involved in this analysis are mentioned in Chapter 10.

Information Processing Strategies During Retrieval

The next level of analysis is an identification of the mental strategies that *can* be used to perform the identified information processing tasks. An important part of this analysis will include a determination of the resource requirements of the individual strategies in order to judge how each of them matches the set of subjective preferences and process criteria that will guide the user's choice. Strategies developed and employed by readers in different situations are good indicators of their preferences in their various search environments. These will not necessarily be relevant in future environments, since search behavior of course will be influenced by the retrieval tools available to support the search. Nevertheless, information about search strategies in present day systems is very valuable as a background for the formulation of possible strategies and the criteria underlying user preferences and for design of support in new information systems.

Users visit the library to select documents that have a reasonable chance of satisfying their needs. With the intermediary's assistance, this document selection can be made in essentially two different ways. Since the choice of documents consists of a search for a match between a multilevel means–ends mapping of contents of the book stock and the corresponding attributes of user needs (as shown in Fig. 9.6), it can only be done by a single person. Either the user or the intermediary can perform this search. The user knows their need and has a mental model of their own world, but has the problem of formulating their needs and intentions in terms of the library domain—a necessary operational step. If the reader searches by themselves, they must procure information from the intermediary about the library system, the characteristics of the book stock and its contents. On the other hand, the intermediary has

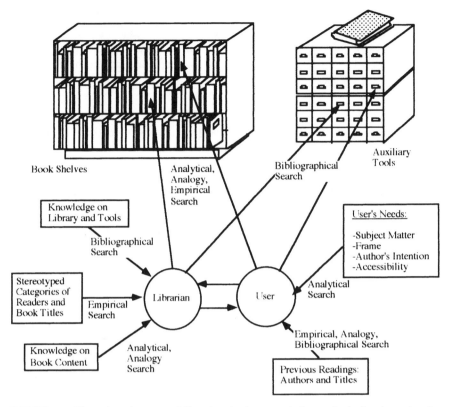

FIGURE 9.6. The roles of users and librarians in the negotiation to find a book serving the needs of a user.

a mental model of the library domain and possesses knowledge about the library system, the character of the information in stock, and its contents, but must obtain information about the user's domain, their needs, and task related intentions in order to undertake the search for a match.

Several different search strategies have been identified that represent different ways in which users and intermediaries categorize information during information retrieval in libraries. Figure 9.7 summarizes these as well as typical user-intermediary roles (Pejtersen, 1979a, 1982, 1984, 1988).

Analytical Search Strategy. When explicitly formulated information about the user's intentions and information needs is communicated, the librarian can compare the user's domain with the library domain and suggest a selection. The user probes their own need and communicates their intentions to the librarian, or the librarian examines the user's need by means of clarifying questions and by repeatedly inviting the user to explicate their information needs.

The User	The Librarian
Bibliographic Search with Instrumental Assistance	
- Explore book-stock - Compare book content and needs - Select and decide	- Assist in physical search - Assist in use of aux. tools - Explain arrangement and tools
Bibliographic Search with Verification Assistance	
-As above	- Communication of information about contents of books selected by user.
Analytical Search	

- Communication of information about
  needs
- Acceptance or rejection of proposed
  books
	- Explore user's needs
	- Compare to contents of books in stock.
	- Select book
	- Proposal for user consideration

Empirical Search	

- Transmission of implicit and explicit
  signs
- Acceptance or rejection of proposed
  books
	- Initiates signs by questions and
	  proposals
	- Classification of users according to
	  verbal/visual cues
	- Association to stereotype categories of
	  authors and titles

Search by Analogy	
- Identifies a book he/she liked to read	- Identifies another book with 'similar' content
Check Routine	

- Request for information on book
  contents
- Acceptance/rejection
	- Communication of information about
	  contents of selected books

FIGURE 9.7. The allocation of roles in different search strategies; as identified from user–librarian negotiations.

The analytical search is a knowledge-based search in a network of relations between document attributes and attributes of user needs in order to achieve a match between the two. The analytical search involves a complicated data treatment at several information processing levels: analysis of users' need, translation of needs into book contents in matching retrieval terms, comparison of match, situation judgment based on users' relevance feedback, and frequent iterations among these four processes with a combination of induction and deduction of users' information needs depending on the current accessibility of information about users' needs and book content.

The types of mental model involved consider the user's domain in terms of a semantic network of user intentions and information needs as well as the library domain, particularly with respect to the classification and indexing methods applied. In fiction, however, no such institutionalized methods are available, and the librarian's mental model activated during this process must be based solely on knowledge about the users' world and a corresponding list of categorical attributes of book contents derived from user formulations and queries, which are repeatedly probed to identify a user's need. Information exchanged among librarians and users during retrieval negotiations is perceived and interpreted at a symbolic level to develop, support, and revise the librarian's current model of the actual user's need.

In principle, the success of this strategy depends on classification and indexing methods that match categorical attributes of book contents derived from users' task and domain related intentions. However, when applied today in fiction retrieval where no such methods are available, the success depends entirely on the amount and type of the individual librarian's long- and short-term memory resources and deep knowledge of book contents and user goals and needs. That is, a skilled intermediary is capable of bridging the traditional incompatibilities between users' task–problem related classification criteria and those criteria generally employed in retrieval systems. Problems occur when these capabilities are not present and/or when users' needs are vaguely anticipated and cannot be expressed in explicit terms. This is a rational problem solving strategy that skilled intermediaries typically depart from, as do most users. As the most common reaction to its frequent shortcomings, they prefer purely empirical routines based on associations from typical user categories to a repertoire of books.

Empirical Search Strategy. This strategy is based upon the intermediary's purely empirical classification of users into typical categories that are associated with a repertoire of typical sets of genres and book titles, the contents of which the librarian usually does not know in depth. The users can express their needs in many different ways—in library terms as well as in terms of needs—but the statements are not conceived as a starting point for an analysis of the need. The librarian's mental model of the user domain instead consists of user stereotypes correlated with a few defining book attributes. Based on years of accumulated experiences with users' reading tastes and book characterizations, the intermediary "recognizes" the user's need from a combination of behavior, language, appearance, sex, age, question formulation, and apparently other implicitly used visual and verbal features in the situation, which the recorded material does not inform us about. This information is perceived as signs that characterize the user in relation to the intermediary's typical categories. Information processing is reduced to a simple categorization of users and books, and the way the intermediary comments on the individual books indicates that sets of titles are unconsciously classified in a manner that is not exclusive, but is in accordance with various overlapping types of "user behavior." When searching empirically, librarians therefore compare match by checking whether they operate within the right set of stereotypes–book attributes by simply asking for users' approval of author–title suggestions.

The success of this strategy depends both on the extent to which librarians have developed user stereotypes and gained expertise and skills in user classification, and on the repertoire and size of the matching book samples. Usually, skilled intermediaries have reduced the vast number of unknown factors when searching by creating large sets of user-classified titles–authors. When we make judgements from users' reactions to book proposals, the empirical strategy often seems to represent a sufficient search and it is a very efficient shortcut to attaining results for both users and intermediaries. However, browsing the shelf is the primary source to support librarians' memory and associations of title–author collections. Thus problems can occur when the intermediary's set of titles is too limited for well-read users who are too familiar with the librarian's repertoire—or when a user's appearance is matched to a stereotype to which the user does not actually belong. The most common reaction to problems of this kind is to make associative leaps to other categories of books or simply free associations, the origins of which cannot be traced in verbal records.

The Browsing Strategy. The user's need during browsing is usually vaguely defined and the user and intermediary intuitively explore the book stock and retrieval tools for good ideas. This approach is often manifested by sequences of author suggestions in alphabetical order, following the shelving arrangement of fiction. The information processes involved is an intuitive process of scanning the environment. Little is formally planned but the two participants hope for a spontaneous recognition of relevant concepts and familiar cues and signs for action. There is no explicit mental model of the user's domain and intentions activated during browsing. This search strategy primarily works on the user's and librarian's tacit knowledge of the library and user domain and looks for a match with users' previous experiences. In such situations, many intuitive judgments are to be expected together with the use of a number of simple rules that reduce a complex evaluation problem to more simple judgments. All kinds of pragmatic values, associations, and experiences are involved to recognize relevant documents.

Such a strategy is appropriate in unfamiliar domains when no explicit characterization of the specific information to be retrieved is available. However, it is also useful within familiar knowledge domains in situations when the need is implicit or not known or simply vaguely defined.

Sometimes the browsing strategy, which is less effective in terms of getting a rapid output corresponding to a specific search question, is chosen because factors other than a quick solution have a high priority. Such values include enjoying the scanning process itself because of new associations and ideas, improved comprehension, and learning of the concepts of a domain and its variety of information features.

The browsing approach can be a separate strategy but is also likely to occur as a sequence before and during other search strategies. Since this strategy usually demands few mental resources and meets few constraints, a switch to browsing will often take place in any kind of search as a means for associative exploration, formulation, and even revision of anticipated needs. Even when a need is known, support is needed to phrase the need in a language compatible with the terminology of a domain as well as in terms compatible with the conventions of the information retrieval system. Analytical support tools such as a thesaurus and classification systems, are the usual aids for this task and they will often be browsed for this purpose.

The success of the browsing strategy depends on the availability of aids that meet twofold demands. One is a very information rich environment with a great variety of information sources with different types of representations of information that support associations and cueing to users' context. The other is concerned with the need for some kind of an organization of information matching users' knowledge domain into operational subsets for browsing.

Search by Analogy. In the search by analogy, a specific book mentioned by the user serves as the basis for retrieval of new documents with features identical with those of the user's example. "I want something similar to book X" is the usual question. The search by analogy is based on a typical, previously successful sample from the user's reading repertoire that triggers librarians' associations to patterns of book attributes. The librarian's mental model consists of patterns of book attributes and involves a comparison of match and a judging of relevance to a particular user's need. These sets of book attributes will be derived from intuitive judgments of similarity from past instances of search by analogy. Empirical experience rather than formal information is used to identify common attributes of individual exemplars in a category. Typically, the search by analogy will merge into either the analytical or the empirical strategy.

When the search by analogy merges into the analytical strategy, the search takes place in a network of relations between attributes of documents in order to achieve as complete a match as possible with the user's model book. The model book can be the focus for the analysis of users' needs via a check of document features along different dimensions in order to identify which of the features are representative of the user's need. The information flow during retrieval negotiations then probes the user's concept of similarity by checking attributes of the user's model document and determining the selectivity of the user's expectations to similarity match. This is then followed by a search for a specified, selective number of attributes in order to achieve a partial match with the most significant attributes of the model book as identified by the user.

When the search by analogy merges into the empirical strategy, the example book is perceived as part of the user characteristics constituting user stereotypes, and the intermediary associates to sets of books classified according to user stereotypes. The information flow is then identical with that of the empirical strategy where the user's acceptance–rejection of the intermediary's suggestions of associated authors–titles proves or disproves the similarity match of the intermediary's association from model book to a similar category of books.

Bibliographical Strategy. When searching for bibliographical information, the user probes the content of the book stock and selects books themselves. The user typically asks for instrumental assistance from the librarian with questions that refer to the identification and location of specific books. Both user and intermediary communicate in library terms and they are occupied with the user's need in terms of authors–titles, physical characteristics, and location of desired books in the library. The intermediary's model of the topography of library tools and other physical equipment controls their assistance in the user's physical search in the library in connection with verifying users' incomplete references and instructing in the organization of the library, the management of information, auxiliary tools, and shelving arrangements. Generally speaking, the intermediaries transfer their knowledge of the library system and its tools to the user whose need and task intentions, on the contrary, are not communicated to the librarian but exists only implicitly as the user's own criteria for book selection. Information processing evolves around searching bibliographical tools for specific items and comparisons of match with the user's data about books that are being traced.

This strategy is a highly skilled routine task where the user's and the intermediary's mutual understanding of the task seems quite straightforward. The closing act in this encounter is either the finding of the wanted book or the intermediary's proposal for a book reservation.

Exceptions occur when users refer to book attributes such as color and picture of the front cover, the weight and size of books, which in some cases is the only information the user is able to recall about a known item and that cannot be searched for in traditional library systems. To a large extent, the retrieval tools match the strategy and, hence, the information flow between user and intermediary usually focuses on the cognitive task of comparing the user's, often inadequate, model of the library domain with the actual conditions in the domain.

The intermediary plays two different roles that activate two variations of the bibliographical strategy. One is with instrumental assistance as described above; another is bibliographical searches with verification assistance. This variety of the bibliographical strategy shares all the previously mentioned features. The distinction occurs when the user has selected specific items that might satisfy their need but is uncertain about the contents of the selected books. The user requests the librarian to assist by providing information about book contents. Although the user's need is only expressed implicitly through their criteria for a title–author selection and, therefore, is not communicated explicitly, the intermediary should be in a position to characterize book contents in terms compatible with users' needs. This requires a shift in mental model from the topography of tools in libraries to a model of the semantic network of users' domain and intentions and symbolic perception and processing of information. The information flow is both instrumentally directed as well as directed towards the verification of the attributes of the contents of selected documents.

Check Routine and Relevance Feedback. When the intermediary selects books on behalf of the user during the analytical and empirical searches and during a search by analogy, a checking subroutine is initiated by either the user or the intermediary. The user checks the match between their own perception of need and the intermediary's perception of the user's need by requesting information about the content of suggested documents. Alternatively, the intermediary takes the initiative to feedback by giving supplementary remarks about book contents. The check routine thus functions as a repeated feedback used for mutual judgments of the intermediary's comprehension of the user's world and for judgment of the relevance of retrieved items. Information processes evolve around frequent judgment sessions involving several, different mental models of the user domain and correspondingly various ways of interpretation of information. The success of the check routine depends entirely on having access to book contents corresponding to domain dependent user intentions and needs. Without this access, as is the present situation, information is deduced from casual, secondary sources such as the physical appearance of books, book flaps, and pictures on the cover, which are used as value judgments of its contents. Skilled librarians developed a number of routines for participating in check and feedback verification consisting of a stereotyped social interaction that compensates for missing information about book content, such as reference to quality of books as judged by other users, reviews, and media popularity.

Frequency and Shift of Strategies. Generally speaking, shifts of strategies usually occur when users express several needs for which different strategies are appropriate or when a problem arises. This could be related to the library domain or to shortcomings in information processing during retrieval negotiations, the mental model applied, the interpretation of information, or the inadequacies of the available resources necessary for the cognitive strategies to be fully applied.

Such cognitive shortcomings and resource inadequacies did not often occur during bibliographical searches, which were frequently used with few shifts to other strategies. Shifts from bibliographical to analytical strategies or to searches by analogy occurred when selected books were on loan or otherwise inaccessible, as well as when users ran out of bibliographical information and had an inadequate mental model of the library domain. This was the most frequent shift of strategy and could lead to the assumption that a user's inquiry for known items in many cases actually represents a strategy for retrieving information rather than a wish for specific authors–titles.

Shortcomings and resource inadequacies occurred more often during empirical searches and searches by analogy that nevertheless were both widely used. Shift to analytical strategies occurred when the librarian's mental model of user stereotype, user domain and intentions, was inadequate or based on an incorrect pattern recognition. Consequently, the associated book categories suggested to the user did not match the need. Users' regression to well-known books initiated by the user with a shift to bibliographical strategies also initiated librarians to shift to analytical strategies.

Analytical searches were not frequently used and shortcomings and resource inadequacies took on the form of shortages of books possessing attributes derived from a systematic exploration of the user's needs. The success of the analytical strategy depends on a thorough knowledge of contents of the book stock, which are systematized according to users' criteria of need, as well as an adequate interview technique. Resources that cannot fulfill the requirement of one strategy may be adequate if another strategy is chosen. Alternation of the analytical strategy with the empirical strategy could solve a resource conflict through selections of books from the librarian's standard repertoire. Problems with adequate resources in analytical searches will be described below. For example, the lack of an on-line retrieval system with representations of document contents based on user intentions to simplify the information processing of complex data appears to be an important reason for the infrequent use of the analytical strategy.

Resource Requirements. The information retrieval situation is complex (see Figs. 9.8 and 9.9). To a large extent, the choice of a given strategy will depend on the match achieved between the requirements of the strategy to knowledge about users' domain, information processing capacity, and the human resources that are available in a given search situation. These demands–resource requirements will change from one

INFORMATION BASIS	SEARCH STRATEGY		
	Bibliographical	Analytical	Empirical
Frame of reference	Organization of library data on publications	User needs, goals, and tasks. Contents of documents	Types of user behavior and needs. Categories of Documents
Information flow	Identification, verification, and location of items	Aspects of user needs; Aspects of documents found	Users' appearance, visual and verbal expression; Documents read

FIGURE 9.8. Knowledge needed for search strategies.

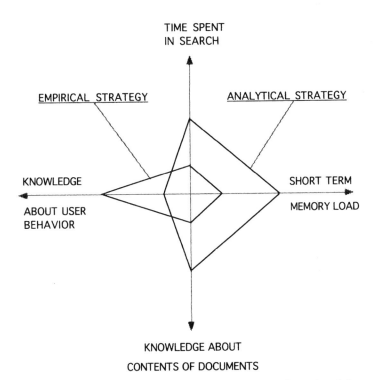

FIGURE 9.9. Comparison of resource requirements of the analystical and the empirical stategies. This figure illustrates how a switch of strategy can resolve a resource–demand conflict.

query to another, from one search environment to another, and from one searcher to another. Resources that cannot fulfill the requirement of one strategy may be adequate if another strategy is chosen. In considering the demands on mental resources, the complexity of the information processing task is important. When going through a list of demands in relation to various strategies, the analytical strategy seems to place the heaviest demands on the searcher's information processing capacity. The demands on short- and long-term memory are higher within this strategy than within other strategies, and there is a greater need for depth of knowledge about document contents and in-depth knowledge about the individual user's domain. In particular, the analysis of the user's world, the probing of need, and so on, put heavy demands on short-term memory. Furthermore, these analyses are very time consuming. However, previous experience with user intentions, types of reading tastes, and habits is not a necessary resource. In comparison, the empirical strategy on the contrary places few demands on memory and information processing capacity, on time resources, and on knowledge about book contents. On the other hand, it requires skill and expertise in the user domain, in the behavior of groups of users and their intentions, and reading tastes and habits in order to recognize proper categories of books.

Strategies for Cultural Mediation. Apart from the mental resource requirements mentioned above, a librarian's choice of search strategy will depend on other context dependent performance criteria, such as a wish to minimize the time spent, optimize the search efforts, improve the user's satisfaction, as well as the social contact between user–librarian, to minimize–maximize the precision or recall of retrieved documents, or to fulfill the optimal goal for work in the library domain, as described previously in terms of the promotion of educational and cultural values for the population, which is the legal basis for library institutions.

The use of value criteria to guide the librarian's search to satisfy a user's request may in some situations be necessary to effectively decrease the degrees of freedom in a specific search. Furthermore, it is usually required in library legislation that librarians support and develop users' cultural activities. Therefore, this will be a very influential criterion for their acceptance of a computerized tool. A review of earlier attempts to develop retrieval tools reveals how professional mediation goals and attitudes can interfere with functional requirements about matching users' needs. While a user's subjective value criteria may either be formulated before accessing a retrieval system or determined during the retrieval process, the value criteria based on the norms of society are determined a priori in the form of traditions and institutionalized cultural values. The mediation of cultural values by librarians depends on their use of their freedom to suggest particular books that match the user's needs as expressed by the user or identified during the librarian's search strategy. The possibility of mediation depends on the librarian's use of those aspects of books that are not specified by the user to influence the user's reading habit in the direction of some higher or alternative value criterion. Since users' needs typically are expressed in terms of topical aspects of book contents, the opportunity for a librarian to introduce new quality aspects to users is frequently present or at least it can be considered during the search process. When it comes to the librarian's task to promote innovation and new ideas, the operational capabilities of the library system in terms of the availability of information about quality and value aspects originating from library traditions, literary history, and the norms of different social groups and dealing with religious, moral, historical, aesthetic, educational aspects may play a fairly important role. This is also the case when users take the initiative to ask for advice and recommendations from the librarian.

Independent of specific literary values, a description of different performance criteria for mediation applied by librarians during information retrieval is needed. The following four categories of mediation processes have been identified.

1. *Passive, Neutral, Mediation.* The librarian retrieves books in accordance with the user's concept of literary values and strives for the best possible match with the user's value concept. The user's view of literary value can be of a subjective character or can be in agreement with generally accepted views of value. In both cases, the intermediary's search is neutral and no attempt to motivate a change in the user's reading interest is made. The manifestation is usually a simple acceptance of the user's request so that the search remains strictly within the frame established by the initial request of the user, whose idea of value is only implicitly present. On the other hand, the librarian may sometimes actively try to make the frame of the request more explicit through a negotiation with the user followed by appropriate proposals. This is the most frequent mediation category.

2. *Active, Neutral, Mediation.* The intermediary seeks information about the user's concept of literary value and explores the degrees of freedom for experiments with alternative value criteria in order to determine potential coincidences between users' value criteria and officially acknowledged paradigms. The intermediary's performance is active and value oriented, but neutral, and he/she chooses books on behalf of the user with one or more sets of generally accepted literary values matching the user's individual value concept. The manifestation is the probation of users' attitudes and a search for books that integrate desired qualities from the user's point of view with a suitable set of supplementary literary qualities. This is the second most frequent category of mediation.

3. *Selective, Objective, Mediation.* The intermediary pursues one selective set of the generally accepted value norms or literary paradigms. The selectivity of quality aspects of the proposed books is based on the intermediary's familiarity with and subjective preference among literary paradigms or current book repertoires as motivated by the user's topical request. The manifestation is a search where the user's subscription to literary paradigm is not probed and thus it is only implicitly known. It can only be indirectly used as a guide for book suggestions all belonging to a particular preference. Thus the librarian's performance is value oriented and he/she chooses on behalf of the user in this third most frequent mediation category.

4. *Selective, Subjective Mediation.* The librarian communicates personal experiences with good fiction of a more subjective character in order to motivate the user and create curiosity about certain book qualities or to participate in a nonprofessional communication with the user. The manifestation is a search where the librarian's

criterion of value is explicitly communicated to the user as comments to book proposals while the user's subscription to literary paradigm is not probed, and thus is only partially or implicitly known. The librarian's performance is value oriented and he/she chooses on behalf of the user on the basis of more personal premises that cannot be traced to coincide consistently with well-established literary norms. Possible conflicts among user and mediation goals did not seem to be a problem. Users' goals had the highest priority as long as they were expressed within the material available in the library.

If deliberately employed, these four categories of mediation of cultural values can be an effective criterion both for choice of search strategy and for choice among retrieved documents that match the user's need. They represent different degrees of application of cultural mediation as a performance criterion—from pure neutrality to being in favor of certain values; from a focus on the user's subjective opinion to a focus on the librarian's subjective opinion; from a pluralistic attitude to a solistic attitude to values. In the active, neutral, and selective objective mediation processes, the librarian generally communicates known literary values that should be accessible in a computerized information system. The librarian should master all four types of mediation and possess skills to shift among them with support from the retrieval system.

Mediation and Search Strategies. The opportunity for mediation depends on the intermediary's identification of the freedom they have to influence the users' choice. In turn, this depends on how specifically the user has identified their own needs. It also depends on how well the intermediary has explored the user's expressed needs in order to identify a freedom to mediate. As a result, the mediating opportunity also depends upon the search strategy employed by the intermediary. There is a natural relation between the various search strategies and the different mediation processes mentioned in the previous section. This relationship is based on the specificity of the user's need in terms of demands for specific documents and in terms of requests for a specific combination of information content in books. The more specific a user need is formulated, the more selective is the search process and the fewer degrees of freedom for cultural mediation. Therefore, during browsing, search by analogy, analytical and empirical strategies, mediation performance is possible to various degrees following the mentioned order: Browsing provides the most possibilities and analytical search the fewest.

In conclusion, it is necessary to include value criteria for mediation from many different perspectives since the success of a retrieval process depends not only on the value perception of the individuals associated with an information system, such as end users, librarians, and indexers, but also on the value perception advocated by society, social groups, and scientific paradigms. The performance criteria related to the promotion of such quality aspects of fiction, and their role in the formulation of proper search strategies in relation to goal–content therefore constitute a very important aspect of the design of auxiliary tools.

Role Allocation

In traditional retrieval sessions, the user delivers a need, the human intermediary transforms this into possible solutions, and the user tries to judge relevance. The simplest version of a computer intermediary would merely receive and respond to user commands and, in turn, deliver information on document contents, on procedural details in the user–intermediary dialogue, and provide more general help information. The user determines need, chooses data bases, selects a search strategy, and makes the final judgment on relevance. Allocation of these activities will depend on the "intelligence" built into the intermediary of the computer system. As intelligent behavior involves extensive knowledge about the domain, the task, and the users in order to take over the control, the degree of adaptivity, and allocation of control need to be determined, as discussed in Chapter 5. As information retrieval calls for such knowledge, retrieval will, even in systems intended for the end users, also take place in cooperation with a librarian during varied stages of a search.

Several factors were identified during field studies that made it reasonable to involve role allocation in design.

1. The democratic management style, which makes reallocation without organizational change a negotiable issue. Internal coordination, problem solving, and planning of activities are organized at democratic meetings with participation of all staff members, as well as at meetings with selective topics attended by those, who have the relevant task expertise that enables them to plan their access to information accordingly.

2. The relative stable domain character of the library domain, which allows long-term planning. In spite of a continuous increase in the amount of items, in new knowledge and concepts, concurrent with change of user needs, the change of principles in retrieval tools adopted to organize data base information (classification systems, indexing policies, etc.) is very slow.

3. The functional criteria applied for role allocation and division of work among librarians. The librarians serve the clients of the library, that is, the users being individual users, or groups of users, users involved in local community projects, or in activities in the local educational system. The service of the users and the library domain knowledge required include all levels of abstraction as seen in Figure 9.2. In larger libraries, the librarians develop a unique expertise in functions of acquisition and retrieval of documents within their preferred areas of the book stock and its associated users. The competence of the librarian as a domain specialist in information retrieval and cultural mediation is not only a main criterion for functional division of work among the staff in a library, it is also a major criterion for the division of tasks among the librarian, the user, and the computer support tool.

Change of Professional Competence. The impact of fiction retrieval tools on the possibility for satisfying professional goals caused considerable concern within the profession. Many resources have therefore been spent on clarification of the nature of the impact of a new system as well as on the formulation of a design strategy that would pay attention to this problem. A key issue appeared to be a need for prediction of the potential deterioration of professional skills during changes in role allocation among users and librarians. The librarians often doubted the feasibility of mapping a fuzzy pattern of user queries onto a likewise fuzzy perception of the content of fictive literature without having to use a too reductive and simplistic representation. In that case, a new system would lead to a simplistic interpretation of users' needs, an impoverishment of their reading experiences and, as well, an impoverishment of the librarian's domain knowledge. In all, this would have a negative consequence on the quality of the work situation.

Thus it was necessary to make certain that the data base information met the competence and skills of the librarians, so that both users and librarians through the use of the system, would increase their knowledge about the document collection. Thus this would ensure that the librarians' skills would not deteriorate as they used the system in collaboration with the users.

Such problems would typically appear in the design cases, when a computer system is going to be introduced into a new domain, in particular in sensitive domains with complex qualitative and value related information. (In addition, this will be increased by new requirements to design as identified by a cognitive domain analysis—and not—as has been the case for much of the computerization of retrieval of scientific and fact literature—to let design be inspired by the previous technology and practice in catalogs, card catalogs, and so on.)

The Librarians' Role and Skills. Even when the context of the search is formalized by a data base system having a classification of books that are compatible with the dimensions of the users' intentions and needs, many books will suit the needs that are only rarely defined by the user in all the aspects relevant for the choice of a book. To guide the specific search process, therefore, a user may need an intermediary to help make decisions and choices among the retrieved documents. In large data bases retrieval of an overwhelmingly number of records is a serious problem—too much information to be processed by a user (Saracevic, 1991). It may then be effective to include novel or individual value criteria in the search process independent of the desirable content and goals of the search to increase the precision of users' need formulation and the selectivity of the search. Computers can overtake the neutral manipulation of the retrieval function of data bases while instructions for judgment of individual values as a performance criterion in various, individual task situations are extremely difficult to design. Such guidance calls for domain expertise, currently updated knowledge on new events and trends, and it is a complex professional skill, which is only fully developed in the total library environment, since it depends exclusively on intuitive judgments by the (human) intermediary. This information is very difficult to describe in the explicit terms needed for computer formalization. Thus, with the advent of more effective retrieval systems the role of intermediaries might ultimately change into the role of more active mediators and consultants.

Access to Change of Data Base Content. Role allocation design also included delegation of access to data base content in order to store new information, to add, change, and remove information in records already in the data base. There is a need for the staff to be able to customize and adapt the information content to the preferred use of material by particular groups of staff. The need for enabling the staff to add and change data base information is in parallel to their practice of developing internal card catalogs and lists of material for use among the staff in areas where information is new, incomplete, or inadequate compared to user queries. Such information would often include characterization of the category of queries and needs of the intended user group, often developed as a cooperation among several librarians as their experience grows. The users with access to data base content will then be professionals, but they are casual users in unfamiliar situations without the expertise of the staff of the organizations that offer the service of standardized data delivery to the library. The interface to support this task in the Book House was then based on principles similar to those used for design of the interface for end users, as described later.

Role Allocation Through the Interface. Most retrieval systems are designed as tools for librarians, and for the benefit of end users some of these have been redesigned mainly by improving the interface to make them accessible to public users with little or no support from librarians. However, design of role allocation also included the possible support of the different needs that may be expressed by an expert, who retrieves in cooperation with an untrained user. The initial analysis of role allocation indicated that librarians, as trained in information retrieval in fact literature through a command based interface, would prefer a traditional, familiar command interface, when supporting users. Access to the BookHouse data base was then provided through selection of an option for writing search commands, but later evaluation tests showed that librarians' preferred to share the graphical user interface.

Organizational Design Aspect

Analysis of the management style and the organization of work in the library domain gave information about the possibility for redesign and change of task content in relation to a new computer system as mentioned above. Another organizational issue covered the related activities in several other coupled work domains, such as bibliographical services and other institutions that supply data services to the library. The author—publisher—bibliographical service domains are essential, because an appropriate information retrieval system accessing documents on the basis of structures of users' needs will call for a new indexing service (see Figure 9.10). Analysis is needed of the stage at which it can be accomplished in the book processing in order to avoid new costly work procedures, which otherwise might decrease the applicability of the system. The planning of data supplies will evidently involve comparison of existing services, which consist of existing data base information or catalog information and possible combinations of existing data supplies with new categories of information. In fact, recent cooperation with the Danish library service center showed how existing data in fiction annotations in catalogs can be reused by reformatting data and then limiting the supply of new data to those dimensions that are unique for the Book-House system.

As the users of public information systems in libraries include children as a group with special needs and limited capabilities, education and school libraries became another essential domain, since teaching library use, information retrieval, and data base development is part of their curriculum.

MEANS-ENDS RELATIONS	PROPERTIES OF THE SYSTEM SELECTED FOR REPRESENTATION			
	USER INTEREST	**LIBRARY**	**AUTHOR**	**PUBLISHER**
Goals and Purposes, Constraints.	Reader's Ultimate Goal; Education, Emotions, Profit, Power, Social career	Society. Mediation of Culture, Public Education; Budget; Political Neutralism; Union Rules	Profit, Literary fame social esteem, Social, Political Campaign, Authors Intentions: support to education, humanistic, cultural, political values	Profit, Social esteem, Power, Influence in Society
Priority Measures; Flow of Information, Values, People, and Money.	Value Criteria Related to Reading Process and/or Product Knowledge, Data, Aesthetical, Psycological, Political experiences	Money flow; money source and application: Books, music, art; Book qualities	Volume of book production, Distribution. Money earned, reception by critics, cultural society, readers	Money flow; money source and application: Books, music, art; Book qualities
General Functions and Activities.	General Topical Interest in Historical, Social, Geographical, Cultural Settings and environments	Serve users; Book purchase and administration; arrange Exhibitions, cultural programmes, employees, finance	Influence, help readers, education, society, increase quality of life, provoke debate and change	Serve customers; Purchase and selling manuscripts, books administration; Distribution, Commercials, employees, finance, arrange book clubs
Physical Processes in Work and Equipment.	Topical interest in Specific Kind of Plot, Subject matter, facts, events	Information acquisition, storage and retrieval, reading books, reviews, other reference materials, tools, talking to users and colleagues	Reading books, talking to colleagues readers, publishers, reading information and experiences, use typing machines, computers, travels	Reading manuscripts, company meetings, talking to authors, literary critics and consultants, book stores, advertising companies, educational system
Appearance, Location, and Configuration of Material Objects.	Reading Ability and Physical Characteristics of books (size, colour, pictures.) and users (Sight, age, sex)	Employees, qualifications, preferences, library inventory, building, tools, equipment and facilities	Books, Films, Libraries, cultural Institutions, family, Colleagues, Tools,	Employees, qualifications, preferences, company inventory, building, tools, equipment and facilities

FIGURE 9.10. The general work domain of a public library, which with a few modifications will have a general validity for all kinds of libraries. The figure illustrates how a local design problem propagates into several different organizations and institutions (cf. Figs. 7.1 and 7.2).

User Characteristics

Field studies and 200 structured interviews with users included questions about age, school grades, the reason for performing information retrieval, and the intended use of the material as a follow-up on recorded negotiations. Users' retrieval behavior was investigated in one particular library through structured interviews. Sample unstructured interviews were performed to get more detailed information, mostly during the prototype evaluation sessions. The general conclusion was that user *profiles* in a public library included a wide selection of the general public with a great variety of demographic characteristics, expertise, professions, educational levels, and tasks including children and adults of both sexes, and of almost any age. Therefore, the overall system design principle cannot be related to one particular problem space or user category.

Some division and characterization of user groups could be made following quite gross distinctions and based on generalizations from particular instances, all of which provided a number of assumptions to be used for design.

Children users were intended to be school pupils ranging from age 6–16 and thus ranging from having started to learn reading to being skilled readers and able to master language. Thus, the greatest range in levels of capabilities were found in this group, in which some of the users would not be able to read the description of books in the data base—nor the retrieved books. They were still considered as important target users since they were frequent library users, and introduction to computer based information retrieval is an important skill that needs to be acquired at an early age. Next, they were known to be able to play computer games, which means that they would not be computer illiterate, and would then eventually, as they learned to read, shift from playing with the system to seriously using it for retrieval. For the same reason they would not be able to write and spell their search needs through a keyboard.

The adult user group would range from 16–70 and very little could generally be said about their visual and other capabilities—except that some would be computer illiterate and that sensory motor problems might occur in use of the mouse by elderly users. Users' terminology and vocabulary was generally a common, natural language with no specialized vocabulary or terminology. It was a general pattern that most library users would be casual users, that their knowledge about the library domain and information retrieval would differ to such an extent that no a priori knowledge could be assumed as a design basis. They were all assumed to have performed their fiction searches primarily by browsing shelves or making their request to a librarian, which would then eventually result in a cooperative shelf-browsing. Design could then be based on familiarity with this strategy, while no knowledge about use of classification shelve arrangements or catalogs was expected. Similarly, no special knowledge about the fiction domain, literary history, or criticism was assumed to be present in library users. It was explicitly decided that professionals with expertise in literary text analysis were not the primary target users; their needs could be fulfilled otherwise.

Although their tasks differed, two typical categories of use of fiction appeared: First, across all demographic characteristics fiction was used as a leisure activity, both for emotional experiences and as inspired by a current leisure activity or problem. Second, by elementary school, high school, and college students it was used in the class assignments, often to support writing.

Interviews with users during field studies were successfully supplemented with discussions with users involved in the various prototyping phases, which, among other findings, also gave more "sensitive" information about the importance of paying attention to emotional reactions caused by different social, racial, and cultural backgrounds among users, for instance among immigrants.

Chapter 10

BookHouse Design: Data Base and User Dialogue

INTRODUCTION

Based on the results of the field studies described in Chapter 9, an information retrieval system was designed for a public library and evaluated in actual use (Pejtersen, 1989, 1992a,b, 1993a,b). The field study was planned according to the principles described in Chapters 2–4, the design followed the principles for ecological system design described in Chapters 5 and 7, and the system was evaluated according to Chapter 8. In this and the following Chapter 11, the design is presented with reference to the design maps discussed in Chapter 7. (Actually, many iterations took place between the design and evaluation of the BookHouse system and the explicit formulation of the framework described in the previous chapters).

The present chapter covers the design of the data base and the retrieval functions of the system. The field studies clearly place the library work domain in the left-hand side of Map 1 of Figure 7.5. Therefore, after considering the task situation of document retrieval in the library, no intrinsic system goal or functionality can be used by the designer to organize the knowledge base and the users' navigation. These features must be based on the structure of users' needs and their preferred search strategies as they are described in Chapter 9. To be implemented in a computer system, however, the results of the field studies must be formalized in terms of a formal classification and indexing system to be used for data base design, and a set of retrieval functions that correspond to the search strategies in a consistent way. This is illustrated by the transformations in the upper-left-hand corner of Map 4 (of Fig. 7.10), shown for the library system in Figure 11.1.

The computer algorithms implementing the data base and retrieval functions in turn are controlling an ecological interface, and thus serve to make the structure and content of the knowledge base transparent to the user. This is discussed in Chapter 11.

BOOKHOUSE DATA BASE STRUCTURE

Given the above description of the library domain, it follows that no physical, logical, or other formal constraints will determine the task sequence to be applied during retrieval. The retrieval domain is a loosely coupled world of document items with no explicit tight coupling among each item. Coupling exists among items within an authorship and across authorships through citations, references to other documents and classification schemes, and indexing languages. No coupling exists between the retrieval functions and the semantics of the user's search domain, that is, the book contents, except for the semantic relationship among terms imposed by the formal Boolean logic. This logic can be applied at the user's discretion. Thus any document selection may be correct, and only the user's need and task perspective will provide evaluative constraints.

The aim of the design of a data base for the BookHouse is to create an organization that reflects the users' needs and task perspective. This is done from an analysis of users' request and ways of characterizing document content during negotiations in the library. To create such a mapping, special efforts must be taken to generate an internal structure in the data base that is (can be) homomorphic with the end user's own need formulations. These need formulations can be either from work or leisure related reading. An important design goal will then be to secure that the constraints and goals implied in the data base design will not be in conflict during use of the system by various groups of users. (Pejtersen, 1986a).

The various levels of representation of book contents and of user needs, as shown in Figure 9.5 should be considered for design of the data base and used as the dimensions within which information has to be classified. How fine-grained the categories and subcategories of such a classification scheme should be depends on the application. It may be different in a public library, a research library, or a company library.

Such a classification scheme will be different from traditional library schemes since it was derived from the dimensions of users' expressed needs. Therefore, a discussion of the related classification and indexing principles will be useful.

Classification and Indexing Principles

A distinction between classification and indexing principles is necessary when designing data bases for information retrieval. We will briefly consider these principles for the traditional library system and for the system proposed here.

The term *classification* refers to the task of locating a given book within a conceptual framework, a classification system, for shelving and later retrieval. Different schemes have been in use for factual and fictive literature. For *factual* literature, a hierarchically ordered, exclusive classification scheme is used (such as the Dewey Decimal Classification scheme (DDC) and the Universal Decimal Classification scheme (UDC). Classification is then based on a standardized, hierarchical list of topics. The lack of correspondence with users' needs that cannot be expressed within a one-dimensional hierarchy has led to proposals of other classification concepts such

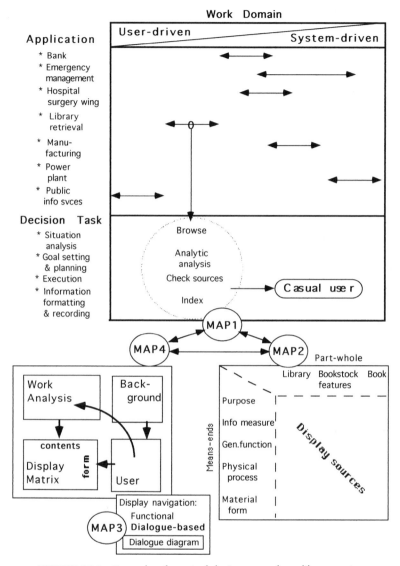

FIGURE 10.1. Example of a set of design maps for a library system.

as the faceted classification principles (Ranganathan, 1967). In this system, facets are not locked by a rigid enumeration in hierarchical schemes, but can be combined quite freely to express any relationship among concepts.

Fiction has traditionally been classified in terms of authors, titles, and countries without relation to the needs of a particular user. Some efforts have been spent on more pragmatic classification in terms of fiction genres. The exclusive classification schemes are needed for precise shelving and retrieval of the individual documents.

Another classification scheme is necessary to relate a particular document to the dimensions of a user's needs that serves to translate from expressed user needs to the shelving system. Such a system is described below.

The process of *indexing* refers to the process of analyzing individual concepts treated in a particular document and determining the attributes by which it should be represented for later retrieval. For a document of factual content, this involves the analysis of content by agencies, with reference to a standardized list of terms in a thesaurus that is based on rules for concept identification and choice of vocabulary. Otherwise the task is simply left with the author supplying the required list of keywords.

The indexing of documents for the classification scheme applied for the Book-House is complex (described in more detail below) because it (1) involves identification and grading of the content of a book along every dimension of a user's potential reading need, and (2) it involves neutral identification and description of concepts in the fiction domain, which is characterized by fuzzy, complex, and evaluative conceptual relationships, which call for strong control measures in the indexing in order to achieve a characterization that will receive general agreement.

An important issue for system design is that the focus needs to be shifted from the information retrieval task to the indexing task. This shift is necessary to identify the practice and rules needed to be developed to support the tasks of classification and indexing in accordance with the structure of users' needs. As this is not the main focus of this book, the following is only a short description of some of the solutions that were found to convert users' domain concepts and queries into an operational knowledge data base. Some of the laboratory experiments with librarians and users are summarized in Chapter 12.

The Fiction Classification Scheme

The classification scheme based on empirical field studies of users' needs used in the BookHouse data base has five independent, facetlike main categories called "dimensions" (see Fig. 10.2). These dimensions closely reflect the levels of abstraction of the representation of users' needs and book contents as shown in Figure 9.5. They range from reference to the authors' goals, their value criteria, their choice of book content, and their level of communication with the reader to the publisher's choice of physical representation of the document. Note that the order of the dimensions within the classification scheme (Fig. 10.2) reflects the order of importance that was identified in laboratory tests to be most effective in support of the indexing task. Consequently, it is different from the order defined by the conceptual level of abstraction (Fig. 9.5). Also, the order of presentation in the interface may be different, being chosen from studies of the natural sequence of queries of the average user of a particular category of users as identified in field studies (see Fig. 10.3). In the children's data base, the appropriate reading age will be at the top of the presentation.

The classification scheme described below differs significantly from the exclusive classification schemes identifying a particular document with one and only one location in the formal system. In the present classification system, a document is

THE AMP CLASSIFICATION SCHEME FOR FICTION	
DIMENSION 1: Bibliographical Data	
Title, author, illustrator, translator, editor, publisher, pages, year of 1st edition, year of classified edition, cover of book, serial, illustrated, screen version, title country and year of original edition, etc..	Supplemented by color and pictures of cover of book, name and age of main characters.
DIMENSION 2: Subject Matter	
a. action and course of events b. psychlogical description c. social relations	The subject content of the novel: What the story is about, action and plot.
DIMENSION 3: Frame	
a. time: past, present, future b. place: geographical, social environment, profession	The setting in time and place chosen by the author as the scenario of his work.
DIMENSION 4: Author's Intention	
a. emotional experience b. cognition and information	The author's attitude towards the subject and subscription to literary school, style and values. The set of ideas, values and emotions which the author communicates to his readers.
DIMENSION 5: Accessibility	
a. readability b. physical characteristics c. literary form e. main characters f. age of main characters	The level of communication in terms of those properties which facilitate or inhibit communication, such as content, form and language, typography etc.

FIGURE 10.2. The multidimensional classification scheme for fiction based on a formalization of the needs expressed in user queries. Levels are represented in an order, which best match the decision task requirements of the indexers for effective analysis of document content.

classified with reference to its location at all five dimensions and, therefore, a high resolution in the classification can be obtained by only a few categories within each dimension due to the large number of possible combinations. Each dimension characterizes fiction literature according to its own set of criteria and is subdivided into a few broad categories that reflect the structure embedded in the users' needs and made explicit in their queries. Because fiction collections and their user communities vary a great deal, no attempt has been made to provide a priori subdivisions below the level of categories. Agencies using the system should individually develop subclasses to suit their particular needs.

The dimensions and categories are not all mutually exclusive but supplement each other since all characteristics may be found in the same book but with different dominance. When a book is classified, it is simultaneously placed in every dimension and category in which it would be reasonable to find, depending on the weight placed on its different aspects by its author. There is no consistent logical order in the established categories according to generic and part–whole principles; e.g., one class is not logically dependent on the other classes in a tree structured hierarchical order and classes may overlap. The dimensions and classes are coordinate and do not comprise uniform subjects, nor do they belong to one well-defined group. This makes it easier to meet users' needs that often are multidimensional, and thus are not centered around one single dimension of the book.

Specific user queries and reading capabilities may change as conditions in society, culture, and education develop and new types of literature are published. New manifestations of authors' and readers' intentions and goals, new choices of general and specific content, and new ways of representing books will appear. Therefore, use of specific concepts in user queries are not used to establish categories. But it is assumed that the abstract dimensions referring to various categories and means–ends levels of book characteristics will still remain. It is, however, an open ended, pragmatic scheme to which in principle new dimensions and categories can easily be added in case new levels of user intentions and needs appear in the future. New dimensions can be added at the will of the user community without ruining any formal logical order. (Pejtersen, 1984).

The reader might argue, quite rightly, that it is incorrect to call this system a "classification system." With its nonexclusive, multiple-location structure, and the lack of a standard notation, it certainly differs greatly from traditional library classifications. It would be just as incorrect to use the label "indexing system" in its traditional sense, since book features are not listed alphabetically but are organized into meaningful groups. In conclusion, the one-sided stressing of a tree structured hierarchical relation found in some traditional classification systems has been discarded so as to take advantage of a combination of some of the functional aspects of fiction, which is determined by its readers' various levels of means–ends intentions. The faceted classification principles closely resemble this approach (cf. Ranganathan 1967).

BookHouse Indexing Language

The BookHouse does not have a full text data base but contains a short representation of the document content based on the classification scheme and a short annotation of content containing keywords for retrieval within each category of the classification dimension. Development of an indexing language was based on a combination of international, linguistic thesaurus standards (Austin et al., 1986) and the pragmatic aspects of the domain language employed by the user group—as an iteration among top-down and bottom-up approaches to the establishment of rules to achieve a consistent practice in the choice of vocabulary. It was influenced by the recordings of users' actual choice of language, domain concepts, and vocabulary, when they were evaluating document content, performing relevance feedback, and formulating and revising a need. The choice of terminology and indexing vocabulary was to be comprehensible to the different age levels of users, being both children and adults according to the assumption that users were primarily lay people.

Special attention was paid to the level of *exhaustivity* of the analysis of books as user requirements varied within each dimension. For each dimension it was necessary to develop special requirements to exhaustivity of the analysis and description of content. The exhaustivity of the annotation in each record was also a balance between the trade-off of providing sufficient information within all dimensions about the content for the user's decision about relevance—without revealing too much about the course of events in a book and, in particular, its ending. Therefore, the indexing included only the most dominant and characteristic features of a book and not minor aspects.

Another concern was the *specificity* of the language needed to represent book contents in a language level compatible with users' formulation of their needs and queries. In a number of laboratory experiments users and librarians found a specific indexing language favorable for recognition of needs they had not been able to formulate in specific terms. It was then decided to index with a specificity corresponding to the level chosen by the author of the book. Generally, this resulted in a more specific level than that expressed in users' queries in field studies (see Fig. 10.3). This was done to communicate to the reader as precisely as possible the level chosen by the authors for writing their story for comparison of match with need—and to support recognition of more specific concepts to be searched by users. It was therefore a deliberate policy to employ a general vocabulary meeting most frequent user queries as well as a specific vocabulary meeting authors' levels of writing in order to meet any level of language and concepts chosen by potential users. At the same time, this would enrich the vocabulary of search and access terms.

To retain some control of indexers' choice of a consistent vocabulary in keywords

LISA ALTHER: KINFLICKS	
Front Page Colors:	White, red and black
Front Page Pictures:	Faces
Subject Matter:	A woman's visit to her mother's sickbed and her revival of her youth, student days and marriage. Her experience of her mother's death.
Time and Place: **Setting**:	1960ties. USA. Tennessee. Southern States. Middle-class. High school. Feminist.
Cognition/Information:	Realistic description of the American society and of a woman's love affairs, her development and identity problems. The relationship between mothers and daughters. Feminist perspective
Emotional experience:	Humoristic.
Literary form:	Novel. Related in first and third person. Feminist novel. Developmental novel.
Readability:	Average.
Typography:	Normal.

FIGURE 10.3. Representation of a book as shown to the user for acceptance. The order of presentation of the dimensions of the classification scheme is here matched to the importance for user recognition.

and book annotations, a thesaurus was developed by a "bottom-up" method, empirically using terms employed in document descriptions as indexing progressed, and building a *controlled vocabulary* of hierarchical and associated relationships.

Library Goals and Subjective Values

A major reason that has hitherto inhibited the development of a retrieval tool with subject access to fiction has been the complexity of the artistic fiction domain. Another reason for this problem became clear from field studies of the present situation combined with historical studies on research in library and information science, and studies of curricula in library education in different parts of the world. This reason is related to the goal of work in the library domain, which involves a quality criterion that can only be met by substantial subject matter expertise and, therefore, gives the library intermediaries a clear, professional identity. Hence, the combination of the complex nature of a fiction book collection and the social and organizational domain constraints relating a cultural mediation task to the information retrieval task is a very important factor to consider as an explicit design parameter, if the system is going to be accepted by the professional staff—even if it primarily aims at the public end users themselves.

Basically, data base design should focus on the development of retrieval systems that reflect users' task related information needs and their subjective value criteria. At the same time it should be flexible enough to allow the search to be controlled by professionals' well-established paradigms or traditional value criteria, which are embedded in society and in the goals to be pursued by the institution.

This was achieved by focusing on the highest means–ends levels of authors' intention and goals and the underlying value criteria as expressed in the writing of a book (Fig. 9.5). Such goals and value criteria will differ among groups of users as well as among individual users, and while there exists some general consensus about the authors' goals and values in judging the quality of a book, its actual implementation in a specific document will be the object of subjective judgments and disagreements—even within specialized groups and environments. Hence, for users and librarians to be able to operate with value criteria and quality judgments during retrieval, the annotation of the cognition–information aspect of the author's intention must provide rich information about subscription to paradigms and affiliation with religious, cultural, and political movements.

As indexers may have different and subjective value criteria, a policy was developed that informed indexers to be neutral and eliminate their own preferences in identifying the author's intentions and value opinions. At the same time the indexing should reflect as many viewpoints as possible to meet the many potential user perspectives. To support this task a dynamically growing list was developed of definitions and enumerations of possible paradigms like political schools and ideologies, religion, philosophical, and literary trends—as well as subjective emotional experiences to be gained during the reading process. The controlled vocabulary in the BookHouse data base then included relative terms, evaluative, subjective, and vague concepts in addition to more content specific terms—the meaning of which can be deduced by users largely through their position in the context of an extensive topical content description.

Cultural Mediation

The classification was implemented in the data base to allow users unconstrained access to all dimensions of the scheme and the indexing vocabulary expressing the meaning and book content of each dimension. The user (librarian or end user) can enter the system anywhere without first having to place the subject in some super-ordinate class and then go on searching for a subordinate class. In effect, any relation between designations and subjects may be expressed.

A search for a book on science fiction may cover a need for a book about future technology (subject matter), an exciting book (emotional experience), a book about criticism of pollution in modern society (cognition–information), or an easy book (readability). When retrieving book examples along these four dimensions, it is possible to identify the actual user need. However, when the user's need is only expressed within a few dimensions, a large number of relevant documents may be found. This leaves the librarian (and end user) with the possibility of including a value and quality criterion by searching along the dimension of the author's intention of cognition–information. In this way the system supports searches meeting a passive, *neutral mediation* of the book stock, as well as searches meeting *active neutral* and *objective mediation* strategies. In the same way, the system supports most of the identified search strategies and, in particular, the *analytical strategy*, by providing a flexible structure of the user's problem space as the foundation for the data base.

NAVIGATION DIALOGUE IN SEARCH STRATEGIES

Results from the analysis of search strategies applied by users in libraries were used to design the retrieval functionality, the search dialogue, and support of users' navigation in the data base. This involved the design of the retrieval algorithms, formulation of the access to keywords as search terms, and the display of book content in a retrieved set.

Several important design requirements were concluded from the field studies because several strategies with different characteristics were found.

Support of Shift of Search Strategy

Frequent shifts among strategies were observed and the system should give the user free access to shift of strategy independent of the current stage of the search. The option to have different responses from the data base to search questions in terms of new search sets with different content according to the strategy applied will help the user to explore the content of the data base. When several search strategies are applied, users' information needs vary dynamically and will be revised according to the response of the system. Therefore, multiple and changing representations of the book contents are needed during a dialogue. The design trade-off is then to make possible a free and flexible interaction as found in field studies with no constraints on the possibility for breaking a retrieval sequence according to subjective preferences. At the same time it is important to ensure that the user will not get lost by providing adequate support of navigation in the dialogue functions.

Reduction of Mental Effort

The analysis of information processing strategies showed a great difference in the mental resources required by the different strategies and in the precision of search. Since mental workload is an influential criterion for the user's choice of strategy, an important design consideration is, when possible, to let the computer take over those search processes that are particularly resource demanding.

The analyses showed that the analytical strategy is particularly resource demanding. At the same time, this strategy is very effective with respect to precision and, therefore, special attention is needed to provide a support that is adequate to make the strategy preferable to the user during search. One way to make sure that each strategy requires an equal amount of user expertise is to match the interface displays to the users' capabilities as described in the next paragraph.

Several criteria were used for allocating roles to the computer and the user: Heuristics based on the studies of user requirements within the various search strategies were used for automation of both Boolean and probabilistic retrieval methods. The emphasis was to let the computer take over the planning of formal combinations of search terms and leave the user resources to focus on need situation analysis and exploration and on the subsequent evaluation of the retrieved set of documents. Consequently, only that part of the functionality that is less resource demanding and relevant at a given stage in the user system dialogue is displayed to the user for active control of retrieval processes.

Retrieval Functions in Search Strategies

Most retrieval systems are functionally based on an implementation of formal Boolean methodology, which is well known to be without correspondence with users' natural, common sense logic. Many studies have shown that the Boolean methodology cannot be effectively used by untrained users, and ways to make on-line searching evolve beyond the inherent limitations of current approaches to Boolean keyword searching, probabilistic retrieval, fuzzy set methods, and so on, are continuously being investigated as alternatives (Thompson and Croft, 1989).

No coupling exists between the functions of the existing retrieval systems and the semantics of the users' conceptual domain. Several, alternative computer functions and retrieval algorithms are needed to serve the user's operation on the data base and the transformation of queries into resulting sets of retrieved documents. Retrieval functions must introduce a proper relationship between the organization of the data base and the users' different formulation of needs as expressed in their choice of search strategy. Each strategy will require its own set of retrieval functionality and heuristics.

In the design of the BookHouse system it is hypothesized that a user's choice or preference for search strategy reflects a particular representation (or mental model) of the user's private–work domain and of the organization of the data base. The aim of the design is to match these two models through proper tailoring of the search dialogue, the retrieval functionality, and the interface. In other words, to match the organization of the knowledge base, as viewed by the shifting user strategy, and the users' conception of their private–work domains.

The resulting allocation between users and the computer of the functions involved in the different search strategies is as follows:

1. Bibliographical Strategy

User: Selects bibliographical data and compares book content with need to verify match of document with need. *User in control.*

Computer: Retrieves and displays keywords with bibliographical data and physical book charateristics for identification of known item. Retrieves known item based on Boolean best match technique. Supports verification of identified item by displaying book content at all levels of abstraction (goals, topical content, accessibility level and physical characteristics, publication, and local library arrangement). Supports location of known item by shelf information.

2. Analytical Strategy

User: Formulates and communicates information need through selection of keywords on goals and values, topical frames, subject matter, and level of accessibility of book content. Compares keywords and content of retrieved documents with need, and accepts and rejects relevance of retrieved documents. *User in control.*

Computer: Retrieves and displays data base classification structures, related keywords at relevant levels of abstraction [goals, topical content, and accessibility level (physical characteristics)] to support the user's exploration of data base domain for need formulation. Combines keywords and retrieves documents automatically based on Boolean best match technique. Supports comparison of match with need by displaying book content at relevant levels of abstraction. (goals, topical content, and accessibility level). Supports relevance feedback through display of Boolean search terms and selectable keywords in book content.

3. Empirical Strategy

User: Communicates characteristic demographic and personal characteristics to the system. Accepts and rejects documents proposed by the system based on comparison of match of books proposed by the system with user need. *User and system share control.*

Computer: Collects, analyzes, and classifies user characteristics to build a model of the user based on user stereotypes. Adapts documents to user profile and retrieves documents with AI and probabilistic partial match techniques that associate user stereotypes with classification and indexing of document contents. (Not yet implemented in the BookHouse.)

4. Search by Analogy

User: Communicates information on need through selection of a model document. Accepts and rejects documents proposed by the system based on comparison of content of retrieved books with model book and with user need. *User in control.*

Computer: Retrieves documents adapted to user's model book with probabilistic partial match techniques based on default selected levels of document content (goals, topical content). Supports comparison of match with model book by display of book content at relevant levels of abstraction (goals, topical content, and accessibility level).

5. Browsing Strategy

User: Selects subset of data base information within field of interest and explores possible book candidates in the data base. Formulates need based on document content. *User in control.*

Computer: Displays randomly retrieved book content at relevant levels of abstraction (goals, topical content, and accessibility level) to support recognition of need and comparison of match.

BOOKHOUSE SEARCH DIALOGUE

The description of the search dialogue will follow the navigational paths as described in Figure 10.4.

The user group of the BookHouse system includes children as well as adults. The fiction material for children differs from fiction for adults. The analyses showed that the concepts and queries used by children and by adult users were different to the degree that made it necessary to develop different data bases. Briefly, two main differences were observed: The frequency of queries within each of the levels in Figure 10.2 differed and information about the lowest level representing the accessibility of documents needed to be tailored to user age, school grades, reading situation and use of material, and so on. Consequently, the data base information on children's books contain additional accessibility information, and indexing language in the description of children's document content has similarly been adapted to children's levels of formulating queries as identified in negotiations with librarians. Field observations also showed how the teenage user groups would retrieve information in the adults' departments and adults would browse shelves with literature for the youth in the children's departments. Correspondingly, many libraries have overlapping shelve arrangement to accommodate this user behavior. The fiction retrieval system was then designed with three different data bases covering two different domains for two user groups with different ages and cognitive capabilities and language levels.

In the following sections, the search dialogue is reviewed with reference to Figure 10.4.

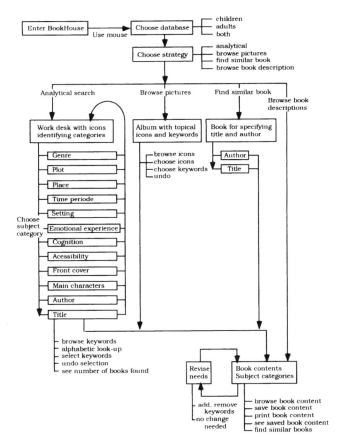

FIGURE 10.4. A map of the functionality of the BookHouse.

Select Data Base

After entering the system, the user is asked to select the appropriate data base. There are three possibilities (1) the children's book base, covering literature for children up to about 16 years of age, (2) the adults' book base, or (3) the total book base covering both children's and adults' books. A single mouse selection is required.

Select Search Strategies and Specify Need

At this point in the dialogue, the user must decide how the search for books will be carried out. Based on the results of the work analysis stated earlier, the BookHouse provides four alternatives: (1) an analytical search, (2) a search by analogy, (3) browsing in pictures, and (4) browsing in book descriptions. The origin of the browsing picture strategy was the discovery in field studies of users' frequent use of the visual appearance of pictures on the book cover as a source of information for judging the content of the book and its value in relation to the user's need.

Analytical Search. The dimensions of the user's needs can be explored interactively and systematically when selecting the analytical strategy. Need aspects can be compared with document aspects within 13 different dimensions corresponding to identified user needs and the classification system employed in the data base. The user can select one of these at a time so as to get access to the textual listings of the particular set of search terms for the selected dimension. The user can then step through this alphabetical list in various ways. They can look for a specific keyword by paging through these keywords. The user can select the first letter to look up the keyword to speed up the process if a known search term is desired. The user can also combine several keywords from the same or different dimensions. The planning of the Boolean combination is done automatically by the computer system based on the multifaceted classification scheme and a number of heuristics on users' needs for combinations of dimensions of book content as identified in user–librarian negotiations. The Boolean AND combination is used to combine categories of the data base structure for analytical subject searches; either combined from keywords in alphabetical lists or from keywords in book descriptions. Boolean AND or OR combinations are used among keywords depending on the dimension. The set of selected terms forms the user's current search profile, which they can revise by deleting terms, adding terms, making a NOT operation, or asking the system to find books similar to any book displayed on the screen. The same retrieval operations can be conducted by clicking on keywords embedded in the book description. Thereafter the user can see descriptions of the retrieved set of books, which are structured according to the classification system so that the user can directly verify the correspondence between their selected terms and the actual contents of a book.

Shift to a search by analogy can be done by asking the system to find books similar to any book displayed on the screen.

Search by Analogy. In the search by analogy the user gets access to a title or author index for the books in the data base. After selection of one of these, known by the user to be relevant, the system will automatically attempt to find other books in the collection that include as many of the same classified and indexed attributes of the model book as possible. A description of the 10 most similar books will be presented, one at a time in decreasing order of relevance. Probabilistic similarity algorithms (Pejtersen, 1991b) are used to search for semantically related documents within classification dimensions, categories, and keywords of the model book. Seven of the thirteen dimensions (plot, setting, place, time, emotional experience, cognition, and genre) of the multifaceted classification scheme are utilized in this calculation of similarity and used as the basis for a weighting of all books in the collection.

The user's current search profile is formed by the selected author–title. The user can shift to the analytical strategy by making a Boolean AND or NOT operation on keywords embedded in the book description or, by repeatedly asking the system to find books similar to any book displayed on the screen. The system calculates collection similarity on-line after each request for "find similar books."

The question of how to aid the user by weighting dimensions of the data base structure to create a model book on the basis of what the system could retrieve to be partially matching and similar books and display suggestions for the user's consideration and evaluation was investigated. The approach to assessing value by weighting is premised on the assumption that the value of an information source is related to the domain attributes that users found valuable in interactions with librarians. See the evaluation chapter (Chapter 12) for experiments to achieve a retrieval functionality and search dialogue appropriate for support and implementation of search by analogy.

Browse Through Book Descriptions. When browsing through the books of the data base for possible matches between intuitive current need and the available items, the system shifts immediately after clicking on this strategy to a book description. A choice of a browsing strategy indicates that the user does not know the specific address of a good book, but would prefer to wander around town until a good, familiar, or interesting item is discovered. Books are retrieved randomly from the data base after each request for browsing books. The user can then continue to step through other randomly chosen descriptions structured according to the classification scheme and continue verification and check of the result. The set of books selected by the system is the user's current search profile and a shift from the browsing strategy to the analytical strategy can be performed by selecting keywords displayed in the context of the book description and a Boolean AND or NOT operation will be performed according to the user's preference. Shift to a search by analogy can be done by asking the system to find books similar to any book displayed on the screen.

Browse Through Pictures. The browsing strategy also includes an iconic version where the user can browse through pages of small pictures describing book content to recognize a need, a so-called picture association thesaurus (the construction of these icons is described in the section on evaluation of the system). The user gets a quick bird's-eye view of what the books in the data base are all about since a single picture communicates many different facets of meaning and, in fact, includes several keywords from many dimensions of the classification scheme.

Keywords associated with icons are combined with a Boolean OR operator and represent a kind of associative semantic network. Each small network represented by an icon can be combined with another network represented by another icon with a Boolean AND operator.

As a result, the user is able to perform a complicated Boolean search and retrieve a small subset of books with the little effort it takes to select a couple of pictures and thus quickly reduce the large number of unrelated documents to a meaningful, related subset. This is an informationally economic way of supporting searches, since a corresponding textual keyword specification in analytical searches is feasible, but tedious, and it would require several access sequences to one dimension at a time in order to select and combine the same number of desired terms.

When using a traditional thesaurus with a textual display of the data base content, the user can freely select and combine any term related to the entry term. In the picture association thesaurus, all keywords related to an icon are searched default simultaneously when an icon is selected. However, if the user does not "associate in agreement" with the a priori meaning of the given icon, they can at any time revise the meaning of an icon, omit any term(s) related to the icon, and thus determine for themselves its associative meaning. When a picture is selected, a set of books containing the keywords associated with the icon is retrieved. The user can then see book descriptions structured according to the classification scheme for comparison of match with the user's need. Shift to the search by analogy is possible by asking for similar books, as well as shift to the analytical strategy. Keywords associated with the selected icon forms the user's current search profile and they can revise by deleting terms, adding terms, or making a NOT operation. The same retrieval operations can be conducted by clicking on keywords embedded in the book description.

Comparison of Match and Relevance Feedback. The final result of any search is an open book description (see Fig. 11.12). When asking to see retrieved books, a description of the first book of the current set is displayed on the screen. The number of books in the current set and the number of each book is displayed. The user can browse through the retrieved set of books, one after the other in both directions or larger leaps can be taken to continue verification and check of the result. At this stage, the user evaluates and, if necessary, revises the search criteria. Then they reexamine the search results and process eventual candidates. If the result is positive and a match is found the user can save interesting candidates for later browsing and/or request that hard copies of interesting descriptions be printed (e.g., for use later in finding the books on the shelves). One can also see previously saved book descriptions. To increase the number of documents that match the user's need, the user can initiate a search by analogy and a new set of documents is automatically retrieved and displayed.

If the result is negative and no match or only a partial match is found, a host of options are available to revise the search and to shift strategy. The user can repetitively execute and control the search through a number of search tools for relevance feedback independently of the search strategy selected. It has been an important criterion to allow the user to shift among all strategies at any time, but particular emphasis has been given to enable analytical operations in terms of a feedback decision, thus avoiding the traditional claim for a feedforward planning of a need formulation and analytical formulation of a search statement. The user can modify the current need by adding or removing search terms from the current search profile, and select keywords in the current book description that the user has access to with the mouse aided by tools for Boolean operations and probabilistic retrieval. Any planning decision can be immediately followed by an action and the system responds instantaneously to each action with feedback in terms of a new search result.

Thus each change generates an automatic search, the number of books found is displayed, and the first book description in the new set is displayed. At the present time, only the current book set and the saved set of interesting candidates is retained by the system. When shifting from one strategy to another, the search string from a previous strategy will remain on the screen and new terms can be added, which were chosen from another strategy. Again, heuristics are implemented to make such combinations from different search strategies logic from the user's point of view.

Shift to the search by analogy will initiate a search based on a current book description, shift to the dimensions of the classification scheme will give access to selection of (different) keywords that will automatically be combined with the current search terms; shift for selection of yet another strategy will either initiate a new search or combine the current search terms with those retrieved by the strategy. All of this depends upon the characteristics of the strategy.

Empirical Strategy. Field studies showed how skilled librarians could retrieve books based on their experience with those groups of user needs that over time were reasonably stable and occurred frequently. Patterns of user and book characteristics evolved during years of experience in information retrieval. At the present stage of its development, the BookHouse does not base its search heuristics–feedback on anything other than the user's age, that is, the category of children–adults, the means–ends level of a specific query, the strategy choice, and the user's response in terms of relevance feedback. For example, no information on clientele characteristics, book use statistics, specific user age, sex, background, and "reading style" is used to color the search process.

Bibliographical Strategy. The bibliographical strategy can be accessed through author–title, which can be searched as two dimensions within the analytical strategy and within the search by analogy, when the user specifies author–title of the model book. More data can be searched from a current book description, but traditional bibliographical data are limited, which correspond to the bibliographical data that is brought by the users to the librarians' during the negotiations in field studies. For the same reason they are supplied with new data on color illustrations, types of series, colors of cover, cover pictures, names and age of the principal characters, institutions, and so on. In the BookHouse experiments performed before implementation of strategies it turned out that laymen users' with no experience in library use found no need to distinguish the bibliographical strategy explicitly from analytical content searches, since content would always be the final goal of a search. Some strategy experiments are summarized in Chapter 12 on the evaluation of the system.

Conclusion

This description has only dealt with a "normal" dialogue trace. Of course, the BookHouse makes it possible for users at any time to make iterative shifts between saved, relevant books and a currently working book set, abandon a strategy, current searches, select a new data base, and so on. The system is designed to serve as a "systematic supplier of alternatives" structured so as to enable the user to cope with the complexities and strain of the cognitive decision of specifying needs, evaluating outcomes, respecifying needs, and reevaluating results—all by themselves. The characteristics of the target users will influence their possible choice of dialogue as it will influence their subjective preferences for strategy, since some strategies will require more cognitive resources and expertise than other strategies. Children in particular are likely to try to learn the system functionality through trial and error. A number of alternatives was assumed to best support this approach, each alternative had different requirements to both time and the number of actions to be taken before a resulting book description appeared on the screen.

Chapter 11

BookHouse Design: Interface Displays

INTRODUCTION

To introduce the discussion in the present chapter, a brief review is given of the paths followed in the design of the ecological interface of the BookHouse through use of the maps discussed in Chapter 7 (see Fig. 11.1).

A library constitutes a work domain at the left-hand side of the domain map of Figure 7.5. From the point of view of a user who is looking for information to satisfy personal or professional needs, no invariant structure is to be found in the domain of search. As discussed in the previous chapters, the representations of the individual book can be shaped to match the dimensions of user needs and the functionality of the retrieval system can be arranged so as to allow users to apply any of the relevant strategies. However, when a retrieval system is intended for use by casual users from a wide variety of educational and social backgrounds, guidance of the search cannot be based on an explicit representation of system functionality shown in Figure 10.4.

In terms of the domain analysis, no overall goal and functionality serving to structure the related knowledge base will be embedded in the domain as viewed by the user (see Fig. 2.10). The intentionality and coordination of functions depend on the user. Therefore, an overall structure of the knowledge base that can facilitate the user's navigation within the system should be introduced by an interface system based on analogies familiar to the user from other activities. These activities are indicated by the transformation paths in design Map 4 of Figure 7.12. The resulting representations are shown in Figure 10.1.

To sum up: (1) the basic content of the interface is defined by the data base and by the dimensions of the classification system applied, (2) analysis of the relevant retrieval strategies gives principles for the selection of the information that should be included in the individual displays, and (3) the overall structure of the interface system depends on the analogy chosen to make its structure visible to the user.

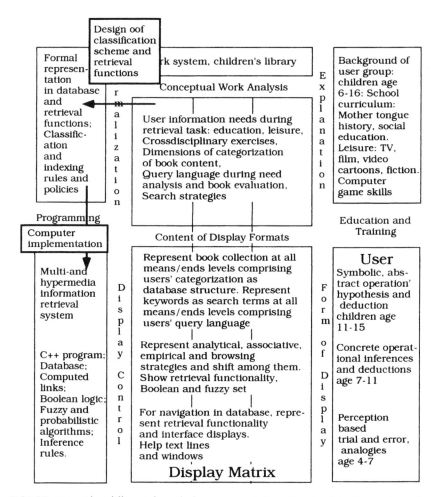

FIGURE 11.1. The different knowledge representations involved in the design of the BookHouse system with reference to Map 4 of Figure 7.10.

The organization of the knowledge base of the BookHouse and the related visualization is shown schematically in Figure 11.2. The overall structure reflects a division of the knowledge base into data base sections corresponding to different user groups and needs. This is visualized by a store house metaphor with several rooms. At the functional level, the knowledge base is structured according to users' search strategies and the interaction is based on iconic representation of book characteristics and search tools. At the lowest level, indexing of books reflects user queries and identification of items relevant to a user. This is based on textual representation for identification of attributes and associative relations, and tools for multi-attribute searching and matching. A collection of items (e.g., books, characterized by the dimensions of the reading needs of the general public) are found at this lowest level.

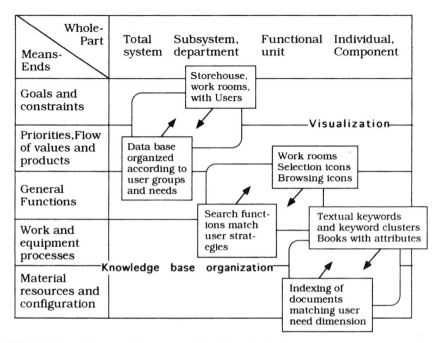

FIGURE 11.2. Schematic illustration of the knowledge-base organization of the Book-House and the related visualization.

The displays used at the three levels are discussed in detail below, but first the intended user population will be described.

USER CHARACTERISTICS

The target users of the BookHouse system were intended to be library customers from about the age of starting school (6 years) to the age of retirement (~ 70 years old). They were assumed to be casual and novice users (i.e., inexperienced in computers and information retrieval techniques). This delimitation was used to determine their various cognitive levels of perception of information involved in skill, rule, and knowledge based operations. Special precaution must be taken with respect to children users.

Piaget's (Piaget, 1958) work suggests that children's capabilities for structural thinking and abstract reasoning can be divided into several developmental stages. Three developmental stages are of relevance in the design context of the BookHouse for young users from age 6–16: (1) the *intuitive* perceptually based stage at which children from 4–7 think in terms of analogies and rely on, sometimes systematic, trial and error approaches; (2) the *concrete operational* stage at which children from 7–11 possess an understanding of logical relations and can experiment in a systematic way with some of the operational structures needed for classification; and (3) the *operational* stage at which children from 11–15 can think in terms of abstract relations and classes and make exhaustive hypotheses and deductions completely independent of perceptually based intuitions. Independent of discussions of the specific details of Piaget's work (Boden, 1980), it is obvious that the capability for making appropriate deductive inferences based on category membership, the understanding of formal logic structures, and the use of well-defined categories will not be available skills in young children. Similarly, their reading capabilities will vary as will their vocabulary, all of which will influence the visual coding and choice of form in the interface.

Children are one important user category and it follows from the discussion above that an ability to use knowledge-based analysis of needs and book contents and of system functionality cannot be assumed. Even adult, casual users are known to identify their reading needs only in implicit and intuitive terms and are not able to verbalize their needs. It is, therefore, a basic design requirement to create an ecological interface that faithfully maps and visualizes book contents and system functionality in a way that is easily recognizable to users.

For this purpose, common life and everyday familiar forms, texts, or objects are well suited for both children and adult end users, whose capabilities, knowledge background, and work context differ to such an extent that it is not possible to find one common background. Then commonalities in and across cultural, educational, national, historical, religious, sexual, and age related knowledge will have to be combined with forms, text, or objects that will most likely be associated with the domain information and the task and its associated actions that the system is going to support. The best forms of icons to be used for user actions on the interface are found when display elements can be immediately recognized as manipulative daily life objects and, at the same time, associate to the library retrieval task.

ICONIC INTERFACE DESIGN

The approach taken to meet this aim in the BookHouse system is an extensive use of configural displays with an iconic representation of the display content. An icon is a graphical object on the display screen which, if selected, will result in the execution of a given operation connected with the information retrieval process itself. This means that an icon represents a concept of relevance for the users' need and, as well, an opportunity for action.

It is well known that icons are faster to "read" compared to text since one icon can communicate complex messages and evoke rich connotations. They are easier to learn than text; they are easier to remember and to recognize once we know which one is advantageous, since recognition is easier than recall. Visually, we can process a far larger amount of information than with any other form.

In designing icons, there are various representational problems to be dealt with. The important aspects to consider are (1) the features of the action to be "iconized" and (2) how these features will be represented. It is important to realize that the interpretation of the content of the icon will be dependent on the current context from which it is viewed, and the comprehension of their form will be dependent on user characteristics and level of expertise. In any single situation, any particular user will choose a given way of responding that will depend on a host of factors, such as need, available time, training, and in particular in our case, the repertoire of heuristics and skills. How a user interprets an icon in a given situation depends on their intentions–experience at the given moment. The challenge for the designer is to provide a match between the context of the information retrieval task, displayed as icons, and the context of the intentions and needs of users' problem space, so that the user perceives the icons in the intended fashion (Agger and Jensen, 1989; Pejtersen, 1991a,b; Pejtersen and Goodstein, 1990).

One way of achieving meaningfulness and suggestiveness would be to study the conventional use and repertoire of icons for directing action in the context of potential users' work environment and everyday life. To ensure a consistent, systematic design of icons based on this analysis, a frame of reference containing the identified scenarios, together with a selection of examples of related icons, should be developed. Such a study in relation to the design of icons for the BookHouse indicated that the design–interpretation of the links between action intentions–formulations and icons can be structured in a systematic way referring to the action scenarios:

1. If *State*—2. Then *Actor*—3. Does *Act*—4. With *object*—5. To reach goal *state*

The visual *content* chosen for an icon may refer to any of these elements (see Fig. 11.3). The design of icons was based on the items in this action scenario. Each icon

---

1. State of affairs needing attention

Everyday context: Traffic signs, a locomotive informs about crossing trains. Be careful.
BookHouse context: Books put aside as interesting candidates for later perusal: "Pile of books" icon

2. The actors who should care

Everyday context: Icon: Lady's and gents on rest room doors
BookHouse context: Adults and children in search of books matching their age. Icon: Adults and children searching the book shelves

3. Action to perform

Everyday context: Icon: Man walking on stairs.
BookHouse context: Put aside interesting candidates of books for later perusal. Icon: Hand putting a book aside on a table

4. Object to use

Everyday context: Garbage can icon.
BookHouse context: Delete a search term. Rubber icon

5. Target State to reach

Everyday context: "Light on" icon.
BookHouse context: Get more books. Icon: Book case full of books

6. State to avoid

Everyday context: "Flame" icon informs about open fire to be avoided.
BookHouse context: None

7. Symbolic reference

Everyday context: "Hammer and Sickle " icon informs about Socialism.
BookHouse context: Books dealing with religion. Icon: The bible

8. Pure Convention

Everyday context: Red light at street crossing.
BookHouse context: Keywords displayed in red, black text

FIGURE 11.3. Design of icons can be based on different reference to items in an action scenario.

was determined on the basis of several sample testing of users' and librarians' interpretation of their meaning during the design process.

In addition to the referent in the action scenario, the pictorial *form* of a sign is chosen to enhance fast learning and recognition. This pictorial form can be based on a pictogram (a realistic reproduction), an analogy, or an arbitrary representation (Lodding, 1983). The possibilities are enormous and there are no rules or guidelines for making the best (or avoiding the worst) selection.

In the following, the form of the individual icon will be characterized as (1) symbols, the meaning of which is determined by a reference to population stereotypes; (2) metaphors, that is, representations with objects, persons, actions, states, and concepts, the meaning of which is determined by an analogy between the current work domain and another familiar context; and (3) pictograms, that is, pictures of objects or persons having a literal meaning that is determined by their resemblance with the real thing. The strength of similarity of the picture with the real device rather than with an analogy is in focus here.

Icons used for the BookHouse have several different functions. One is symbolic, that is, to visualize characteristics of books in a way that matches user needs, another is to act as signs that guide the use of search tools, and, finally, icons present affordance for action. We return to this below.

VISUAL DESIGN OF THE BOOKHOUSE DISPLAYS

As discussed in Chapter 7 and illustrated by the map of Figure 7.12, the visual design of interfaces for the casual user must be based on a conception of mental models that the user has adopted during other familiar activities and, if necessary, that can be easily modified for use in the retrieval task. It will be ineffective to try to force mental models of system functionality and domain characteristics on the user that are considered "rational" by the designer. Consequently, the visual design of the Book-House interface is based on a pictorial representation of concepts and objects that are familiar to the general public.

Discussion of the design of the visual form of the displays shown in Figures 11.4–11.12 will be described with reference to the structure of the knowledge base shown in Figure 11.2. That is, the overall organization of the visual representation reflect (1) the storehouse metaphor used to structure the knowledge base (Figures 11.4–11.7), while the intermediate, functional level is organized according to (2) the characteristics of the search strategies (Figs. 11.8–11.9). Finally, the lowest, most detailed level of representation in the displays serves (3) to visualize the individual items of the knowledge base (the books) and the individual options for action (by command icons; see Figs. 11.11, 11.12).

Organization of the Knowledge Base

The functionality of the BookHouse system consists of data base, retrieval algorithms, and navigational routes, each with a great number of access commands and manipulation devices. Since information retrieval will consist of several different iterations and interactions between the user and the system, the system functionality will play a dominant role because it will change frequently, as will the domain content, which may confuse the user. The many navigational routes and free interaction between any combination of these is too complex to display for the user as it appears in Figure 10.4. Therefore, a representation is needed of the navigational routes in the search dialogue and system functionality at a general level, which makes the various routes and retrieval techniques appear in a uniform manner. In the same way, several interface displays with different content related to various combinations of these manipulation devices will meet the user. Although it is important during the design stage to reduce this number to a minimum, the number necessary in an information retrieval system will be too large to overview and learn by (in particular, inexperienced) users, especially if the system design aims at providing the end user a great deal of control over a rich interaction functionality. A trade-off is necessary between the *multiplicity* in functionality needed to ensure user control and easy learning according to subjective preferences and strategy choices and *simplicity* in representation of a complex functionality to the user. Consequently, it is necessary to design an interface, which furnishes the user with a mental model of the dialogue and functions of the computer system and gives a semantic context for navigation in several interface displays.

The BookHouse Metaphor. The use of the BookHouse metaphor serves to give an invariant structure to the knowledge base, as viewed by the user, according to the search dialogue developed in the previous chapter. Since no overall goals or priorities can be embedded in the system, but depend on the particular user, a global structure of the knowledge base reflects subsets relevant to categories of users having different needs and represented by different rooms in the house (cf. Map 2 in Fig. 7.7). This gives a structure for the navigation that is easily learned and remembered by the user (see Figs. 11.4–11.6). The BookHouse metaphor is a functional analogy to a local library and relates to the population stereotype of a pancake house known from Grimm's fairy tales, familiar to children and adults, and easily understood and remembered, (see Fig. 11.4). The user "walks" through rooms with different arrangements of books and people.

The storehouse metaphor of the BookHouse supports the user's navigation through the varied functionality of the system during their planning and control of the search. It gives a familiar context for the identification of tools to use for the operational actions to be taken. It exploits the flexible display capabilities of computers to relate both information in and about the data base, as well as the various means for communicating with the data base to a location in a virtual space. The many dimensions and facilities are allocated locations in appropriate rooms or sections of rooms within this storehouse and thus advantage is taken of the mnemotechnic trick of Semonides (500 BC), which builds on the fact that items are easier to remember when arranged in a familiar topography. As George Miller (Miller, 1968) has phrased it, information is a question of "where." This approach supports the user's memory of where in the BookHouse the various options and information items are

FIGURE 11.4. A BookHouse metaphor is chosen for the fiction retrieval system in support of users' memory and navigation by locating information about the functionality and content of the system in a coherent and familiar spatial representation. The BookHouse access is through the open doors with a click on the mouse.

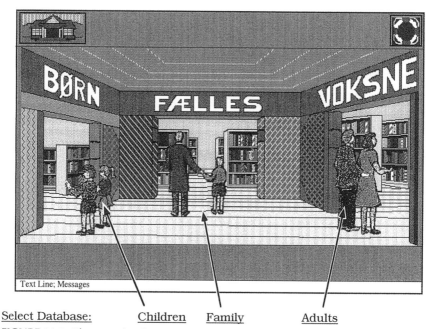

Select Database: Children Family Adults

FIGURE 11.5. The room for choice of data base in the BookHouse. In the upper left corner of the screen the rooms visited by the user will be displayed concurrently as the search progresses.

Start again   Select other Data Base   Help

Select: Browse Pictures   Select: Browse Book Descriptions
Select: Search by Analogy   Select: Analytical Search

FIGURE 11.6. The room for choice of search strategy. The display is a metaphor with a functional analogy to users' activities in libraries. From the left, search by analogy, browsing pictures, analytical searches, browsing book descriptions are signs indicating action possibilities through icons referring to persons, who should care. When the user selects a strategy, the system supports the search accordingly by a number of heuristics.

located. It facilitates the navigation of the user so that items can be remembered in given physical locations that one can then retraverse in order to retrieve a given item and/or freely browse in order to gain an overview.

In addition, support of the "cognitive momentum" (Woods, 1984) is given by indication of the displays previously used during a session at the top edge of each display by icons serving as command icons for easy backtracking. (see Fig. 11.7). Users can always keep track of where they are in the BookHouse, which path they have taken, their current position in the dialogue, and the available next steps in the dialogue. In effect, they can then trace their way back if they so desire (e.g., in order to change data base or strategy). Once a room has been visited, it always appears on the top menu line together with the other rooms of the house that have been visited. The user category that has visited the room will be represented by the age of the persons appearing in the display rooms to indicate in which section of the data base the user is searching (see Figs. 11.4–11.6 and compare with Fig. 13.7).

Animation is used in relation to the movement of the mouse cursor on the screen making different categories of users appear on the screen. Animation of adults and

	- Shift to another strategy; continue or abandon current search
	- Continue analytical subject searches and choose a dimension
	- Shift to another data base and abandon current searches

FIGURE 11.7. Displays for support of navigation in the data base and shift of search strategies. It directs the user's attention to the user's current "mode" and their previous choices. At the same time it displays alternative action possibilities through the functional analogy of a metaphor.

children or both walking around in the house was used to support the identification and selection of appropriate domain information (children's data base, adults data base and data base for both) as well as to support the impression of information, location, and actions relevant for a particular user group during navigation in the system.

Then, very efficient spatial navigation can take place by moving the mouse followed by a selective clicking with the mouse button. This will directly reinforce the development of manual skills involving the mouse to cope with the resulting spatial-temporal working space in a direct manipulation mode (i.e., an efficient automated interaction with the entities on the visible surface).

Room for Selection of Data Base. As it appears from the field studies, children, adults, and families searching for books apply different attributes for their search. In the BookHouse, the metaphor of "rooms" in the storehouse is used to distinguish different categorization and attributes of the general book stock and thus to give a global structure to the knowledge base. In effect, each book is to be found in different rooms (data bases) by different attributes, depending on the user's choice of "room" (see Fig. 11.5). Depending on the choice made, the attributes of the books applied for the search will be matched to the related user group. This will be visualized by the age of the agents appearing in the next room, the strategy selection room.

Room for Selection of Search Strategy. At the more detailed, functional level, the content of the individual displays represent the requirements of the various search strategies.

Depending on the previous choice, the display for choice of search strategy shows people, children, adults, or both—busy searching for books in different ways. (see Fig. 11.6 for an example). The user can thus select one of four different search strategies. The person at the right is browsing through books taken randomly from the shelves; another at the extreme left is looking through the shelves for a book similar to one already read and currently in the user's mind. At the next position, a user is browsing through icons reflecting book contents similar to users looking at pictures on book covers when looking for good books. The fourth person is sitting at a table analyzing and combining information, which graphically displays the total set of semantic options in the system and defines the boundaries to be followed during situation analysis and goal setting. Strategies are displayed as combinations of people with books, book shelves, pictures, or book attributes performing as signs for action through icons referring to persons who are involved.

The direct manipulation feature allows the user to click directly on the figure executing the same strategy that the user is interested in. A selection here leads the user to a new area where the required set of supportive icons are available for carrying out the chosen search strategy. Then one of three different displays will be fetched showing children, adults, or both depending on the user's previous data base choice.

Functional Organization of Search

At the functional level, the knowledge base is structured according to search strategies by giving access to further choice of work room. The visual design must take into account that several different user strategies will be applied when exploring and retrieving domain information to make decisions about information needs and document relevance. Information must be provided by the system to support all the provided strategies. When searching with a selected strategy, the value of the information from the information system relates to the degree to which it supports a choice of a proper search strategy in a given situation. Each strategy has specific requirements regarding the information available during a search, and the value of information displayed depends on the degree to which the messages in form and content match these requirements. To support different mental strategies, the content of the data base should be organized in different semantic networks (e.g., as attributes organized in hierarchical, generic structures for analytical searches and as associative relationships for browsing and search by analogy).

Work Room for Analytical Search in Domain Structure. The work room display makes available to the user a multidimensional information space in terms directly compatible with (intuitive) user needs arranged to permit a spatially based exploration "dialogue" for a satisfactory match (see the display in Fig. 11.8).

In this work room, the abstract attributes of items in the data base are consistently represented by familiar physical objects in spatial arrangements in order to reinforce the development of manual skills in a spatial working space. On and around the table are icons representing the different dimensions of the classification structure used to classify books in the system. They give a pictorial link to the semantics of the dimension they represent and are also signs for action in order to proceed further. The majority are metaphors representing an analogy to objects familiar to the general user. For example, the world globe, representing the geographical setting of the book, the clock icon that denotes the time dimension, the glasses, that refer to the accessibility of books, and the authors writing their stories with the old technology of a writing machine; Some icons represent an *analogy* to previous technology in the library like the card catalogs for author–title searches and genre searches. Others refer to the state

Choose Other Strategy

Specify Environment
(Social or Work Place)

Specify Plot   Specify Actors

Specify Author Intention

Specify Time

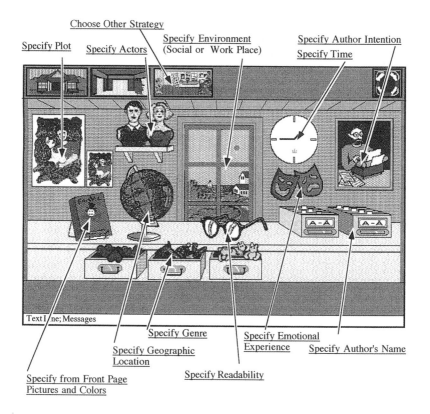

Text Line; Messages

Specify Genre

Specify Geographic
Location

Specify Emotional
Experience   Specify Author's Name

Specify Readability

Specify from Front Page
Pictures and Colors

FIGURE 11.8. The iconic display of the means–ends representation of the work domain in terms of a multifaceted classification scheme based on an empirical analysis of users' search questions in libraries. It supports the user's analytical situation analysis and reasoning, domain exploration, and planning and execution of a search as well. Most icons are metaphors referring to object to use with analogous functions, others are symbols. When the user clicks on an icon, keywords are displayed and the user can specify need by clicking on one or more keywords. A subsequent search is then performed.

to reach, like the theater masks, as a metaphor for the emotional experience provided by books. The posters on the wall are symbols referring to stories in books. The view from the window is a symbol of the geographical, environmental or social setting in books, while the action of looking out of the window for book settings is a functional analogy. The form of these icons is chosen as objects in houses having a functional meaning familiar to the user population.

In this way, the structure of the data base is immediately made transparent to the user and supports their analytical situation analysis, domain exploration, and execution of a search. This is important since it cannot be expected that casual users, who may not even be able to express their reading needs explicitly, can actually apply knowledge-based situation analysis. Analytical search strategies are, however, very effective. The aim of the BookHouse interface design is explicitly to support this strategy, but in a more implicit, intuitive form. As mentioned previously, during field studies of cognitive decision tasks, users will meet particular difficulties in the situation of need analysis and formulation; hence, special efforts must be dedicated to the interface design for support of the situation of need analysis to help overcome users' cognitive limitations.

The solution chosen for support of the analytical search strategy is, therefore, the work room described above with symbolic icons and metaphors identifying the potential dimensions of a user's needs and acting as command icons for control of the search. These symbolic icons are based on an analog reference to the dimensions of user needs by means of figures representing concepts familiar to the general user. In this way, the user is prompted to browse and consider all relevant aspects of the present need even if they are not able to explicitly formulate them verbally. In this way an analytic search process is embedded in a recognition process enabled through tacit knowledge. When users resort to this kind of browsing strategies, they rely on natural, highly effective recognition skills, and scan their information environment for recognition of cues to tacit, unconscious cognitive processes. One of the objectives for interface design is therefore to support this approach actively by probing preattentive, unconscious cognitive processes to improve the use and awareness of implicit knowledge more efficiently.

Then, once a need is recognized, no analytical activity is needed for planning a search—another cognitive task that poses problems for casual users: A selection of the icons of interest will activate an immediate search.

Photo Album for Browsing in Associative Domain Structures. Another way to support domain exploration and intuitive searches as identified in the work study, is through a picture association thesaurus with icons that can be selected to provide a quick birds-eye view of the semantic content of the data base and subsequent recognition of need situation and goal setting (see, e.g., Fig. 11.9). When an icon is selected a search is performed on keywords associated with each icon. To serve this purpose, each icon represents a semantic, associatively related network with a large number of keywords. Each network represented by an icon can then be combined with another network represented by another icon as the user browses, associates, and recognizes more icons of relevance to the need. However, apart from being an effective representation of an associatively related semantic network, the rich connotations of icons can provide an unexpected aesthetic and emotional experience, and give rise to potentially new perspectives and associations on vaguely perceived needs.

Pictures are implemented as a thesaurus aid in the form of a "photo album" metaphor with 6 pages of small icons, with 18 icons on each page, and a text line at the bottom of the screen, which informs the user of the cluster of meanings associated with each icon, whenever the user touches the icon with the mouse. The user can then step through several pages in the picture album by turning the red corners of the pages.

Most of these pictures are designed as symbols with reference to users' conceptual levels of prototypical concepts and needs; some are metaphors and a few are pictograms. The richness and variety of forms of these icons can be used in knowledge-

FIGURE 11.9. The display for browsing icons is designed for exploration and recognition of needs. It is intended to probe the user's tacit knowledge and to support intuitive recognition of concepts within the user's current need situations. Six pages of icons in the metaphor of a photoalbum represent the book content of the data base. These icons are symbols that represent a semantic network of associatively related keywords belonging to several different dimensions of the structure in the user domain. The red corners of the picture album are used to page forward and backwards.

based reasoning but are typically designed to support immediate recognition and signs for action. The effectiveness of these icons in supporting prediction and symbolic reasoning depends on the user's familiarity with their content and form. Since pictures are well known to evoke rich connotations and to be more multidimensional in their communicative means than words, they will perform more efficiently than the text as a support for browsing strategies.

The approach to constructing these icons was to use index terms as the basis for imagining and drawing appropriate pictorial analogies. In order to organize keywords and pictures in an associatively based semantic network, several types of word association studies were conducted. A multiple choice association test was conducted among library users to determine the associative meaning of each icon based on users' associations between keywords and icons. The purpose was to reach a graphical, functional form that originated from the system users' context and was recognized as familiar concepts. The form of each picture is a symbol chosen from population stereotypes to be intuitively recognized by the user as relevant to the current need situation, and once recognized, to be spontaneously related to a subsequent search for a match.

When the meanings of these icons were tested during the design phase by three groups of end users (children, adults, and librarians), it turned out that children and adults sometimes associated differently. Therefore the meanings of some icons in the children's data base are different from icons in the data base for adults. Thus, when the user chooses an icon in the children's base, a Boolean search can be initiated with different and often more subject terms than with the same icon in the adults' data base. Only icons that had terms with at least 50% consensus were implemented. When less than 50% of the subjects agreed on the meaning of an icon, it was redrawn. When more terms had been selected as being meaningful for a given icon, only those terms with 50% or more were assigned to the icon. The results gained from these experiments were used to design, implement, and evaluate the picture association thesaurus for browsing in the BookHouse system (see Chapter 12 on the evaluation of the BookHouse for more detailed information).

Display for Search by Analogy. Yet another way to circumvent the need for explicit analysis is to find a book by search for a book "similar" to the book the user has read with pleasure. The user can specify the title and author by selecting from a list presented on request from a menu option on the front cover of a book. For this strategy only the lowest level of information in terms of author–title information is displayed (Fig. 11.10). After selection of an author or title, the system will automatically attempt to find other books in the collection that include as many of the same attributes of the model book as possible within the classification system.

Text Line; Messages

FIGURE 11.10. Display of a model book; for selection of author–title, when searching for something similar to a known book. The book metaphor display refers to an action to perform, and supports the execution of a search procedure connected to the strategy of search by analogy.

Presentation of Individual Books

The lowest level of the knowledge base includes the representation of the characteristics of the individual books so as to match the search strategies. As discussed in the previous chapter, this requires a classification of the individual book by the attributes of the classification system and an indexing to derive the keywords as search terms, all of which are used in the retrieval processes. The retrieval functionality then needs to match these attributes and keywords systematically to the semantic network of keywords used in the picture browsing, the template derived from the model book in search by analogy, and to the dimensions represented in the analytical work room. All these processes are invisible to the user and controlled by the strategy displays.

However, an explicit, verbal representation of keywords and of the individual books are required for the selection of keywords to meet a specific user need, and for the final comparison of retrieved books with the user's need. A set of displays are available for this purpose.

Displays With Keywords for Analytical Subject Access. During an analytical search the execution and control of the search activity is based on keywords. In the work room display, the user can select one icon at a time so as to get access to an open book with keywords containing the textual listing of selectable keywords in alphabetical order derived from the dimensions of the classification scheme (Fig. 11.11). The user can explore terms and compare match with need and then can combine search terms from the same or different dimensions. After clicking a keyword, the user can see book descriptions structured according to the dimensions of users' needs. The book metaphor display refers to an object to use and it is a sign for action, encouraging the user to make the analogy of reading and turning pages by using the red corners of the book.

Book Content Display. The analytical strategy displays of pure classification categories and single keywords related to categories without display of their relationship with the content of books will provide a sufficient basis for need formulation and selection of relevant category and keyword. A display of the domain structure with related book content is needed for a comparison of the user's goal as it is formulated within the domain structure. Also, the actual content of retrieved documents is described within the same structure, for the user to be able to validate the match between the actual content of each document and the user's need. Similarly, a display with exhaustive information of book content is needed to enable the user to perform relevance feedback on the content of each book (see Fig. 11.12).

This is particularly relevant in the information retrieval task, as a user will rarely be able to define a need and select accordingly within all the dimensions of a display containing the domain structure. Each retrieved document will contain the part of the structure selected by the user, but it will also contain much additional information. Furthermore, each retrieved book will differ and the content will change dynamically,

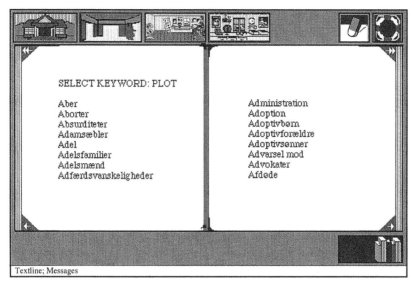

FIGURE 11.11. The list of controlled keywords; displayed in red in an open book format, which is a metaphor strongly emphasizing the analogy to the retrieval task in the context of the library domain. The display support planning and execution of an analytical search by the alphabetical list of terms to be combined and searched. The rubber icon assists by the action undo and delete undesired or regretfully selected terms, and the book case shows the result. The user can specify need by clicking at one or more keywords and a subsequent Boolean search is performed. The red corners page one book or several books forward and backwards.

which makes it difficult for users to compare the degree of relevance of each item in comparison with the other items of a retrieved set. A display explicitly showing the overall structure and content at a finer grained level will help this comparison of similarities and differences in content.

Then, as mentioned earlier, to help users recognize content in comparison with need, the dimensions included in the structuring of the book content will differ depending on the user's choice of search strategy. The system tracks the user's choice of search strategy and choice of classification dimension. Next it adapts the display content in accordance with the user's underlying current mental model of the domain. The difference in the representation of the levels of content is particularly related to the lowest level in the users' means–ends model, as described in the section on strategies identified in field studies.

Verification of the actual state of the search with the desired goal state is effected by showing the users' choice of keywords in a search string on the screen in the sequence in which they have been chosen. These keywords are displayed in red in separation from the content of retrieved documents, but they are also displayed in red in the content of the book description within the adequate dimension of the classifica-

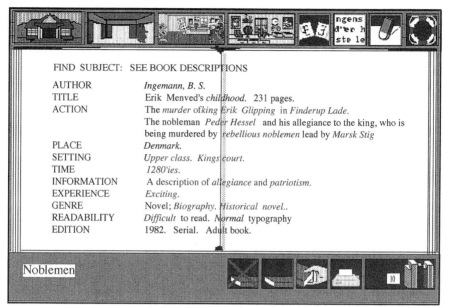

FIND SUBJECT: SEE BOOK DESCRIPTIONS

AUTHOR	*Ingemann, B. S.*
TITLE	Erik Menved's *childhood.* 231 pages.
ACTION	The *murder of king Erik Glipping* in *Finderup Lade.*
	The nobleman *Peder Hessel* and his allegiance to the king, who is being murdered by *rebellious noblemen* lead by *Marsk Stig*
PLACE	*Denmark.*
SETTING	*Upper class. Kings court.*
TIME	*1280'ies.*
INFORMATION	A description of *allegiance* and *patriotism.*
EXPERIENCE	*Exciting.*
GENRE	Novel; *Biography. Historical novel..*
READABILITY	*Difficult* to read. *Normal* typography
EDITION	1982. Serial. Adult book.

Noblemen

FIGURE 11.12. An open book metaphor display with a description of a retrieved document. It illustrates the multidimensional information contents of a book analyzed, represented, and made accessible according to the structure identified in the users' domain. The user can step through other book descriptions and continue verification and check of the result. This display will be presented as the final output after browsing books, browsing icons, analytical keyword searches, and search by analogy. Controlled, selectable keywords are displayed in red in the context of the book description. Below the book description, the user's current search term is seen.

tion scheme. In both cases these keywords can be selected or removed, and thus used in a feedback process involving a judgment of the relevance of the content of a document.

The open book display has red corners with arrows pointing to the left and right in case more than one page of information is available. The upper corners have double arrows to indicate that larger leaps can be taken, while one arrow indicates a step that takes one page at a time. The book metaphor display refers to an object to use and is a sign encouraging the user to make the analogy of reading and turning pages by using the red corner of the book.

Display for Strategy of Browsing Books. A choice of the browsing strategy allows direct and unspecified access to book content. In this strategy display no distinction is made between the cognitive decision of need situation analysis and the comparison of match of a retrieved set with user need. An open book with book descriptions structured according to the dimensions of users' needs is displayed in Figure 11.12.

The book content is selectively chosen to match this strategy by focusing on the topical content levels of information. Red keywords are available in the description for the user to begin an analytical search as well as a shift to search for similar books and thus initiate a relevance feedback process.

Display for Browsing Books Put Aside. Books considered to represent a match can be put aside for later consideration during browsing their content. An open book with yellow pages is displayed to help the user associate to "old" book pages and thus help to distinguish books in a current set from those put aside when switching among these books. The levels of book content displayed to the user when browsing saved books will depend on the strategy that was employed by the user when the book was retrieved and put aside. Red keywords are available in the description for the user to begin an analytical search, as well as a shift to search for books similar to those put aside and thus initiate a feedback process for judgment of the relevance of retrieved documents.

Navigation Revisited

In addition to the guidance given the user's navigation in the system by the house and room metaphor, some detailed guidance of navigation is necessary for the manipulation of the books retrieved during search, that is, for the bottom line part of the dialogue map of Figure 10.4.

For support at this interaction level, the design is aimed at direct manipulation guided by an easy formation of cue–action sets. For this, the BookHouse makes extensive use of a kind of "menus" that are essentially displayed as command icons in standardized positions on the screen. Only those that are relevant for decision tasks at a given stage in the user-system dialogue are visible.

Icons to be perceived as signs and cue information regarding alternatives for proceeding in a search need to be designed with a unique, unambiguous form communicating to the user a specific action possibility to govern the user's choice. Some examples of action icons are given in the following sections with reference to Figures 11.13 and 11.14. The form of icons is designed to ease naive and casual users' ability to understand, learn, and distinguish among the numerous possibilities for control actions at the lowest level of a search. Metaphors and pictograms are chosen to support analogical reasoning determined by a familiar context for the general, public novice user. The content of icons refer to *objects* to use, current *state of affairs*, and *states to reach*.

It should be noted that according to the ecological design principles and the use of an analogy to an environment familiar to the casual user, each icon will be a reference to concepts related to the users' needs, to the book contents, or to the search process, and at the same time it should afford relevant actions (see the discussion of cognitive resources and direct perception in a natural environment in Chapters 4 and 5).

The library context makes book metaphors in displays perceptually strong as

	Browse through the set by taking books out of the book case. The book metaphor attracts the user's attention to the *state* of the search
	Save interesting candidates for later browsing. The book metaphor attracts the user's attention to the *object* to use
	See previously saved books. The book metaphor attracts the user's attention to the *state* of the search
	Get rid of previously saved books. The book metaphor attracts the user's attention to the *state* to reach

FIGURE 11.13. Book metaphor action icons; with reference to objects to use or the state of the search in order to indicate alternative action possibilities for rule-based planning and evaluation of the search.

iconic signs for action possibilities. In particular, it is desirable to have the ability to display an action and its results by analogy to familiar routines like reading books, taking books in and out of book cases, putting books in piles, both with a subsequent increase and decrease of stacks. The form of the book icon is both the professional object of the domain and its associated retrieval target and a familiar object from the user's background.

	Delete search terms from the current search profile. The metaphor refers to object to use.
	Print book description for later use in finding books on the shelves. The pictogram refers to object to use.
	Help is a metaphor that refers to object to use.
	Select new search terms from the book description and modify the need by adding to the current search profile. The metaphor refers to the state to reach.
	Modify the current search profile by removing red search terms in the current book description and a Boolean NOT is automatically performed. The metaphor refers to state to reach.
	Find books similar to any current book on the screen. The metaphor refers to state to reach.

FIGURE 11.14. A selective repertoire of command icons; to be perceived as signs for action. The icons are selectable for execution and control of the search whenever a book description is on the screen. The user can repetitively select among these options including shift of strategy without entering a previous room. The metaphor and pictogram are cues for action referring to *objects* to use and *states* to reach and supports rule based activity.

Object to Use. The execution needed to put books aside in a contemporary pile on a table is done by an icon in the form of a *hand holding a book*, indicating the action the user is supposed to take. Command icons can be used to modify the current need by adding or removing search terms from the current search profile by means of an *eraser icon*, which deletes terms clicked with the mouse. The *Printer icon* serves the purpose of making hard copies of interesting descriptions for use later in finding the books on the shelves. Request Help by taking the *Life belt icon* gives the user both a context-dependent explanation of where the user is at the moment, as well as more general information about the system facilities (see Fig. 11.13).

State of Affairs. The visualization of the outcome of a search in terms of the number of retrieved sets of documents, in addition to the content of retrieved documents are related to the user's goals. When a search is begun the initial state is "no documents." After the search is over, the final state is the number of documents in the retrieved set. The desired number of documents in a set will be defined on occasion by the individual user according to the task situation. The number of books retrieved as a result of each new choice of keyword will then be displayed concurrently to enable users to verify a match, as users may have preferences with respect to the number of books. The feedback to change of numbers in a set is given through numbers and is visually given through an increase or decrease of a pile of books, which appear on the screen as soon as a set is retrieved. The same feature adheres to the number of books put aside for later browsing.

The number of retrieved books in any current set is displayed in a *book case* to inform about the state of a search. By clicking on this icon descriptions of retrieved books are displayed for comparison of match with user need. An icon with *books in a pile on a table* gives information about the number of saved books. By selecting this icon the user can get descriptions of previously searched and saved books and compare match with need at any stages of a search.

State to Reach. State to reach refers both to the *target state* in the relevant part of the system and to the user's reading need. The *minus icon* can be used to operate a Boolean NOT search on a not wanted keyword. Feedback will be shown to the user by inserting the keyword in the user's search string with a minus on the term. The same control of a Boolean AND search can be obtained through use of the *red keywords*, which when selected generates the number of books found and the first book description in the new set. An icon with a *cross on a book pile* will remove books put aside. Find similar books will, if the icon with the *equation* sign has been selected, as a start use any current book description on the screen as the basis for finding other similar books in the selected book data base (see Fig. 11.13).

CHOICE OF SOFTWARE CONCEPT

So far we have considered the transformation at the center and right-hand columns of Map 4 in Figure 11.1. Here we will review the representations used in the left-hand column showing the formalized computer representations.

The choice of technological and software solutions are not the concern of this book, but it will be relevant to mention the hypermedia software concept as particularly relevant for ecological information retrieval and storage systems. Hypermedia systems are based on the concept of free and flexible navigation in semantic relationships enabling design of relationships tailored to match the semantics of the problem space with the semantics of the users' cognitive worlds. It allows our suggested design of a multidimensional "resource envelope" for multiple levels of activities instead of support of a particular normative interaction procedure. The strength of hypermedia systems for representing information is the ability of links to explicitly represent flexible semantic structures and nonlinear access to information. As argued previously, in many modern work domains it is becoming increasingly evident that users need a rich semantic network that corresponds to their preferences within the work domain constraints.

Hyper (media) is a term used in information systems applying the cognitive concept of nonlinear and nonsequential structures (i.e., problem solving based on multidimensional relationships connected by associations), while media is a generic term for multimodal information entities communicated by computers (text, video, sound, film, graphics, and animation). Any subcategory of hypermedia computer systems can exist, such as hypertext, hypervideo, hyperfilms, and any combination of these "hyper categories" are called hypermedia systems. Multimedia systems apply multimodal information, but these are not necessarily implemented within the hyperstructure that characterize and distinguish hypersystems from other systems.

The hypermedia concept (although not the term) originated from Bush (1945), Nyce and Kahn (1992) with the claim for a design based on the idea that multiple associative structures are possible for the same information content. He suggested an application for storage and retrieval of information content of libraries as the users' personal tool for locating and representing knowledge so that research workers could avoid being bogged down by the mass of new knowledge and could support selection of information, i.e., research findings in a different way than those usually applied by formal, hierarchical library classification and indexing schemes. Personal, individual associations were the basic principle to be applied in access and use of records instead of principles generally applied for the whole library collection.

In the terminology usually employed software components of hypermedia programs basically consist of *nodes* that are the single data base entities containing the information content to be processed, stored, and retrieved. The content of a node may be collections of multimodal information, or single pieces of text, graphics, pictures, film, video, sounds, and so on. *Links* are relations among nodes and connection to the attributes and substance of node contents. Typically, links allow associative interaction and free navigation through keyword searching or browsing in the semantic network subject to the users' own discretionary choice, thus creating users' navigational *paths*.

Design of User Control in Hypermedia

Three classes of hypermedia systems have been developed so far as *active* and *passive* systems and a combination of both with different degrees of user control.

Active–dynamic systems aim at letting the users create their own semantic networks by adding content (nodes) and adding–changing new links. Thus, this develops new relationships in the network according to their own goals and subjective preferences. Users may create new links, remove alternate links and add new information and develop their knowledge base.

At the opposite end, *passive–static* systems are designed with preestablished links in the network and without user access to node contents. Inbetween are systems where a combination of active–passive is found (Streitz, 1990). What kind of system is most appropriate in a given design will depend on the total work system when character-ized within the cognitive framework, and the degree of user control will to a large extent be determined by the requirements of the domain and the capabilities of its users. For the library system described a combination was appropriate with a static data base structure for the information retrieval task and links computed to relate data base content to retrieval algorithms and coded search strategies, which cannot be changed by the user. The information storage task obviously will be conducted through access to dynamic change of the content of the data base system within the stable domain structure for retrieval, but without access to change of the structure and the computed hypermedia network links.

As hypermedia systems allow great interactivity, and as interaction depends solely on user activity and responsibility users must be mentally active while interacting with information (Jonasson and Grabinger, 1990).

The degree of user control is high and the cognitive load on users will increase as will the users' experience of disorientation and getting lost in hyperspace (Conklin, 1987), particularly in systems with active dynamic user interactivity and a large amount of user control, both in use and in redesign of the network and content of the system. Basic design issues are raised to solve the conflict originating in the trade-off between user control of a rich semantic network and limited cognitive resources.

The creation of meaningful semantic *network structures* is essential to support users' awareness of domain information without loosing control of goal related decision activities. A stable domain structure is necessary to avoid cognitive over-load, since association based actions are situation dependent. Even expert users will not be able to recall their interaction with the system in a variety of situations, unless the same situation is repeated, which is not often the case in complex problem solving. Obviously, in the mind of casual users in unfamiliar situations that are untrained in the domain, association based behavior will not occur, since associative behavior is based on recognition of a given situation. A means–ends domain structure will provide the boundaries of the domain invariants to support user awareness (Pejtersen, 1993a,b).

The creation of links is creation of users' action possibilities in the domain structure. Design of computed links as coded user search strategies in the means–ends networks of the domain is essential to avoid a sense of user disorientation in the information network. If no constraints are put on strategy shifts and strategy combina-tions, the user will have multiple choice of subjectively defined navigational paths even with computed links. The choice of a metaphor will then be necessary to support location and navigation.

The creation of an ecological interface with visualization of the domain structure as links to activate domain structures and boundaries will map the semantics of the domain in the interface to the users' capacity for cognitive control. The suggestion of such design considerations for hypermedia systems will extend the design focus from the general cognitive characteristic of the users' way of accessing and processing information to their interplay with the task and domain characteristics.

Limitations due to premature software tool kits with limited capacity and speed for design of a large information retrieval system is one reason why hypermedia often only work as the interface to a data base that is adding multimedia representation and associative interaction and retrieval from the data base. But hypermedia designs can be implemented in almost any kind of software environment. The hyperstructure existed in application programs before the hypersoftware was marketed. The software chosen to achieve hyperinteraction structures is of minor importance. The concept relates originally to humans' way of thinking and processing information, and later on it was implemented in dedicated software tools. The BookHouse is an example where the hyperstructure was derived from the analysis of users' search strategies and frequent, associative shifts among these strategies while interacting with the multidimensional domain information.

Finally, choice of technological platforms and concepts deal with many issues. The following points are of particular interest in this context: (1) the choice among a stand alone system or in networks; (2) an integrated part of other, local computerized library systems and data delivery services; or (3) the end users' IR system as a front end interfaced with a centralized data base system.

HELP AND TRAINING

Finally, we consider the direct user interaction with the software. When designing an ecological interface it is presupposed that the emphasis on displaying the affordances, the signs for action, will encourage users to explore the content boundaries of the domain and the functional action possibilities of the system. At the same time, information richness based on the multidimensional framework for domain analysis combined with display of the affordances of several types of system–domain invariants will enable users to learn about the domain and the system functionality with minimal instruction and little practice.

Then, the overall aim of advocating design of ecological interfaces in library systems is that they are intended to (1) help the user to understand the domain content in the data base without previous training and thus to support problem solving in unfamiliar situations, which occur frequently for casual library users, by making the boundaries of the data base information on book content explicit and visible in the interface; (2) ease and speed up the user's learning of the data base content and its associated retrieval functionality by making the transition from a novice to an expert smooth and fast; (3) support experiments and learning by doing through trial and error by making several search dialogues and search strategies available to the user; (4) support expert users development of new search strategies, new task sequences, new dialogues, and new solutions to their retrieval problems and needs (which may be solutions unpredicted by the designers). Design trade-off: Adequate information for experts, to continue their increase of expertise, but not too much for novices to be able to learn easily.

Help texts are not "voluntarily" applied by users in a problematic situation. The basic assumption is that when learning by use of an ecological interface, structure and meaning given to the individual display is dependent on the interplay between user, domain and task characteristics. Similar approaches are found in a discussion of interaction in hypermedia systems for education (Fischer and Mandl, 1989).

Special considerations had to be adopted for the design of the allocation of aids intended to be given through the ecological interface displays, in interface help texts and users' guides. Several different forms of aids may be needed in relation to (1) the various levels of the means–ends structure of the domain that can be activated in retrieval and classification tasks and (2) covering the action possibilities communicated by the objects on the screen and the functionality of the system in relation to users' goals at the individual stages of a search. Help with respect to the first point has to take into account the relationship with the conventional guides for information retrieval, classification, and indexing; most commonly the use of the thesaurus for support of retrieval query formulation as well as for control of indexing vocabulary during storage of data base information.

BookHouse Design: Help Texts

When the user explores the screen with the mouse a text line at the bottom follows the mouse movements and gives the user two kinds of messages: If the mouse is not in an area with selectable objects, the text appears: Move the mouse and see what you can do. When the mouse points at a selectable object, a square is drawn around the icon to evoke the user's attention and to indicate mouse sensitive limits of the object. A text appears with a message telling the user to press the mouse button and with functional information on what will happen next. These help texts serve two purposes: (1) It supports interactive, self-acting learning by doing during exploration of information on the screen. (2) It supports first time users in understanding the meaning of each single icon. Prototype testing with users showed that not all users will be able to interpret the meaning of icons and predict a unique retrieval action associated with an icon. If the user needs help beyond the level of execution of action the Life belt icon gives the user help windows with both a context-dependent explanation of where the user is at the moment, as well as more general information about the system facilities.

In addition, there are always help facilities available that attempt to give a context-dependent explanation of where the user came from, where the user is at the present moment, what the user can do, a brief explanation of how to do it, and what happens if an option is chosen. In addition to the context dependent help, the user can continue to ask for help, and from an index of help texts choose any text to get more information about the system facilities.

As mentioned, each icon has an associated short textual explanation that appears at the bottom of the screen whenever the mouse passes over the sensitive area on the screen "owned" by the particular icon. This was felt to be necessary—especially to help first-time users, since the icon can have multiple meanings and can be ambiguous to a user. Thus the top and bottom rows of, for example, Figure 11.8 illustrates some of these icons and their fixed locations on the screen; the dedicated text area is at the lower left. Figures 11.13 and 11.14 depict these command icons in more detail.

Finally, a written folder is available on search options.

Chapter 12

BookHouse Evaluation

EVALUATION EXPERIMENTS

Empirical evaluation of the concepts underlying the BookHouse design, as well as the means chosen for implementation, has involved considerable experimentation. This experimentation was conducted in the laboratory during the design phase and in libraries to validate the final product. To set the stage for a discussion of the evaluation with reference to Chapter 8, some preliminary remarks will be necessary.

The design process that unfolded during the development of the BookHouse system is a very clear example of design as a creativity-resource trade-off. Many degrees of freedom remained for the designer when they were given the results of the various field studies and the order to design a computer-based tool for the retrieval of fiction by the casual user of a public library without requiring any special instruction. These had to be resolved in trade-offs between available programming tools and resources (the hypertext software was not available at the time of design); the established practices in the (Danish) library institutions (including indexing practices); the uncertainty about user acceptance of new concepts (such as retrieving books by visual browsing in iconic representations); and so on. As a result, the design process resulted in a long sequence of iterations among design ideas, prototyping of part solutions, and experimental evaluation of ideas, hypotheses, and solutions. Happily, the sequence of experiments performed turned out to be a very good illustration of the need to consider the conditions underlying experimental designs at the various boundaries between the user and the ultimate work context in the library. The sequence of experiments illustrated in Figure 12.1 maps the BookHouse design and evaluation experiments onto Figure 8.3. For further details see Goodstein and Pejtersen (1989), Pejtersen (1986a,b,c,d), Pejtersen et al. (1987), and Morehead et al. (1984).

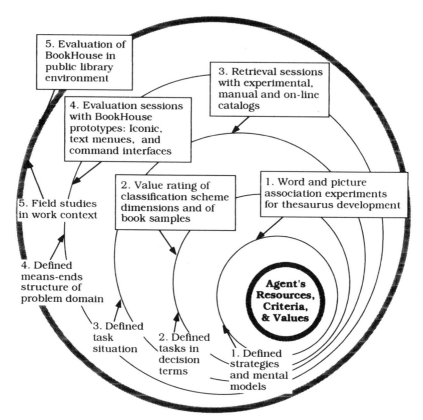

FIGURE 12.1. Mapping of the design and evaluation experiments involved in the Book-House design onto the framework for specification of boundaries of experiments (see Figs. 8.2 and 8.3).

Experiments at Boundary 1: Association Processes

One sequence of experiments was aimed at gaining needed design data on a boundary close to the user. These focused attention on an evaluation of users' visual and verbal association processes as a basis for the design of interfaces to support browsing.

Information retrieval is embedded in the general cognitive control of the state of affairs in the user's work environment. This control greatly depends on tacit knowledge, that is, on intuitive reactions of users in a familiar or unfamiliar context. Thus, there is a distinction between the preconscious, recognition based cognitive processing of this tacit knowledge, that cannot be accessed, and the (verbalizable) conscious, analytic cognitive processes used in higher level performance to draw inferences and solve problems. Since recognition processes enabled through tacit knowledge play such a major part in human information processing, interfaces in "intelligent" information systems ought to exploit these human skills to aid in browsing through complex information. When users resort to browsing strategies, they rely on natural, highly effective recognition skills and scan their information environment for recognition of cues to tacit, unconscious cognitive processes. One of the objectives for interface design is therefore to actively support this approach by probing preattentive, unconscious cognitive processes in order to improve the utilization of implicit knowledge more efficiently. This kind of interface support is also discussed in Pejtersen (1991).

Experiment 1. The hypothesis studied in one experiment was that, if a thesaurus was based on associative semantics derived from the users' own tacit, episodic knowledge bases instead of dictionary definitions and denotatively definable relations, then the thesaurus could be used to trigger recognition and recall of stored, hitherto hidden, conceptual relationships of relevance to the user's need formulation. In other words, a network of word associations obtained from user groups is comparable to a semantic network and could thus be usefully incorporated into a thesaurus to support browsing. Such a network will also encompass the users' points of view, their previous experience, and various other pragmatic factors of which the user may not be aware, but which contribute to the recognition of an information need for which traditional retrieval methods are not adequate or effective.

Word association experiments have gained an accepted status as one way of probing the preconscious levels of our inner minds. These word associations appear to be one method for gaining information about eventual connections between a stimulus word and a response word that, in some sense, is important to the subject. It also seems reasonable to expect that a given group of people with similar goals and interests, such as readers of fiction, will demonstrate similarities and show common patterns of responses to stimulus words.

To test this hypothesis, a session of experiments was conducted that limited this complex problem to the design of a thesaurus to be displayed to the user in a bibliographical data base. This consisted of purely associative relationships that originated from empirical data concerning associative responses from potential users of the data base system to the controlled vocabulary of the data base. A *discrete continued association* test was chosen (Cramer, 1968). By comparing the distribution of associations to any two keywords, quantifiable judgments were made of how similar or dissimilar any two keywords were with respect to their associative meaning. As a result, a thesaurus was constructed based only on the relationships in mentally connected, that is, associated, subject terms in its vocabulary. In order to evaluate this aid for browsing, an interface to a data base was composed of a text-based menu with access to three major facilities of (1) browsing through an associative thesaurus, (2) browsing through a conventional thesaurus, and (3) browsing through a combination of these two aids. These were evaluated with a combination of rating scales, interviews, and thinking aloud methods. The word association thesaurus was deemed to be in good correspondence with the subjects' perception of the semantic relationships among concepts. The thesaurus is an aid with unanticipated references between concepts and, according to the users' evaluation, was found to help them probe their own need through associations and new ideas. Thus it helped to extend, clarify, and explore the limits of their field of interest (Pejtersen, 1991). A comparison of subjects' choice of conventionally, hierarchically related terms, and choice of terms derived purely from the word association tests showed a higher percentage of terms based on word associations in the initial stages of browsing and need formulation. But as browsing progressed, the percentage of these terms decreased. However, even in the final need formulations, the majority of terms belonged to the word association thesaurus.

Experiment 2. Experiment 1 indicated that aids based on connotative, empirically derived relations between terms can be a fruitful supplement to traditional aids. The word association method was then used to design a Picture Association Thesaurus in which associative relationships of terms represented by icons would improve the associative arousal within the personal context of the users' need. The hypothesis used was that since pictures are known to evoke rich connotations and to be more multidimensional in their communicative means than words, they will perform better as a support for browsing strategies. In order to organize keywords and pictures in an associatively based semantic network, several types of word association studies were conducted.

The aim of the first *free, continuous* picture association experiment was primarily to determine whether it was feasible for designers to associate to pictures from keywords chosen from the different dimensions of the fiction classification scheme used for the BookHouse (see Fig. 10.2) and create *pictorial associative responses* to stimulus words instead of the usual verbal expressions of associations. Based on 450 of the most frequently used terms in the data base, 96 icons were drawn with emphasis on prototypical, commonplace phenomena. Any form could—and was—used by the designers, who tried to act as potential system users with their experience and knowledge of symbols and books. The icons utilized (a) symbols such as the "hammer and sickle" for books about communism, "hearts" for love; (b) conventional *signs* like those from traffic; and (c) *pictograms*, such as an automobile, for books about cars and car accidents. Unique and original symbols were invented to signify abstract concepts, such as a man split into two parts and colors to signify identity problems and narcissism.

The primary aim of the *second* picture–word association experiment was to determine whether a sufficient consensus could be achieved between designers' conceptions of associative relationships between pictures and keywords and those of different groups of potential users of the system. A *controlled, multiple choice word association* experiment was carried out in which different groups of users (children and adults) had to select one or more words associated with a picture. Subjects had five word choices for each icon. Three of these were the terms from the index that the designers had used as the basis for drawing the icon. In some cases, the fourth was a check term not at all related to the intended message(s) of the icon. The fifth gave the subject the opportunity to answer that none of the terms matched the icon. They were instructed to react spontaneously to the stimulus pictures and to respond intuitively when they selected those words that they thought were associated with each picture.

Since the strength and effectiveness of pictures for browsing rely on attaining a multiple and associatively rich meaning in each picture, the analysis of data from the word association test included an analysis of each individual picture in order to determine the frequency and distribution of the subjects' choice of associated relationships betweem picture and term. Users were able to associate most of the pictures with one or more of the terms used as the design basis for creating the pictures. Highly distributed responses confirmed a strong semantic relationship among terms associated with an icon and this was the common pattern (Fig. 12.2). In some cases, pictures and terms had semantically little in common given the clear choice of one term out of the four possibilities. Difficulties arose when the pictures were too detailed to communicate a relatively abstract notion. The results gained from these experiments were used to design, implement, and evaluate the Picture Association Thesaurus for browsing in the BookHouse system.

Experiments at Boundary 2: Effectiveness of Classification

The invariant structure of the classification scheme used in the data base is a static framework that constrains the majority of user–system interactions and operations. In some cases, such an inflexible approach may be inappropriate because information has to be retrieved and displayed in a manner not directly compatible with the static structure. For instance, the user–system interaction will depend on the user's choice of strategy and mental model of the domain, or it can depend on a particular individual's value structure as well as their previously acquired knowledge. Thus a more flexible, individualized situation-dependent model of users' intentions and values can lead to a more flexible interaction. In many instances, the usefulness of information will be user dependent and can also be time dependent. To have models of users' intentions, which could be adaptable and derived for individual users, would be advantageous. A step in this direction was taken in experiments to derive a value

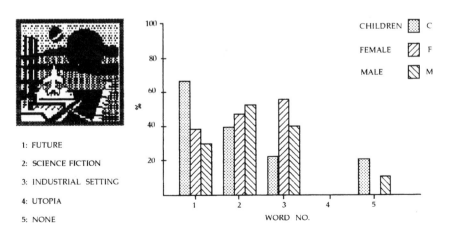

FIGURE 12.2. The figure shows the reaction of different user groups to an icon as identified in a controlled multiple association test.

structure for a fiction data base that could be viewed as an individualized framework within which a user could be aided in their analytical search for information.

Nine subjects performed a search in the data base on a topic of their own choice and at the end of a search they rated the importance of the dimensions and categories of the classification scheme followed by a rating of the features in the most highly rated dimensions and categories. The user was then shown the "model book" to see if it provided a reasonable representation of their need. The system next suggested a set of books, that most closely matched the model book, which the subjects were to evaluate and rate. Additional feature ratings were added to the model book and incorporated into the value calculation for the next suggestions by the system.

Data from the experiment was used to obtain an estimate of the dimensions and categories of the classification scheme (Fig. 10.1), which were most important to the retrievers. Correlations among user ratings of dimensions, categories, and the product of dimensions and categories were calculated to obtain a general characterization of users' value structures. Without interpreting each correlation individually, it was apparent that some dimensions and categories of the information classification scheme were more sensitive to varying user needs than others. For instance, the data indicated that subject ratings of the Frame dimension were independent of other dimension ratings. Ratings of the three categories within the Frame dimension (time, place, and setting) were, however, highly correlated.

Fine-grained interpretations of the correlations found would of course be influenced by characteristics of the actual subject population. For example, Typography might be an important consideration for elderly readers but not for younger readers. Ratings such as those performed in the fiction data base using a diverse subject population allows the creation of homogeneous user groups as well as stereotypical needs for a particular group. No effort was made to perform this type of analysis since a relatively small number of subjects was used in this experiment. Of more direct importance for this experiment is the simple fact that user ratings along different dimensions and categories are highly correlated. In the future, an understanding of these relationships could be incorporated into the value function and/or be used as the basis for inferring an individual's value structure. In this particular application, the value function allowed subjects to formulate a need and search on a relatively large number of features. However, results indicated that their evaluation of the information retrieved was most likely based on a smaller number of those features or the interrelationships between them (Morehead et al. 1984). Results from these experiments were employed in the formulation of the "search by analogy" facility in the BookHouse.

Experiments at Boundary 3: Performance in Retrieval

At this level of evaluation, efforts were concentrated on testing retrieval including the efficacy of the indexing policy employed (see Chapter 10 for further information on the development of this policy).

In particular, the aim was to study whether:

- the conversion of the early field data analyses to the proposed classification scheme would function as a complete model of the users' problem space for determining the information content in the data base.
- the policies chosen for indexing to solve the most crucial problems for concept identification in books, content and form in records, access terms, vocabulary control, and retrieval functions were efficient to provide the proper information to support users' decision tasks.
- these policies could be implemented in an actual document collection and accessed by well-known manual and computerized types of retrieval systems.

Retrieval Experiments. Early fiction schemes were difficult to use in practice for a number of reasons:

1. The design of new systems, as well as modifications of existing schemes drew heavily upon the designers' personal sets of values and perception of fiction.
2. Genre classes were regarded as true and adequate reflections of users' needs.
3. These systems did not reflect the typically complex nature of fiction requests where the subject matter–genre is only one among several possible and equally relevant criteria.
4. The fundamental incompatibility between the multidimensional character of users' needs and one-dimensional shelving of books was not understood even though systems were intended for shelving as well as for retrieval purposes.

Manual Tools. Three different manual tools were developed containing a test collection of documents varying from 200 to 400 books and, subsequently, three different data bases were developed as test beds. The manual forms were chosen first, since this was familiar to searchers. A systematic catalog was compiled on cards, an alphabetical index in a book format, and a combined systematic–alphabetic tool in a book format. The purpose was to investigate the potential need for a more specific approach to fiction similar to indexing of fact literature. A number of fact systems were examined for their suitability in governing specific concept identification, vocabulary control, and a standard citation order. Most of them involved problems, simply because they had been developed for nonfiction. These tools were tested under laboratory conditions by librarians, students, and heterogeneous groups of users from the general population in a number of different experiments focusing on a comparison of tools that adopted indexing policies similar to fact literature with tools based on special rules devised for fiction. Users' preferences and understanding of the classification scheme, the phrases and sentences, the access terms in the thesaurus, and so on, gave important information for the indexing of the 3500 books in the BookHouse data base. (Pejtersen and Austin, 1984).

Computerized Tools. Cards or printed catalogs can be efficient for small collections, but providing access to an entire library's fiction stock calls for the use of computerized facilities. Information cannot be retrieved, combined, and displayed with manual tools in a sufficiently flexible manner so as to support the various retrieval strategies users employ with their accompanying needs for suitable dialogue and display facilities. The possibilities for implementing different strategies in on-line fiction retrieval and the evaluation of their match with users' mental models underlying these strategies were investigated in a computer based environment.

The strategies chosen for experimentation were the analytical strategy, the search by analogy, and the browsing strategy (see Chapter 9). The theory behind the choice of these strategies is the distinction between, on the one hand, problem solving based on the tacit dimensions of human knowledge (i.e., the skill of knowing and acting intuitively without being able to realize and formulate explicitly) and, on the other hand, problem solving based on a conscious, analytical, and explicit specification of needs. Thus the first relates to supporting the use of browsing strategies, while the latter has more to do with analytical strategies.

Four different data bases were developed in which each record represented a book and contained its bibliographic data, its "classification–annotation," its access points similar to the systematic or the alphabetic catalog and, finally, the specific indexing terms that were assigned to each book. Ordinary free-text searching on all the data in the records, as well as traditional Boolean searching on indexed terms, were possible in the retrieval tests, but the main focus was on the analytical strategy, the search by analogy, and the browsing strategy. In these experiments, users were constrained to use one of the three strategies.

Computer-Aided Analytical Search. The feasibility of a combinatorial retrieval in the adopted classification structure that reflected the user's own categorizations was explored, as was the influence of an on-line environment on the effectiveness of the chosen indexing policy.

Fiction requests often contain direct or indirect clues to the categorical context of the requested features. To restrict the search regarding a desired term to the relevant category might therefore speed up retrieval and increase precision. There are types of requests for which this "systematic" approach might be of advantage. If, for instance, a user wants a novel about war with a focus on a serious, factual analysis, one could restrict the search for the subject term "war" to the category "cognition–information" and avoid the retrieval of novels that deal with war from a nonrelevant viewpoint (e.g., those in the category "emotional experiences" and "action"). A large number of on-line Boolean data base searches using the same kind of experimental setting as that used for testing the manual tools, were conducted employing the analytical strategy. Results were promising and revealed that due to the efficiency of the analytical strategy with its high speed and precision, combined with high indexing specificity, the size of the data bases quickly became too small for testing this approach. This experience influenced the size of the BookHouse data base. Furthermore, it appeared that the data base structure based on the adopted classification scheme could not be fully exploited during analytical searches due to users' unfamiliarity with Boolean logic, which had to be used for combining terms from retrieval dimensions and categories. This task lead to an overload of mental resources and, as a consequence, it was decided to insert the Boolean operators automatically in the BookHouse as users select dimensions and terms.

Search by Analogy. Similarly, the search by analogy was tested in a number of laboratory searches constrained to the use of this strategy. Recorded user–librarian negotiations indicated that a considerable amount of fiction inquiries take the form (directly or implied): I have read book X and want something similar. For the sake of retaining control, the librarian needs to discuss book X with the reader in order to identify specific features that the user valued and found interesting. These features are then expressed using terms from the indexing language and relevant books found in a Boolean search. Alternatively, if the "model" book is already in the system, its details can be projected on the screen, the terms assigned to it can be discussed with the user, and keywords for the subsequent search can then be selected directly. Books retrieved in this way will share some features with the "model" but they will also differ from it and from each other in possessing additional features that enable the user to select some and reject others (or they may suggest ideas for new combinations of features).

The subjects were acting as an intermediary between a user and the experimental data base. They were instructed to retrieve books from the data base similar to a title suggested by the user that could serve as a model together with a description of the features rated highest by the user. This search template was tested in a dialogue with a user about features of a book read by the user and displayed on the screen. This formulation process was performed on-line and was followed by a Boolean search. As with the analytical strategy, the results were promising and showed that searches by analogy often turned out to overlap with the analytical strategy. This lead to the idea of changing the role allocation among system and user. The user identified a model book on the basis of which the system could calculate similarity, retrieve similar books, and display suggestions for the user's consideration. Thus on-line searching was able to evolve beyond the inherent limitations of current approaches to keyword searching in a model book.

Experiments at Boundary 4: Evaluation of BookHouse Functions

The first experiments in laboratory settings, as seen in Figure 12.1, were constrained to separate aspects of the users' interaction with a new system to retain some control over the selective evaluations of particular functions. Later on, a shift to a more unconstrained environment for evaluating retrieval performance had to be taken to check performance and end users' satisfaction with a complete system. However, this had to be done progressively, first by involving library staff and end users in an evolving prototyping process leading to the introduction of the "final" version as a tool in the ordinary daily library routine.

The BookHouse concept involves two significant innovations compared to the existing systems. One is the new classification scheme for fiction, the other is the introduction of the ecological, transparent interface concepts. To distinguish between the effects of these changes then becomes a major question for the BookHouse evaluation. In addition, several librarians were concerned that an iconic interface would lack precision in search and, consequently, experiments with the command-based "BookMachine" in a library setting would resolve this question (see the next section).

Consequently, three different prototypes were developed for the BookHouse data base, each with approximately the same system functionality but with different interfaces: The Command Interface, The BookMachine, and the BookHouse, which reflected differing degrees of similarity with traditional interfaces for data base retrieval. As the name states, the Command Interface supports traditional searching by the direct writing of commands, keywords, and Boolean operators—including commands for search strategies. The BookMachine is a front end system connected to a main frame computer with a menu driven, text based interface with mouse-selectable buttons for choosing options. The BookHouse is a stand alone version with an icon-text based interface. For the purposes of comparison of the three versions, and for the development of the experimental setting to be used in the final BookHouse testing, the experiment was performed as a stepwise introduction and evaluation lasting for almost a year.

Experiments at Boundary 5: Evaluation in Library

The overall goal of this phase, which took place in a public library, was to build and evaluate a user-friendly computer system for assisting users in finding fiction from an information system that (1) could be accessed by the general public and (2) provided the books asked for to the user's satisfaction. This evaluation was carried out at the Hjortespring public library in a town outside of Copenhagen. Apart from evaluation and comparison of three different interfaces, each evaluation sequence was planned to focus heavily on data collection related to individual system features. When evaluating the BookMachine, the focus was on indexing—the classification and representation of books in the data base—as well as the retrieval functions. When evaluating the BookHouse, the focus was primarily on the interface and the navigation possibilities available in the user-system dialogue. The evaluation of the Command Interface aimed at a combination of these two, since it was intended to be an interaction designed for specialists and trained users.

The evaluation cycle for each of the systems consisted of a thorough verification of functionality, a set of interviews before the system was installed, system installation and training, detailed interviews, observations and logging during and/or just after system use and, finally, later feedback on satisfaction with the retrieved books. In addition, running contact with the librarians was maintained to exchange overall impressions and adjust the interaction with users when necessary.

Field Evaluation of the BookMachine

Thus, before the BookMachine (BM) text version was released for public testing, it was subjected to an exhaustive functional verification to ascertain that the system could meet the functionality requirements that had been imposed (i.e., stated concisely, that the design was right in that the results met the specifications). Both system designers and librarians at Hjortespring participated—the latter did so as part of their training in the system. The subsequent evaluation using the general public was thus an attempt to validate whether or not the BM, and later the BookHouse (BH), systems actually were the right designs for supporting library users in finding good fiction.

Before the BM system was installed, a questionnaire was prepared and distributed to both the Hjortespring librarians and the public to gain an insight into their experience in connection with finding good fiction. Queries were made about: their reading habits, the problems they had, the types of help that they utilized in order to find books, their thoughts on better aids for assisting them in searching for literature, and their experience with computers. In addition, a group interview with the Hjortespring staff was held regarding the intentions behind the project, as well as the basis for the system and its functionality.

When the BM had been installed and the appropriate connections to the data base in the host computer were established, the system was opened to the public. There were three terminals in the library—one in the children's section, one in the adults' section and one near the librarians' desk. All three were available for public use. Each consisted of a computer with a color display, keyboard, mouse and printer.

Several methods were used to evaluate users' use and response to the BM: (a) a questionnaire, (b) on-line logging of all dialogue events (mouse clicks, etc.), (c) observation, and (d) interviewing by the librarians who (e) also kept a logbook with reports of user responses, system behavior, and so on. It should be noted that users of the terminal had the opportunity of bypassing the BM and later the BH facilities so that they could write their search terms in the conventional way using the standard command language of the Command Interface. Thus there were two questionnaires: One for those who preferred the command mode and one for the BM users.

Some redundancy was achieved by having both questionnaires and logging. The use of multiple methods offered some possibilities for checking what users did against what they said they did. In a realistic work situation, it is difficult to keep complete check of every detail in users' responses and their use of various data collection tools. It was decided that use of several, different methods would provide more reliable results—especially, if every method gave the same result. This actually happened in the case of the BH.

In order to test user satisfaction with the books that the BM had helped to find, a separate questionnaire was enclosed with each book by the library clerk who requested that the user fill it out after reading the book and then return it to the library together with the book. This questionnaire asked about: user satisfaction with the book, whether the book corresponded to the subject that was sought, whether the book content corresponded to the description displayed on the display screen, and which search strategies the user had used to find the book. It should be mentioned that librarians maintained contact with users in order to give assistance and to observe and interview users while they searched for books; they also kept a logbook.

The experience and result of this evaluation work was used in the design and planning of the evaluation of the final BH system.

Field Evaluation of the BookHouse

The BM evaluation cycle was repeated with the BH. However, two changes were introduced. Experience indicated that a more concentrated contact with users would be beneficial in order to get more qualitative opinions on the effectivity and in particular, the acceptability of the system, especially from adults—and to ensure a consistent communication between the individual librarians with users that later could be systematized. In this connection, an observation form was prepared that the librarians could use to guide their conversations with users and their introduction in use of the BH, if desired by users.

In addition, querying users about the BH displays and icons was enhanced by programming an on-line questionnaire that appeared automatically on the screen when the user ended their session with the system. The questionnaire adapted to the individual user's navigation trajectory and displayed those icons that the user had met and employed during a search. It also contained questions about the understandability of the Picture Association Thesaurus with focus on the usefulness of this thesaurus in need recognition and searching. The questionnaire displayed one example from the Picture Association Thesaurus to all users independently of the user's individual search path. BookHouse users were asked to evaluate the associative relationship between the message of the icon and the description of the book content, which was the result of a selection of this icon. The purpose was to evaluate this aid not only as a tool in need formulation, but also as a means for retrieving relevant documents, which after all was the final purpose of a search. Answers to the 25 displayed questions were mouse clicked, and thus were included in the automatic on-line logging. Data gathering concluded with a group interview with the Hjortespring librarians on their overall evaluation of the system.

In connection with the evaluation, considerable effort was devoted to publicizing and demonstrating the systems to the public, to librarians, and to colleagues in Denmark and elsewhere—in order to get a broader feedback for discussions of possible generalizations of the system design to other non fictional domains. Word of mouth channels also helped to attract users to participate. A brochure for distribution in the library was prepared for each prototype as an introduction to using the system.

The evaluation effort indicated an overwhelmingly positive response to and acceptance of the systems by both children and adults. The BH was found to be an especially helpful and pleasurable aid for finding good fiction. The cognitive engineering approach to uncovering the basic information processing underlying reader–librarian negotiations gave the necessary foundation for specifying system functionality and its pictorial representation and for carrying out the subsequent evaluation.

Evaluation Results. The BH was evaluated at a public library in the town of Hjortespring over a 6-month period. The experimental subjects were the normal everyday users of the library and consisted of both children (from 6 to 16 years) and adults (from 17 to over 70 years). The evaluation program was carefully planned and consisted of both an analytical and an empirical phase. The analytical portion concentrated on the functionality of the system and could build on the earlier studies of users' search behavior, which was uncovered in field studies, as well as in numerous design experiments. The empirical validation was then necessary (e.g., to see how—whether users would accept this functionality and the iconic, direct manipulation metaphor based interface as supportive of their efforts to retrieve relevant fiction).

Evaluation Goals and Measures. The goals of the evaluation are listed in Figure 12.3 under five headings. The dependent variables are expressed as "the degree to which" various design and/or system usage subgoals were or were not realized—and their degree of success and effect on work. First, what was being evaluated was a prototypical implementation of a design concept for a system aimed at ultimately giving the general public a useful product for supporting its need for retrieving relevant information (in this case fiction). Second, this evaluation with unconstrained

Book House Evaluation Goals
1. Evaluation of the Readability of Interface
Preference for "writing" search profiles vs. "selecting" search words/pictures with a mouse.
The associated text descriptors support of the first-time users and ease users' acceptance and ultimate recognition of icon meanings.
The success of the iconic approach when used in combination with text.
2. Evaluation of the Understandability of Interface
The chosen metaphor for the BOOKHOUSE is understood and accepted.
Users' possible difficulties in understanding the systems.
The utilization of the provided "help" facilities.
3. Evaluation of the Usability of the System
The correspondence of the dimensions in the classification scheme to users' needs. Their relevance as expressions of need as well as categories in the description of books.
The level of specificity and exhaustivity of keywords.
Relevance and understandability of indexers' representation of content.
Users' and indexers' agreement on keywords and book contents.
Users' utilization and preference of the provided search facilities and search strategies.
The support of icons for book searches, description of book contents and as a catalyst for finding interesting and relevant literature.
4. Evaluation of Users' Acceptability of the System
Users like the system and consider it pleasurable to use.
Preferences for "writing" search profiles vs. "selecting" search words/pictures with a mouse.
A text-only retrieval system (BOOKMACHINE) is preferred to a picture and text retrieval system (BOOKHOUSE) or vice-versa (including comparative details).
Use of the two systems and opinions in relation to age (children and adults) and sex.
5. Evaluation of Its Impact on Work Context
Influence on user-librarian interaction
Increase or decrease of work activity
Influence on book loans

FIGURE 12.3. Goals of the evaluation. The dependent variables are expressed as the degree to which various design and/or system usage subgoals were or were not utilized, and their degree of success and effect on work. The five evaluation goals correspond to the evaluation issues described in Chapter 8 (Figs. 8.1 and 8.2).

conditions was necessary to confirm or disprove the experimental data and analysis results from earlier evaluations of separate parts of system design and constrained aspects of the user–system interaction as described in Figure 12.1.

The goals of the BH system are reflected almost directly in the items for evaluation given on Figure 12.1. Of particular importance is the invariant data base structure needed to represent users' information needs in the work space, the system functionality needed to allow navigation among a repertoire of search strategies, and the content and form of the iconic displays needed to ease understanding of the system.

The work situation to be studied was the equivalent of a typical user–librarian negotiation where the former attempts to verbally express a need for a suitable book to the librarian who then uses their expertise to match the request as formulated with one or more appropriate selections from the book stacks. With the BookHouse (and its text-based predecessor, the BookMachine), the user sits down at the terminal and communicates their need directly to the computerized intermediary, which then searches through the data base and fetches and displays possible book candidates for the user to consider. However, the final task of finding the desired books on the shelves remains unchanged as are other functions, such as reservations, or return.

The evaluation was planned to involve as many of Hjortespring's users as would be willing to participate within the available resource and time frames. All information was to be anonymous, and there were no possibilities for following identifiable individual users over a period of time. The important problem with the effects of intermediate variables on user performance was solved by ignoring it or at least by treating it in a greatly simplified form. Users were classified only by sex and age, frequency of use, and whether they were librarians or not. The resulting raw data were first sorted and summarized within these categories. It must be remembered that the BookMachine and the BookHouse are prototypical systems for the general public so that any fine-grain distinctions in user characteristics would not be a primary ingredient in an evaluation of the type performed. A brief summary of results was based on 7100 data logged, 1030 on-line questionnaires, 220 questionnaires, 75 log book observations, 75 deep, structured interviews, and two group interviews, with librarians and with a group of teenage users.

Boundary 1 Issues: Readability of Interface

Size of text and icons were deemed appropriate and easy to read and comprehend, but for some younger children the length and complexity of book descriptions were not appropriate. Most users preferred the "direct manipulation" interaction with icons and text provided by the BookHouse to writing their search requirements in a command language, or the text menus of the BookMachine. But some elderly users had problems with control of the mouse clicking.

Boundary 2 Issues: Understandability of Interface

Users and librarians found the BH system easy to understand and use. Very few had problems in navigation and making analogies to getting books retrieved. All the BH metaphor displays were deemed to be easily understood, except for the search strategy display. The analogy with users' search activities was not easily interpreted by all users. In the same way, command icons were perceived in the intended fashion except for the search by analogy icon, which failed in both form and content. The iconic representation of the classification system worked according to the designers' intentions and was deemed as very helpful. Most users preferred the combination of icons with a help text line to support users' choices of action.

Boundary 3 Issues: Usability of the System

The functionality of the *classification system* was validated; users appreciated the invariant domain structure with the various ways of choosing among search terms. In general, the many alternatives together with the optional possibility for combinations from these alternatives enhanced the usefulness of the classification system. The BookHouse experiment has demonstrated that a highly structured and selective access to content keywords divided into 13 dimensions or facets helps the user to formulate his/her need more precisely and, therefore, leads to a better search result. The different dimensions and their related keywords were judged to be in good agreement, and the dimensions were judged to be relevant to users. The classification system was the most popular aspect of the system. All dimensions of the classification scheme were used, but with varying frequency—as was the case in the investigation of user-librarian negotiations.

The indexing of 3500 books resulted in about 6000 different keywords. The users' use of keywords was widely spread and included very specific, as well as broader terms. There was no significant peaked choice of keywords. The indexing principles and rules used for the concept analysis of documents effectively matched the users' perception of the contents of the books.

In the same way, evaluation of *search strategies* showed that all four strategies were used by both children and adults. The user is offered a choice of four different routes to explore the contents of the data base. This choice made the user feel that the system was flexible. The option of choice and shift between different strategies made help texts superfluous. When bogged down in a search, users found it easier to shift to another strategy than to consult the help texts. Shift of strategy replaced help texts. The analytical search strategy was the most popular of the BH strategies. In comparison, in our investigation of manual searches, the analytical search was the least popular and the most difficult to conduct. Thus, an important finding of this evaluation is that users like to be able to perform analytical subject searches if the system gives the user enough support. This apparently happened in the case of the BookHouse—where the invariant data base structure for analytical searches was displayed as icons. This facility helped the users recognize facets of their need instead of forcing them to recall them by introspection and reformulation in order to match verbal formulations to the data base contents.

Other changes were seen in preference of strategies. Browsing through books was the least popular strategy in the BookHouse and the search by analogy was the second least popular—in contrast to manual searches where these strategies were among the most frequently used. Browsing through pictures was the second most popular, but there is no basis for a comparison with traditional searches. In manual searches, the use of pictures on the cover of the book is close to the picture browsing option—but this use was not investigated in detail. The alternative search technique, picture browsing, gave unexpected (and gratifying) results. The results confirmed that users were able to make a direct mapping from a single selected example from the Picture Association Thesaurus to a fine grained book description.

However, the browsing pictures strategy in the BookHouse was actually used for a variety of purposes: for browsing and getting new ideas for reading, which was the original reason for including this facility, but also for analytical subject searches and for bibliographical searches (i.e., a search for a specific author or title). In the same way, the search by analogy was also used to find a specific author–title. The conclusion is that users do not distinguish formally between strategies but perceive them as alternative options or routes for any type of information need. Each strategy option was designed to be the optimal way of satisfying different needs, but users' subjective preferences for ways of searching greatly influenced their choice of strategy. Users were inventive and used the flexibility of the system to create individual shortcuts to satisfy their needs.

Boundary 4 Issues: Acceptability of the System

Ninety five percent of the public library users, who participated in the evaluation, and all librarians, liked working with the system and preferred this to any other system, including those traditionally used for on-line retrieval of fact literature. It was accepted in the information retrieval task situation, and it was considered to be a useful tool for the daily work with retrieval of fiction and a good alternative to browsing the shelves.

A group of teenage users were followed for 5 days over a 5-week period. The project leader watched their way of using the system while progressing from novice to expert. Their development of search methods, their preferences, and understanding of the system were investigated. They preferred the system to browsing the shelves— their usual way of searching. They were a bit reluctant to use the browsing through pictures option but, at the end of the period, when they were skilled in "revise search" tactics, it became one of their favorite search methods—both for browsing, subject searches, author searches, and so on. Teenage users related that they learned to understand books better from the description of the books in the data base. The classification system widened their perspective on the books, helped them to understand the books, and gave them new ideas for making queries and for searching for new topics. The librarians related the same experience of seeing new book aspects and new ways of reading–using books.

Boundary 5 Issues: Impact on Work Context

Librarians preferred the BH for mainly two reasons. First, once the dialogue was learned, the librarian had developed sensory motor skills for searching the system—

with very little demands on mental skills, while manipulating the system. For expert users, all icons will, when learned, be perceived as signs (i.e., cues for action). Hence, resources were left for a more thorough dialogue with the user, for analyzing the user's need while displaying keywords and book descriptions on the screen. Their contact with users was improved, the book descriptions in the system worked as an "intermediary" between users and librarians, and turned out to be a powerful means for communication and exchange of reading experiences.

Second, unsuccessful searches did not occur. The answer "zero books" did not occur as in command searches after (possibly incorrectly) writing a term. All selectable keywords were displayed on the screen and at least one book was available when a keyword was selected. The usual error with writing a nonexistent term that was not in the data base or was represented by a synonym was avoided, and misspelling errors never occurred. The only kind of unsuccessful search was due to the unavailability of a term–book on the screen—and the librarian was not responsible for the contents of the data base.

Other than the normally most popular books began to be borrowed. Before the BookHouse, when shelf browsing was necessary, these books were not borrowed "because they looked boring." Now that the cover of books no longer had the same significance in relevance judgments, the book stock was used more efficiently and in accordance with the librarians' acquisition criteria. This has economic implications for a library's costs for book acquisition. More reservations of novels were made after the BookHouse was introduced, and more users searched subjects themselves. There was a general increase of activity in subject searches of fiction. Librarians arranged successful book exhibitions based on lists of keywords extracted from the data base. It is relevant to point out that, in common with nonfiction areas, it is impossible for librarians to be completely acquainted with the full scope of literature in fiction that eventually could match users' needs. The many thousands of books, which either are not read by librarians or which, from their external appearance, do not appeal to readers, suddenly become as equally relevant as all the "best sellers." This leads to a better utilization of the collection, as well as an introduction of readers to a broader spectrum of writers and their viewpoints, opinions, facts, fantasy, and so on. The ultimate institutional goal for public libraries is to promote education and cultural values. The BookHouse enhances this goal.

In the long run of course, book reservations and other functions should be incorporated in an integrated search and delivery system in order to achieve an optimal automated tool. Indeed, users suggested that fact literature should be incorporated in the BookHouse as well. They saw no difference between fact and fiction retrieval in this kind of system. As a result, the latest version of the BookHouse system has been further developed to also serve retrieval of fact literature.

Chapter 13

Catalog of Annotated Displays

INTRODUCTION

A complete work analysis covering all aspects of the framework discussed in the previous chapters is a time and resource consuming undertaking and will not be possible for the design of every new information system. Instead, it will be necessary to transfer the results of analyses from one work domain to the design of systems for other domains and/or to adopt solutions from previous designs and from other designers' successful results. Thus it becomes necessary to generalize and to transfer solutions from one case to another while adjusting them according to the differences in domain characteristics. We find that the framework presented in this book makes explicit some of the basic dimensions that must be consciously considered for such a transfer and modification of solutions.

We also find it important for a systems designer to have been through a comprehensive work analysis at least once in order to get a full appreciation for the complex relationships between features of a work situation, the characteristics of the various strategies people invent for their tasks, and the criteria they use to choose among the various alternatives. It is our experience that having this background makes it possible to adapt earlier experience to a new domain by means of explorative surveys and selective studies in the new domain.

WTU-SAMPLE CATALOGUE

We have also argued that design to a large degree is guided by the designers' intuition and tacit knowledge. In Chapter 7, we mentioned how the publication of highly illustrated catalogs of elementary functional design solutions influenced the industrial developments of the past. Correspondingly, we suggest that designers collect their own catalogs of successful design solutions within their sphere of interest, organized with reference to their location in the conceptual maps of the design territories discussed in Chapter 7. Following the conclusions of the discussion in Chapter 7, such catalogs should organize exemplars according to their Work domain/Task situation/User characteristics. Hence the term "WTU-sample catalog."

In the course of our interaction with workers, operators, and display designers, we have often felt the need for a condensed format to capture and to communicate in a comprehensible way the many different aspects of the work domain and the work situation for which a particular design is intended. As illustrated by Figure 1.7, missing just one dimension in the complex human–work coupling will likely turn an otherwise promising design into a failure. An example may serve to clarify this point. Diagnosis during disturbances is a critical task in process plants. Since operators often identify the plant state by immediate recognition, it has been suggested on more than one occasion to design displays aimed at easy pattern recognition, such as Coekin's polygon–polar diagram (Fig. 13.2) or Chernoff's faces (Fig. 13.16). Experiments have shown that the displays are well suited for the recognition of familiar patterns and, in particular, to recognize when a pattern is no longer "normal" or "as usual." As a result, the polygon has been used—for example, to monitor a limited set of critical safety parameters in nuclear power plants. However, when talking to operators after these displays are installed and operating, their response is likely to be "a nice display, but we do not actually use it." The reason seems to be that although the polygon display supports recognition of familiar disturbances, it does not support a verification of the diagnosis because it is functionally context-free. Therefore, other information sources are required that can confirm and further substantiate the recognition process. To sum up, the polygon only supports a very specialized cognitive process and its acceptance depends very much on the need for aiding this process separately.

An annotated catalog can direct attention to the various attributes of the situation for which a design should be matched. Whenever an interesting interface design is found, a "prototypical" WTU sample should be formulated by explicitly annotating the defining attributes in terms of **W**ork domain, **T**ask situation, and **U**ser characteristics. In this way, we suggest, the maps of the design territories in which a particular designer operates will be built up of sets of familiar, well-defined landmarks.

When the relationships among different WTU sets are made evident the designers can get an effective intuition about the design territory in the same way as a skilled traveler develops a prototypical map of world capitals and their characteristic localities. Thus the hypothesis underlying the approach is that by giving a designer such a conceptual map, they will obtain an improved awareness of important similarities and differences between information system designs relevant for different WTU combinations and, at the same time, assimilate a working understanding of the influences of all of the dimensions of the framework—both system and human related.

Each prototype in the catalog is intended to be a prime example of a class defined by a WTU-triplet (see Figs. 13.1–13.17). Classes are defined in a multidimensional WTU space describing the properties of the work domain, the task situation, and the competency of the user according to the framework described in the previous chapters. Thus an interface will be characterized by its location in this map because it will be most suitable for certain types of tasks in certain domains for certain user profiles. One possible application would be to abstract from a specific interface design and induce the class represented (i.e., define the boundary conditions of its applicability in terms of a prototypical WTU triplet). Identification of the general class, of which the prototype is an instance, thereby would allow one to generalize design results across work situations which, on the surface at least, might seem to be very different from each other.

Therefore, the first goal in making this design approach realizable is to provide a number of varied examples that can serve as concrete instances to get designers thinking about the conceptual issues that need to be considered in the design of ecological information systems. Hopefully, these can serve as an incitement to

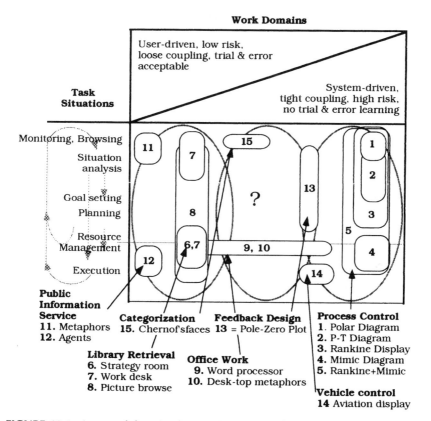

FIGURE 13.1. A map of domain characteristics versus decision task that identifites and characterizes a set of three prototypical WTUs and associated display candidates.

designers to investigate the potential benefits of organizing their own bank of prototypes. The examples of WTU sets are drawn from field studies in process control, manufacturing, hospitals and libraries, and the design of interfaces for libraries and control rooms. Displays based on ecological principles are only available for process control, aircraft piloting, and libraries. The Friend'21 public information service system (Friend'21, 1989) is included to illustrate the use of purely metaphorical systems.

A WHITE SPOT ON THE MAP?

In Figure 13.1, examples across the domain map have been found for intuitive browsing, recognition and monitoring, and for resource management–execution, but not for the functional situation analysis and planning functions. Displays for the monitoring of quantitative information are more or less context-free; the context is attached by labeling the axes and variables and using appropriate scaling. Similarly, browsing aids are in principle neutral until the semantics for the domain are added. As for resource management–execution, there exist an impressive set of general tools with applications in a wide variety of domains; these are often made accessible through some kind of desk-top metaphor.

However, examples for supporting situation analysis and planning will be related to the functionality and intentionality of the particular work system and "ecological" interfaces have only been developed for well-structured process systems and for well-documented task situations, such as the information retrieval task in libraries. We have not been able to find "ecological" displays for analysis and evaluation of work system states for autonomous system users in a constrained environment in the upper, middle part of the map in Figure 13.1, that is, for work domains in which the source of regularity is related to intentional structures, such as laws, regulations, and company strategies. The reason for this situation is that symbolic representations of intentional structures have not as yet been developed. Activities, such as production flow control in just-in-time production systems, patient treatment planning in hospitals, or case handling in public service institutions, are handled by autonomous actors within the constraints posed by regulations, by the intended product–case flow, and by the intentions of colleagues handling the state of affairs up- and down-stream the of flow of work items. For design of ecological interfaces, we need to formalize a description of the different forms of intentional structures and to find suitable symbolic representations. This is an area for further development.*

---

*In the last minute, Kim Vicente has kindly directed our attention to an approach to information system design for this white spot in the domain map, that is actually very similar to our ecological approach. See Smith, 1992. See Fig. 13.7.

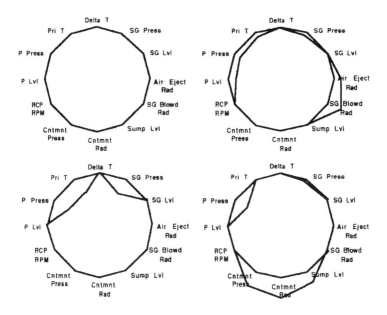

Display Example No. 1

Content: Mutual Relationship within a set of normalized magnitudes of n variables representing the intentional, normal state of system.

Form: Scalar: regular polygon to facilitate immediate perception of changes, based on human perception characteristics, cultural conception of figural goodness.

WTU Characteristics:

Work domain:
Tightly coupled "parts" each having one or a few critical variables reflecting state, performance, etc. Useful mostly for information at the general functional level in the means-ends structure.

Task situation:
Process monitoring: surveillance of need to act. During monitoring the state of affairs, an actor will easily, by fringe consciousness detect the need for intervention. To the trained user, the shape of the distortion will tend to indicate the identity of the abnormal variable(s) and, thus, where to look for further information for diagnosis.

User profile:
For detection of change: no requirements. For perception of clues for further search: Dedicated operator with specialized training

Comments

This display correlates visual patterns with overall system performance by extending from a common geometric center the magnitudes of a set of variables each of which strongly characterizes the functioning of a given system entity. The scaling of the variables is such that "normal" results in a "circle". The polar display is best suited to domains where the coupling or interaction between entities is high so that different system events, faults, diseases, etc. give rise to distinctive, discriminative and predictable deviations from roundness

FIGURE 13.2. A polar display of the operational state of a tightly coupled system for rapid detection of disturbances; suggested by Coekin (1969).

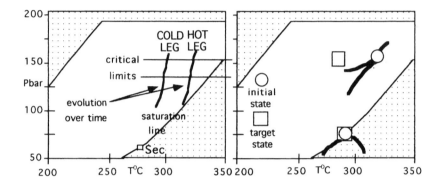

Display Example No. 2

Content: Relationships among physical variables representing actual operation together with boundaries of allowed operation and an indication of a recommended band of operational values.

Form: Diagrammatic, scalar representation of variables and boundaries as defined during system design; based on thermodynamic relations; related to family of two-dimensional contour and line charts.

WTU Characteristics

Work domain: Technical process system governed by physical laws. Useful mostly for information at the general functional level in the means-ends structure.

Task situation: Monitoring and control of physical process; Dynamic displays of e.g., primary-secondary heat transfer relationships over time.

User profile: Dedicated, professional process operators with specialized education and training.

Comments

Object, system, event behavior is predictable and can be described by a relationship over time between two critical variables. This relationship has associated with it certain boundaries, forbidden ranges and/or contours which define levels or limits to the possible and/or allowable values of the variables.

FIGURE 13.3. Symbolic display for monitoring normal operation with reference to the intended normal operation of a technical system and to limiting values.

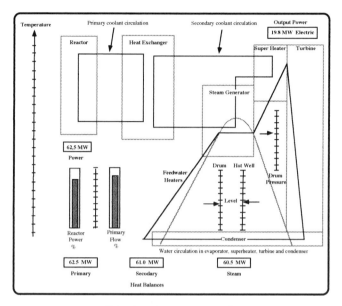

Display Example No. 3

Content: Representation of the relationships among state variables of a physical process and the boundary conditions of acceptable relationships. For critical units, measured variables are shown with reference to intended operational states and limits of acceptable states.

Form: Composite configural pattern. Functional structure of complete industrial plant with representation of major functional units. Overall pattern reflects a particular system anatomy. Configural elements are represented by engineering diagrams and diagrammatic representations of variables and boundaries, based on professional stereotypes for representing physical laws.

WTU Characteristics

Work domain: Technical process system governed by physical laws. In its basic form most useful for information at the general functional level in the means-ends structure. Extra overlays can be used for additional physical functional information.

Task situation: Situation assessment (detection, diagnosis, hazard priority judgment in operation of physical process. Identification of control task during disturbance. Selection of tools for action will require shift to other display.

User profile: Dedicated operators with specialized education and training.

Comments

Experts can develop direct manipulation skills, and know-how rules based on the "expression on the face of the display" at the same time as novices or experts in unfamiliar situation can use the displayed representation as an externalized mental model for reasoning.

FIGURE 13.4. Symbolic display for monitoring normal operation of an entire nuclear power plant (see Figs. 7.14–7.17 for details). Based on display proposed by Lindsay and Staffon (1988) by permission.

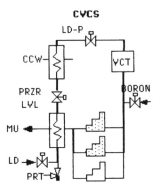

Display Example No. 4

Content: Representation of the physical anatomy of a technical system including objects (means for actions) and their operational availability based on system design, configuration, and running measurements of configuration and availability.

Form: Diagram comprising a physical analog (mimic+ alphanumeric text) based on professional & other established stereotypical symbols from text books, engineering blueprints, etc.

WTU Characteristics

Work domain: Typically tightly coupled system based on technical components which are connected in different functional configurations. Can be used for any system which lends itself to a decomposition into elements represented by an accepted symbology.

Task situation: Resource management, monitoring, planning, and changing the operational configuration for start-up and shut-down of the system and in response to component failures. Most useful for operation at the physical functional level.

User profile: Dedicated operator with specialized education and training.

Comments

Within the limitations of the two-dimensional display space, pictorial analogs of the components are connected together as in the actual system or object. Textural and line width coding (plus color) can denote component status, (in)active paths, etc. Text information can supply performance data. However, in most cases, this display has to be combined with appropriate dialogue facilities and/or displays in order to gain access to more detailed operational status and control information, coordination guidance, etc. regarding the various components on the diagram.

FIGURE 13.5. "Mimic diagram" representing the physical configuration of a technical process plant and the operational state of the components (pumps, valves, tanks, etc.).

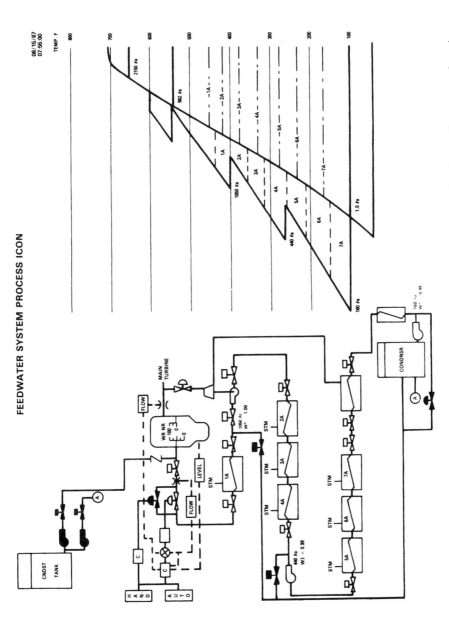

FIGURE 13.6. Symbolic display for operation of the feed-water system of a power plant. The display combines a configural representation of the measure process variables with reference to the intended operation (Rankine symbols) with a mimic diagram for resource management. For annotation see Figures 13.4 and 13.5. The display is designed by Bettracchi, reproduced by permission.

Display Example No. 6

Content: Representation of the set of alternative tools for supporting various strategies for carrying out a particular task.

Form: Iconic matrix with metaphorical representation of alternatives: Room metaphor with pictorial elements referring to situations occurring when the strategies are used in a traditional setting. Based on population stereotypes; user population sampling and testing.

WTU Characteristics

Work domain: Work system offering a limited number of support alternatives.

Task situation: Planning or selection of actions or activities, selection among resources available.

User profile: Autonomous, casual users of varying age, education, and cultural background.

Comments

Pictorial representations of a surrogate for the user mulling over a choice among alternatives. The cartoon-like situations reflect the sense/purpose of each alternate strategy. Any strategy can be selected by pointing the mouse at the person located at the desired alternative. The format is particularly useful when branches in the dialogue are classifiable and visually mappable - in this case the branches correspond to alternate strategies for searching in a database for fiction.

FIGURE 13.7. Metaphoric display for choice of a search strategy in a library information retrieval system. For more details see Figure 11.6. The display is a metaphor with a functional analogy to users' activities in libraries.

Display Example No. 7

Content: Iconic matrix representing the dimensions of the space of attributes characterizing multi-dimensional items in a collection or store. Based on an analysis of the aspects of descriptions relevant for the intended task situation and user group.

Form: Work room metaphor with pictorial representation of objects illustrating the concepts defining the dimensions of the search space; based on population stereotypes identified from user population sampling and testing..

WTU Characteristics

Work domain: A large collection of similar, multidimensional items; books, pictures in museums and libraries, warehouses.

Task situation: Analysis and specification of need for information, formulation of query questions for retrieval of items from large collections; control of search.

User profile: Autonomous, casual and professional users of varying educational and cultural background.

Comments

Symbolic representations based on population sampling, cultural aspects, fantasy. Each icon affords selection as an integral part of the user-system dialogue. Useful when domain attributes or characteristics are classifiable and visually mappable - in this case a database for fiction.

FIGURE 13.8. Metaphoic, symbolic display for formulating reading needs and specifying search terms in a library information retrieval system. It supports the user's analytical situation analysis and resoning, domain exploration, and planning and execution of a search as well. Most icons are metaphors referring to object to be used with analogous functions; others are symbols. For more details, see Figure 11.8.

Display Example No. 8

Content: Representation of population stereotypical semantic network based on the set of concepts, events and objects necessary to map the domain of knowledge to be covered. Representation of the prototypical features of elements in a large collection.

Form: Iconic matrix; pictorial representation of objects or concepts representative for the categories within this set based on user population sampling and testing; based on analysis of the intuitive classification by the target cultural groups.

WTU Characteristics

Work domain: User-paced information retrieval domain; high number of items and relationships. Exhaustive identification and description of elements difficult. Useful for large collections of similar elements in e.g., libraries, museums, warehouses, etc.

Task situation: Intuitive browsing; exploration for potentially useful sources of information matching badly specified needs.

User profile: Casual, typically autonomous user belonging to a defined group (defined by age, profession, nationality. etc.) sharing cultural prototypes.

Comments

Symbolic representations based on population sampling, cultural aspects, fantasy. Each icon affords selection as integral part of the user-system dialogue. Domain attributes or characteristics are classifiable and visually mappable - in this case a database for fiction.

FIGURE 13.9. Symbolic display for intuitive browsing of reading needs in a library retrieval system. Intended to probe the user's tacit knowledge and support intuitive recognition of items of relevance to the user's current need situations. For details see Figure 11.9.

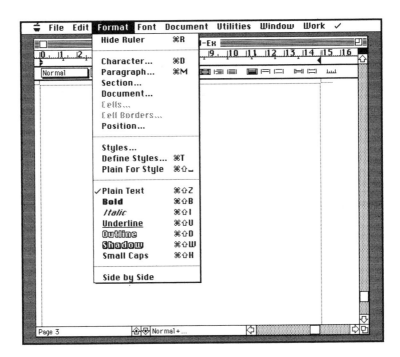

Display Example No. 9

Content: Repertoire of tools for text typing and editing
Form: Desk top metaphors; see Figure 13.11

WTU Characteristics

Work domain: Tool for specialized work process found in a wide variety of work domains.
Task situation: Text typing and editing.
User profile: Autonomous user.

FIGURE 13.10. Desk-top metaphor interface for a text editing system. From Microsoft™ Word™.

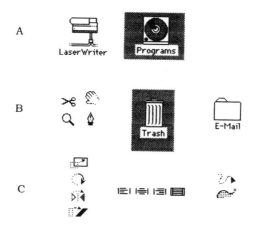

Display Example No. 10

Content: Representation of tools and work items

Form: Metaphors. pictorial icons referring to the appearance of a tool (row A). to the appearance of similar tools in the old technology (row B) or to the effect of the tools (row C).

WTU Characteristics

Work domain: A set of separate tools and work items.

Task situation: Manual operation of separate tools. selection of familiar work items and tools.

User profile: Users familiar with tools and work items from other technology.

FIGURE 13.11. Desk-top metaphors for office work by computers.

Display Example No. 11

Content: Display surface and agent to ask for information.

Form: TV news-reel metaphor.

WTU Characteristics

Work domain: Large sets of databases and other information sources of heterogeneous contents, application and sources.

Task situation: Search for information for a need in a work situation which is unknown at the time of system design.

User profile: Autonomous, casual users of a wide range of ages, educational and cultural backgrounds.

Comments

The display is an example of a larger set of displays based on various metaphors for databases of different contents and formats, such as TV-news-reels, newspapers, albums, catalogs, etc.

FIGURE 13.12. Interface based on the multiple metaphor–multiple agent concept of the Friend'21 project. The figure illustrates the TV-news reel metaphor. The displays in Figures 13.12 and 13.13 are reproduced with permission of the MITI Institute of Personalized Information Environment, Tokyo.

Display Example No. 12

Content: Metaphor for information source and agent to ask for information.
Form: Card-catalogue metaphor.

WTU Characteristics

Work domain: Large sets of data bases and other information sources of heterogeneous contents, application and sources.

Task situation: Search for information for a need in a work situation which is unknown at the time of system design.

User profile: Autonomous, casual users of a wide range of age, education and cultural background.

Comments

The display is an example of a larger set of displays based on various agents that are specialized for various search, selection, retrieval and storage functions on behalf of the user.

FIGURE 13.13. Interface based on the multiple metaphor–multiple agent concept of the Friend'21 project. The figure illustrates the multiple agent concept: A user has asked an agent to search through a data base.

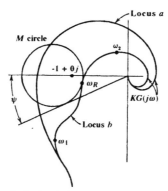

Reshaping locus to provide stability

Display Example No. 13

Content: The location in a complex plane of the roots and zeros of the closed-loop equations of a control system.

Form: Diagrammatic representation of variables.

WTU Characteristics

Work Domain: Closed loop control system design.

Task situation: Analysis of closed-loop stability. Manipulation of the loop gain and location of roots & zeros to ensure system stability.

User profile: Highly specialized professional user.

Comments

The figure is an example of a highly specialized work interface based on scientific notation and diagrammatic traditions.

FIGURE 13.14. A diagrammatic representation of the root and pole loci of the differential equation describing the closed-loop behavior of the control system. For further details see Figure 7.20. Reproduced from Truxal, 1955, with permission by McGraw-Hill Inc.

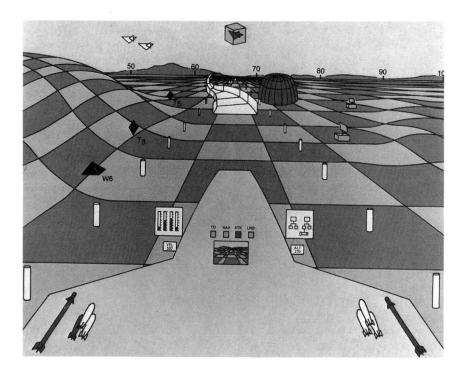

Display Example No. 14

Content: The flight environment of an aircraft with indication of boundaries of safe navigation, identified during design.

Form: Pictorial representation with overlay of constraint boundaries.

WTU Characteristics

Work domain: Piloting a vehicle. In case, an aircraft. Useful also for remote controlled vehicles and robots.

Task situation: Remote manual control of vehicle.

User profile: Dedicated operators with specialized training.

Comments

Aircraft piloting is one of the first domains applying (implicitly and explicitly) displays based on Gibson's direct perception theories. Compare to Gibson's 'field of safe driving' in Figure 1.10.

FIGURE 13.15. A "highway-in-the-sky" display showing the constraints around the flight path as defined by the flight-management system. It shows the recommended path that can be overruled by the pilot. (Courtesy Armstrong Lab, WPAFB, Ohio; reproduced with permission.) For discussion of the "highway-in-the-sky" concept, see Stokes and Wickens, 1988.

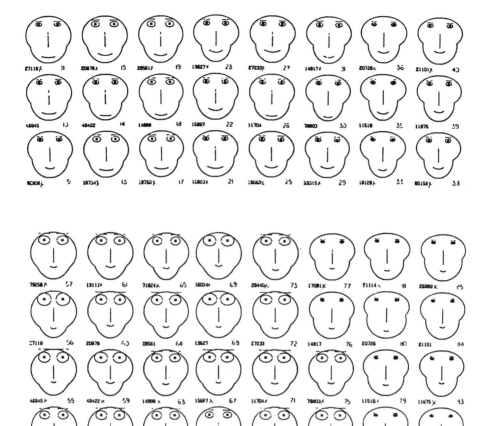

Display Example No. 15

Content: Representation of a set of items with multiple defining attribute.

Form: Pictorial representation of multidimensional category attributes using a human face metaphor.

WTU Characteristics

Work Domain: Large sets or collections of more or less similar items .

Task situation: Categorization and labeling of items in a large collection.

User profile: Autonomous user with specialized training.

Comments

Chernof's face display was proposed for initial categorization and labeling of large set of scientific. e.g., geological samples to draw a practical use from the particular human ability to recognize faces and to interpret face expressions.

FIGURE 13.16. Shows the display based on mapping the multiple attributes of scientific sample items onto various features of a human face. Proposed by Chernof (1971).

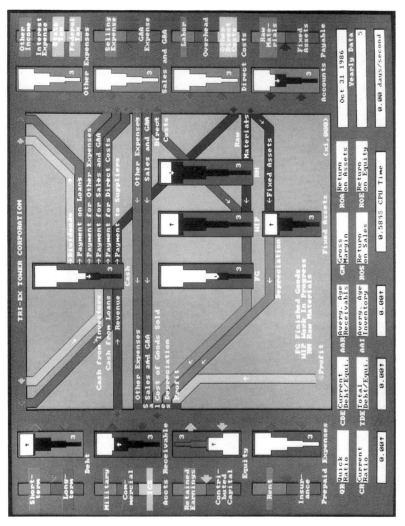

FIGURE 13.17. Last minute example: A symbolic, configural display of the flow of monetary values through a company. The display represents the abstract function and priority measures and is useful at the location of the question mark in the map of figure 13.1. A similar display design has been used for visualization of the flow of energy through a power plant (see Goodstein et al. 1988). The display is described by Smith (1992). © Minard Software, reproduced with permission.

CONCLUSION

A design tool in the form of an annotated catalog can be adopted by the individual designer and customized in several different ways to suit individual needs: It can be implemented as a manual, informal paper-based notebook containing a collection of the designer's associations, new ideas, and ideas from existing, familiar design examples. The catalog can also be implemented as a (company) hypermedia data base with interface examples scanned into the data base with WTU annotations added and keywords assigned for later on-line retrieval of interface examples—for example, from the dimensions of 1) content, 2) form, 3) work domain, 4) task situation, 5) user characteristics, and 6) general comments from the designer. Such comments may include comparisons, relationships, and information about the use of technology and other information useful to represent the particular perspective of the (company) design style and requirements.

A particular issue is to determine the content and form of the textual annotation and the keyword characterization of each display example. Thus, the characterization of the content and the form of a display as well as the WTU characteristics could reflect both the designer's identification of the current use of an interface and the perception of possible use in other similar task domains and user categories. Furthermore, WTU characteristics may refer to the content and form of a total display, and/or the display may be decomposed into elementary units, such as icons, video, text, and so on, which are then characterized within the form-content and the WTU concepts.

The present rudimentary catalog is based on display examples and their annotations as they have emerged during our discussion with designers of similarities and differences of displays for various work domains. As the "white spot" clearly indicates, many more display examples from a variety of successful applications will be required to suggest a useful catalog format and a consistent representation of WTU display annotations that may ultimately serve as a knowledge base to help designers generalize from previous successful examples to a current design task.

References

Addison, F. W., III, and Mack, E. E. (1991): Creating an Environmental Ethics in Corporate America: The Big Stick of Jail Time. South Western Law Journal, 44, 1427–1448.

Agger, S. and Jensen, H. (1989): The BookHouse: Visual Design. Roskilde, Denmark: Risø National Laboratory, Risø-M-2795.

Alexander, C. (1964): Notes on the Synthesis of Form. Cambridge, MA: Harvard University Press.

Alting, L. (1978): Systematic Theory of Manufacturing Environment and Planning B. Vol. 5. pp. 131–156.

Alting, L. (1994): Manufacturing Engineering Processes. New York: Marcel Dekker.

Amalberti, R. and Deblon, F. (1992): Cognitive Modelling of Fighter Aircraft's Process Control. International Journal of Man–Machine Studies; Vol. 36, No. 5 (May), pp. 639–671.

Aoki, M. (1988): A New Paradigm of Work Organization: The Japanese Experience. World Institute for Development Economics Research of the United Nations University. WP 36, February.

Aschenbrenner, K. M., Biehl, B., and Wurm, G. M. (1986): Antiblockiersystem und Verkehrssicherheit: Ein Vergleich der Unfallbelastung von Taxen Mit und Ohne Antiblockiersystem. (Teilbericht von die Bundesanstalt für Strassenwesen zum Forshungsproject 8323: Einfluss der Risikokompenzation aut die Wirkung von Sicherheitsaussnahmen.) Mannheim, Germany. Cited in Wilde, G.S. (1988): Risk Homeostasis Theory and Traffic Accidents: Propositions, Deductions, and Discussion in Recent Reactions. Ergonomics. 31, 441–468.

Ashby, R. (1960): Design for a Brain. London: Chapman & Hall.

Asimov, M. (1962): Introduction to Design. Englewood Cliffs, NJ: Prentice-Hall.

Atherton-Cochrane, P. A. (1981): Tasks Performed by Online Searchers in Presearch Interviews: A Report of the Presearch Project. Syracuse: School of Information Studies.

Austin, D. and Dale, P. (1986): ISO 2788. Guidelines for The Establishment and Development of Monolingual Thesauri., 2nd ed.

Babbage, C. (1826): On a Method of Expressing by Signs the Action of Machinery. Philosophical Transactions of the Royal Society for the Year MDCCCXXVI; Part III; 250–265. London: Royal Society.

Bambrough, R. (1963): The Philosophy of Aristotle. New York: Mentor Books.

Barber, G., DeJong, P., and Hewitt, C. (1983): Semantic Support for Work in Organizations. In Mason, R. E. A. (Ed).: Information Processing 83. New York: Elsevier Science Publishing.

Barfod, A. (1983): Forsøg med Nye Produktionssystemer: En Introduktion (in Danish): Experiments with New Production Systems: An Introduction. Copenhagen: Jernets Arbejdsgiverforening.

Baron, S., Fehrer, R., Pew, R., and Horwitz, P. (1986): Approach to Modelling Supervisory Control of a Nuclear Power Plant. NUREG/CR-2988. ORNL/Sub/81–70523/1.

Bartlett, F. (1932): Remembering. Cambridge, UK: Cambridge University Press.

Bartlett, F. C. (1943): Fatigue Following Highly Skilled Work. Proceedings of the Royal Society, London, 131, 247–257.

Bateson, G., (1979): Mind and Nature. New York: Elsevier-Dutton.

Beer, S. (1966): Decision and Control. Chichester: Wiley.

Belkin, N. (1978): Information Concepts for Information Science. Journal of Documentation, 34, 55–85.

Belkin, N. (1990): The Cognitive Viewpoint in Information Science. Journal of Information Science, 16(1), 11–16.

Belkin, N., Seeger, T., and Wersig, B. (1987): Distributed Expert-based Information Systems: An Interdisciplinary Approach. Information Processing and Management, 23(5), 395–409.

Beltracchi, L. (1984): A Process/Engineering Safeguards Iconic Display, Symposium on New Technologies in Nuclear Power Plant Instrumentation and Control, Washington, DC, 28–30 Nov.

Beltracchi, L. (1987): A Direct Manipulation Interface for Water-based Rankine Cycle Heat Engines. IEEE Transactions on Systems, Man, and Cybernetics, SMC-17, 478–487.

Beltracchi, L. (1989): Energy, Mass, Model-Based Displays and Memory Recall. IEEE Transaction on Nuclear Science, 36, 1367–1382.

Bigelow (1840): see Hindle (1981).

Bisantz, A. M. and Vicente, K. J. (1994): Making the Abstraction Hierarchy Concrete. International Journal of Man–Machine Studies. In press.

Boden, M. A. (1980): Jean Piaget. Penguin Modern Masters. London: Penguin Books.

Boff, K. R., Kaufman, L., and Thomas, J. P. (Eds.) (1986): Handbook of Perception and Human Performance, Vol. II, Cognitive Process and Performance, New York: Wiley (see chapters by Chase, W. G.; Rock, I.; Treisman, A. and Pomerantz, J. R., and Kubovy, M.).

Boff, K. R. and Lincoln, J. E. (Eds.) (1988): User's Guide, Engineering Data Compendium, Human Perception and Performance, H. G. Armstrong Aerospace Research Laboratory, Wright-Patterson Air Force Base, Dayton OH.

Bogetoft P. and Pruzan, P. (1991): Planning with Multiple Criteria. Amsterdam: North-Holland.

Boucher, T. O. and Jafari, M. A. (1990): IDEF0 to Petri Net Transformation with Multiple Routings, Tech. Report, 90–114, Industrial Engineering Deptartment, Rutgers University, New Brunswick, NJ.

Brehmer, B. (1981): Models of Diagnostic Judgments. In Rasmussen, J. and Rouse, W. B. (Eds.): Human Detection and Diagnosis of System Failures. New York: Plenum Press, pp. 231–241.

Brehmer, B. (1992): Human Control of Complex Systems. Acta Psychologica, 81, 211–241.

Brehmer, J., Leplat, J. and Rasmussen, J. (1991): Use of Simulation in the Study of Complex Decision Making. In: Rasmussen, J., Brehmer, B., and Leplat, J. (Eds.): Distributed Decision Making: Cognitive Models for Cooperative Work. London: Wiley.

Broadbent, D., Fitzgerald, P., and Broadbent, M. H. P. (1986): Implicit and Explicit Knowledge in the Control of Complex Systems. British Journal of Psychology, 77, 33–50.

Bruner, J. S, Goodnow, J. J., and Austin, G. A. (1956): A Study of Thinking. New York: Wiley.

Bucciarelli, L. L. (1984): Reflective Practice in Engineering Design. Design Studies, Vol. 5, No. 3.

Bucciarelli, L. L. (1988): An Ethnographic Perspective on Engineering Design. Design Studies, Vol. 9, No. 3, pp. 159–168.

Burk, C. F. and Horton, F. W. (1988): INFOMAP: A Complete Guide to Discovering Corporate Information Resources. Englewodd Cliffs, NJ: Prentice Hall.

Bush, V. (1945): As We May Think. Atlantic Monthly, 176, p. 101–108.

Cacciabue, P. C., Codazzi, A., and Decortis, F. (1990): The Analysis of the Functional Role of Man and Machine in the Control of a Notional Auxiliary Feedwater System. In Rasmussen, J., Brehmer, B., deMontmollin, M., and Leplat, J., (Eds.): Proceedings of the First MOHAWC Workshop on Taxonomy for Analysis of Work Domains, Liège, 15–16 May.

Calabresi, G. (1970): The Cost of Accidents: A Legal and Economic Analysis. New Haven: Yale University Press.

Canter, D., et al. (1986): User Navigation in Complex Database Systems. In Behaviour and Information Technology, Vol. 5 (3), pp. 249–257.

Casey, S. (1993): Set Phasers on Stun. Santa Barbara, CA: Aegean Publishing Co.

Chernoff, H. (1971): The Use of Faces to Represent Points in N-Dimensional Space Graphically. Tech. Report AD 738 473, 48pp. Stanford, CA: Department of Statistics; Stanford University.

Christoffersen, K. Pereklita, A., and Vicente, K. J. (1993): Effect of Expertise on Reasoning Trajectories in an Abstraction Hierarchy: Fault Diagnosis in a Process Control System. Tech. Report. CEL 93-02. Toronto, CA: Department of Industrial Engineering, University of Toronto.

Churchman, C. W. (1971): The Design of Inquiring Systems: Basic Concepts of Systems and Organizations. New York: Basic Books.

Clanton, C. (1983): The Future of Metaphor in Man–Computer Systems. Byte, nr. 12, pp. 263–270.

Coekin, J. A. (1969): A Versatile Presentation of Parameters for Rapid Recognition of System State. In International Symposium on Man–Machine Systems. Cambridge September, 1969. IEE Conference Record No. 69, 58–MMS Vol. 4.

Conklin (1987): Hypertext: An Introduction and Survey. IEEE Computer, 20 (9), 17–41.

Craik, K. J. W. (1943): The Nature of Explanation. Cambridge, UK: Macmillan.

Cramer, P. (1968): Word Association. New York: Academic Press.

Croft, B. (1986): User Specified Domain Knowledge for Document Retrieval.

Crossman, E. R. F. W. and Cook, J. E. (1962): Manual Control of Slow Response Systems. In Edwards, W. and Lees, F. P. (Eds.): The Human Operator in Process Control. London: Taylor and Francis.

Dertouzos, M. L. et al. (1988): Made in America: Regaining the Productive Edge. Cambridge: MIT Press.

Dörner, D. (1987): On the Difficulties People Have in Dealing with Complexity. In Rasmussen, J., Duncan, K., and Leplat, J. (Eds.): New Technology and Human Error. New York: Wiley.

Dörner, D. (1989): Die Logik des Misslingen. Reinbek bei Hamburg: Rowohlt.

Dörner, D., Kreuzig, H. W., Reither, F., and Stäudel, T. (1983): Lohhausen: Vom Umgang mit Unbestimtheit und Komplexität. Bern: Verlag Hans Huber.

Drucker, P. F. (1988): The Coming of a New Organization, Harvard Business Review, January–February, pp. 45–53.

Duncker, K. (1945): On Problem Solving. Psychology Monographs, 58, (5) pp. 1–113.

Dutton, J. M. and Starbuck, W. H. (1971): Finding Charlie's Run-Time Estimator. In Dutton, J. M. and Starbuck, W. H. (Eds.): Computer Simulations of Human Behavior. New York: Wiley.

Eddington, A. (1939): The Philosophy of Physical Science. New York: Cambridge Univ. Press.

Ekner, K.V. (1989): On: 'Preliminary Safety Related Experiences from Establishment of Bicycle Paths in Copenhagen, 1981–83.' Technical Report, In Danish. Copenhagen: Stadsingniørens Direktorat.

Ericsson, K. A. and Simon, H. A. (1984): Protocol Analysis: Verbal Reports as Data. Cambridge, MA: MIT Press.

EWICS (1981, 1986): Guidelines for the Design of Man-Machine Interfaces. Technical Committe 6 on Man-Machine Communication, European Workshop on Industrial Computer Systems. Levels 0–2 published by: Trondheim, Norway: SINTEF; Level 3 published by: Delft, Holland: Technical University.

Feinberg, F. (1965): Action and Responsibility. In Black, M. (Ed.): Philosophy in America. London: George Allen and Unwinn Ltd. Reprinted in: A.R. White (Ed.): The Philosophy of Action. Oxford, UK: Oxford University Press.

Ferguson, E. S. (1977): The Mind's Eye: Nonverbal Thought in Technology, Science, 197, (4306), 827–836.

Fischer, P. M. and Mandl, H. (1990): Towards a Psychophysics of Hypermedia. In Jonassen, D. H. and Mandl, H. (Eds.): Designing Hypermedia for Learning. Heidelberg: Springer-Verlag, pp. 18–25.

Fitzgerald, P. J. (1961): Voluntary and Involuntary Acts. In Guest, A. C. (Ed.): Oxford Essays in Jurisprudence, Oxford, UK: Clarendon Press. Reprinted in White, A. R. (Ed.): The Philosophy of Action. New York: Oxford University Press.

Flach, J. (1990): Control with an eye for perception, precursors to an active psychophysics, Ecological Psychology, 2, 83–11.

Flach, J. M. and Vicente, K. J. (1989): Complexity, Difficulty, Direct Manipulation and Direct Perception. Engineering Psychology Research Laboratory, University of Illinois at Urbana-Champaign, Report EPRL-89–03.

Flach, J., Hancock, P., Caird, J., and Vicente, K. (Eds.). (1994): Ecology of Human–Machine Systems. Hillsdale, NJ: Lawrence Erlbaum.

Flores F., Graves, M., Hartfield, B., and Winograd T. (1988): Computer Systems and the Design of Organizational Interaction. ACM Transactions on Office Systems, Vol. 6, No. 2, April, pp. 153–172.

Friend'21 (1989): Proceedings of the International Symposium on Next Generation Human Interface Technologies, September–November 1991, Tokyo: Institute for Personalized Information Environment.

Fujita, Y. (1991): What Shapes Operator Performance? JAERI Human Factors Meeting, Tokyo, November. To appear in International Journal of Man–Machine Studies: Data, Keyholes for the Hidden World of Operator Characteristics.

Funke, J. (1988): Using Simulation to Study Complex Problem Solving. Simulation and Games, 19 (3), 277–303.

Fussell, J. B. (1987): Prisim—A Computer Program that Enhances Operational Safety. Presented at the Post-Smirt Workshop on Accident Sequence Modeling: Human Actions, System Response, and Intelligent Decision Support. Munich, August.

Gallagher, J. R. and Easter, J. M. (1982): Disturbance Analysis and Surveillance—System Scoping and Feasibility Study: EPRI NP-2240. Palo Calto, CA: Electrical Power Research Institute.

Gibbs, J. W. (1873): Graphical Methods in the Thermodynamics of Fluids. Transactions of the Connecticut Academy, II, 309–343.

Gibson, E. J. (1970): The Development of Perception as an Adaptive Process. American Scientist, 58, 98–107.

Gibson, E. J. (1988): Exploratory Behavior in the Development of Perceiving, Acting and the Acquiring of Knowledge. Annual Review of Psychology, 39, 1–41.

Gibson, J. J. (1950): The Perception of the Visual World. Boston: Houghton-Mifflin.

Gibson, J. J. (1966): The Senses Considered as Perceptual Systems. Boston: Houghton-Mifflin.

Gibson, J. J. (1979): The Ecological Approach to Visual Perception. Boston: Houghton-Mifflin.

Gibson, J. J. and Crooks, L. E. (1938): A Theoretical Field-Analysis of Automobile Driving. The American Journal of Psychology, LI, July (3), 453–471.

Goodstein, L. P. (1985a): Studies of Operator–Computer Cooperation on a Small-scale Nuclear Power Plant Simulator. Roskilde, Denmark: Risø National Laboratory, Risø–M-2522.

Goodstein, L. P. (1985b): Functional Alarming and Information Retrieval. Roskilde, Denmark: Risø National Laboratory, Risø–M-2511.

Goodstein, L. P. (1990): Decision Support Systems—A Survey. Roskilde, Denmark: Risø National Laboratory, Risø–M-2930.

Goodstein, L. P. and Pejtersen A. M. (1989): The BOOK HOUSE. System—Functionality and Evaluation. Roskilde, Denmark: Risø National Laboratory, Risø–M-2793.

Goodstein, L. P. and Rasmussen, J. (1988): Presentation of Process State, Structure and Control. Travail Humain, Vol. 51, No. 1, pp. 19–39.

Gould, J. D. (1988): How to Design Usable Systems. In Helander, M. (Ed.): Handbook of Human–Computer Interaction. Amsterdam: North-Holland, (pp. 757–789.)

Grudin, J. (1989): The Case Against User Interface Consistency. Communications of the ACM, 32, 1164–1173.

Hadamard, J. L. (1945): The Psychology of Invention in the Mathematical Field. Princeton:

Princeton University Press.

Hammond, K. R. (1986): Generalization in Operational Contexts: What Does it Mean? Can It Be Done? IEEE Transactions on Systems, Man, and Cybernetics, SMC-16 (3), 428–433.

Hammond, K. R., Hamm, R. M., Grassia, J., and Pearson, T. (1984): The Relative Efficacy of Intuitive and Rational Cognition: A Second Direct Comparison, Technical Report No. 52, Center for Research On Judgment and Policy; University of Colorado, Boulder.

Hancock-Beaulieu, M. (1989): Evaluating the Impact of an Online Public Library Catalogue on Subject Searching Behaviour at the Catalogue and at the Shelves. Journal of Documentation, 46 (4).

Häntzschel-Clairmont, W. (1912): Die Praxis des Modernen Maschinenbaues. Berlin: Verlag von C. A. Weller.

Hildreth, C. R. (1982): Online Browsing Support Capabilities. ASIS Proceedings, 19, 127–132.

Hildreth, C. R. (Ed.) (1989): The Online Catalogue: Development and Directions. London: Library Associations.

Hildreth, C. R. (1991): Intelligent Interfaces and Retrieval Methods for Subject Searching in Bibliographic Retreival Systems. Washington, DC: K. G. Saur.

Hillier, B. and Leaman, A. (1973): How is Design Possible? A Sketch for a Theory. DMG-DRS Journal, 8, (1), 40–50.

Himmelblau, D. M. (Ed.) (1973): Decomposition of Large-Scale Problems, New York: North-Holland.

Hindle, B. (1981): Emulation and Invention. New York: New York University Press.

HMSO (1987): M V Herald of Free Enterprise. Report of Court No. 8074, Department of Transport, November 1987. London: Her Majesty's Stationary Office.

HMSO (1989): Investigation into the Clapham Junction Railway Accident. The Department of Transport. London: Her Majesty's Stationary Office.

Hogarth, R. (1986): Generalization in Decision Research: The Role of Formal Models, IEEE Transactions on Systems, Man, and Cybernetics, SMC-16 (3) 439–449.

Hollan, J. (1989): Metaphors for Interface Design. Presented at U. S. Nuclear Regulatory Commission's Man–Machine Interface Workshop, 10–12 Jan.

Hollnagel, E., O. M. Pedersen and J. Rasmussen (1981): Notes on Human Performance Analysis. Roskilde, Denmark: Risø National Laboratory. Risø–M-2285.

Hovde, G. (1990): Cognitive Work Analysis: Decision Making in Operation Theater Planning. Working Report (in Danish). Roskilde, Denmark: Risø National Laboratory.

Huber, P. (1991): Galileo's Revenge: Junk Science in the Court Room. New York: Basic Books.

Hubka, V. (1982): Principles of Engineering Design. London: Butterworth Scientific.

Hutchins, E. L., Hollan, J. D., and Norman, D. A. (1986): Direct Manipulation Interfaces. In Norman, D. A. and Draper, S. W. (Eds.): User Centered System Design: New Perspectives on Human–Computer Interaction. Hillsdale, NJ: LEA, (pp. 87–124).

Ingwersen, P. (1986): Cognitive Analysis and the Role of the Intermediary in Information Retrieval. In: Davies, R. (Ed.): Intelligent Information Systems. Chichester: West Sussex, UK: Horwood, 206–237.

Ingwersen, P. (1992): Information Retrieval Interactions. London: Taylor Graham.

James, J. and Sanderson, P. M. (1991): Heuristic and Statistical Support for Protocol Analysis with SHAPA version 2. 01. Behavior Research Methods, Instruments and Computers. 23,

449–460.

Johnson, S. E. (1984): DASS: A Decision Aid Integrating the Safety Parameter Display System and Emergency Functional Recovery Procedures. Tech. Report No. NP 3595, Palo Alto, CA: Electric Power Research Institute.

Johnson, W. G. (1980): MORT Safety Assurance Systems. New York: Marcel-Decker.

Johnson-Laird, P. N. (1983): Mental Models, Cambridge, MA: Cambridge University Press.

Jonassen, D. H. and Grabinger, R. S. (1990): Problems and Issues in Designing Hyper-text/Hypermedia for Learning. In Jonassen, D. H. and Mandl, H. Heidelberg (Eds.): Designing Hypermedia for Learning. Heidelberg: Springer-Verlag. pp. 3–27.

Kaavé, B. (1990): Exploration of the User–System Interaction in Production Control (in Danish). Unpublished MSc dissertation, Copenhagen: Technical University.

Kaplan, R. S. (1989): Management Accounting for Advanced Technological Environments. Science, 245, August, 819–823.

Kaplan, R. S. (1990): The Four-Stage Model of Cost System Design. Management Accounting, February 22–26.

Klein, G. A. (1989): Recognition-Primed Decisions. In Rouse W. B. (Ed.): Advances in Man–Machine System Research, 5, Greenwich, CT: JAI Press, pp. 47–92.

Klein, G., Orasanu, J., Calderwood, R., and Zsambok, C. E., (Eds.) (1993): Decision Making in Action: Models and Methods. Norwood, NJ: Ablex.

Knaeuper, A. and Rouse, W. B. (1985): A Rule-based Model of Human Problem-Solving Behavior In Dynamic Environments. IEEE Transactions on Systems, Man, and Cybernetics, SMC-15, 708–719.

Lanir, Z., Fischoff, B., and Johnson, W. (1988): Military Risk Taking C^3I and the Cognitive Functions of Boldness in War. Journal of Strategic Studies, 11, 96–113.

Leplat, J. and Ramussen, J. (1984): Analysis of Human Errors in Industrial Incidents and Accidents for Improvement of Work Safety. Accident Analysis and Prevention, 16 (2), 77–88.

Leupold (1724–1739): see Ferguson (1977).

Lind, M. (1982): Multilevel Flow Modeling of Process Plants for Diagnosis and Control. Proceedings of International Meeting on Thermal Nuclear Reactor Safety, Chicago, IL.

Lind, M. (1992): A Categorization of Models and its Application for the Analysis of Planning Knowledge. Proceedings of POST ANP'92 Conference on Human Cognitive and Co-operative Activities in Advanced Technological Systems, Kyoto, Japan.

Lind, M. (1993): Multilevel Flow Modeling. AAAI93 Workshop on Reasoning about Function. Washington DC: July 11.

Lindsay, R. W. and Staffon, J. D. (1988): A Model Based Display System for the Experimental Breeder Reactor-II. Paper presented at the Joint Meeting of the American Nuclear Society and the European Nuclear Society, Washington, DC.

Lodding, K. N. (1983): Iconic interfacing. In IEEE Computer Graphics and Applications, Vol. 3, pp. 11–20.

Lopes, L. L. (1986): Aesthetics and the Decision Sciences, IEEE Transactions on Systems, Man, and Cybernetics. Vol. SMC-16, No. 3, pp. 434–438.

Luchins, A. S. (1942): Mechanization in Problem Solving. Psychological Monographs, 54, 1–95.

Mackie, J. L. (1965): Causes and Conditions. American Philosophical Quarterly, 2. 4, 245–255 and 261–264. Reprinted in Sosa, E. (Ed.): Causation and Conditionals, Oxford, UK: Oxford

University Press, 1975.

Markey, K. (1990): Experiences with Online Catalogues in the USA using a Classification System as a Subject Searching Tool. In: Fugman, R. (Ed.): Tools for Knowledge Organization and the Human Interface. Frankfurt/Main, Germany: Indeks Verlag. Vo. 1, p. 35–48.

Maxwell, J. C. (1868): On Governors. Proceedings of the Royal Society London, 16, 270–283.

McNeese, M. D. and Zaff, B. S. (1991): Knowledge as Design: A Methodology for Overcoming Knowledge Acquisition Bottlenecks in Intelligent Interface Design. Proceedings of the Human Factors Society 35th Annual Meeting, 3, pp. 1181–1185. New York: Human Factors Society.

McNeese, M. D., Zaff, B. S., Peio, K. J., Snyder, D. E., Duncan, J. C., and McFarren, M. R. (1990): An Advanced Knowledge and Design Acquisition Methodology: Application for the Pilot's Associate. AAMRL-TR-90–060, Dayton, OH: Armstrong Aerospace Medical Research Laboratory, Wright-Patterson AFB.

McRuer, D. T. and Krendel, E. S. (1957): Dynamic Responses of Human Operators, W-DC-TR-56–524. US-Air Force. Dayton, Ohio: Wright Patterson Air Force Base.

Meister, D. and Farr, D. E. (1967): The Utilization of Human Factors Information by Designers. Human Factors, 9, 71–87.

Mesarovic, M. D. (1970): Multilevel Systems and Concepts in Process Control, Proceedings of the IEEE, Vol. 58, No. 1, New York: The Free Press of Glencoe.

Michaels, C. F. and Carello, C. (1981): Direct Perception, Englewood Cliffs, NJ: Prentice-Hall.

Mihram, D. and Mihram, G. A. (1974): Human Knowledge, the Role of Models, Metaphors and Analogy, International Journal of General Systems, 1, 41–60.

Mihram, G. A. (1972): The Modeling Process, IEEE Transactions on Systems, Man, and Cybernetics, SMC-2, (5), 621–629.

Miller, A. I. (1989): Imagery in Scientific Thought Creating 20th-Century Physics. Boston: Birkhauser.

Miller, G. A. (1968): Psychology and Information. In: American Documentation, Vol. 19, p. 286–289.

Minsky, M. (1975): A Framework for Representing Knowledge. In Winston, P. (Ed.): The Psychology of Computer Vision. New York: McGraw-Hill.

Mitroff, I. I. and Linstone, A. I. (1993): The Unbounded Mind: Breaking the Chains of Traditional Business Thinking. New York: Oxford University Press.

Moray, N. (1987): Intelligent Aids, Mental Models and the Theory of Machine (unpublished manuscript).

Moray, N. (1990): Prologomena to the Study of Strategic Behavior. Engineering Psychology Research Laboratory, University of Illinois at Champagne-Urbana, Technical Report EPRL-90–04.

Moray, N. and Dessouky, M. I. (1990): Human Factors of Strategic Behavior: Scheduling Theory as a Normative Model of Strategic Behavior. Proceedings of the 34th Annual Meeting of the Human Factors Society, Orlando, FL, pp. 596–597.

Moray, N. and Dessouky, M. I. (1993): Taxonomy of Scheduling Systems as a Basis for the Study of Strategic Behavior. Engineering Psychology Research Laboratory, University of Illinois at Champagne-Urbana, Technical Report EPRL-93-02.

Moray, N., Lee, J., Vicente, K., Jones, G. J., Brock, R., Djemil, D., and Rasmussen, J. (1991):

A Performance Indicator of the Effectiveness of Human–Machine Interfaces for Nuclear Power Plants: Preliminary Results.

Moray, N., Lootsteen, P., and Pajak, J. (1986): Acquisition of Process Control Skills. IEEE Transactions on Systems, Man, and Cybernetics, SMC-16, 497–504.

Moray, N. and Rotenberg, I. (1989): Fault Management in Process Control: Eye Movements and Action. Ergonomics, 32, 1319–1342.

Morehead, D. R. Pejtersen, A. M., and Rouse, W. B. (1984): The Value of Information and Computer-aided Information Seeking: Problem Formulation and Application to Fiction Retrieval. Information Processing and Management, 20, (5/6), 583–601.

Morris, N. M., Rouse, W. B., and Fath, J. L. (1985): PLANT: An Experimental Task for the Study of Human Problem Solving in Process Control. IEEE Transactions on Systems, Man, and Cybernetics, SMC-15, 792–798.

Norman, D. A. (1981): Categorization of Action Slips. Psychological Review, 88, 1–15.

Norman, D. A. (1986): Cognitive Engineering. In Norman, D. A. and Draper, S. W. (Eds.): User Centered System Design: New Perspectives on Human–Computer Interaction. Hillsdale, NJ: LEA, (pp. 31–61).

Norman, D. A. (1988): The Psychology of Everyday Things. New York: Basic Books.

Nyce, J. M. and Kahn, P. (1992): A Machine for the Mind. Vannevar Bush's Memex. In From Memex to Hypertext: Vannevar Bush and the Mind's Machine. New York: Academic Press, pp. 38–65.

Paget, M.A. (1988): The Unity of Mistakes. A Phenomenological Interpretation of Medical Work. Philadelphia: Temple University Press.

Parsons, T. (1960): Structure and Process in Modern Societies. New York: The Free Press of Glencoe.

Pedersen, S. A. and Rasmussen, J. (1991): Causal and Diagnostic Reasoning in Medicine and Engineering. Proceedings of workshop on CEC Basic Research Project MOHAWC. May; Stresa, Italy. Roskilde, Dk; Risø National Laboratory.

Pejtersen, A. M. (1974): User-Librarian Negotiations in the Adults' Library. Samples. (in Danish). Copenhagen: The Royal School of Librarianship.

Pejtersen, A. M. (1979a): Investigation of Search Strategies in Fiction Based on an Analysis of 134 User-Librarian Conversations. In Henriksen, T. (Ed.): IPFIS 3. Conference Proceedings. Oslo, Norway: Statens Biblioteks- och Informations Högskole. pp. 107–132.

Pejtersen, A. M. (1979b): The Meaning of Aboutness in Fiction Indexing and Retrieval. Aslib Proceedings, 31, pp. 251–257.

Pejtersen, A. M. (1980a): Design of a Classification Scheme for Fiction Based on an Analysis of Actual User-Librarian Communication, and Use of the Scheme for Control of Librarians' Search Strategies. In Harbo, O. and Kajberg, L. (Eds): Theory and Application of Information Research. London: Mansell, pp. 167–183.

Pejtersen, A. M. (1980b): Classification of Fiction. Historical Survey. Technical Report (in Danish). Copenhagen: The Royal School of Librarianship.

Pejtersen, A. M. (1981a): Theory and Practice of the Librarian's Role as a Mediator in Fictive Literature. Technical report. Copenhagen: The Royal School of Librarianship.

Pejtersen, A. M. (1981b): The Librarian's Role as a Mediator in Fictive Literature. In Friberg, I. (Ed.): The Fourth International Research Forum in Information Science. Conference Proceedings. Borås, Sweden: Bibhotekshögskolan. pp. 178–207.

Pejtersen, A. M. (1982): Search Strategies and Categorization in User Librarian Negotiations. Technical Report. (in Danish), Copenhagen: The Royal School of Librarianship.

Pejtersen, A. M. (1984): Design of a Computer-aided User-System Dialogue Based on an Analysis of Users' Search Behaviour. In Social Science Information Studies, no. 4, pp. 167–183.

Pejtersen, A. M. (1986a): Design and Test of a Database for Fiction Based on an Analysis of Children's Search Behaviour. In Ingwersen, P., Kajberg, L., and Pejtersen, A. M. (Eds.): Information Technology and Information Use. Towards a Unified View of Information and Information Technology. London: Taylor Graham, pp. 125–147.

Pejtersen, A. M. (1986b): Design of Intelligent Retrieval Systems for Libraries Based on Models of Users' Search Strategies. Proceedings of 1986 IEEE International Conference on Systems, Man and Cybernetics. New York: IEEE, Institution of Electrical and Electronic Engineers.

Pejtersen, A. M. (1986c): Implications of Users' Value Perception for the Design of a Bibliographic Retrieval System. In Agrawal J.C. and Zunde, P. (Eds.): Emperical Foundation of Information and Software Science. New York: Plenum Press, pp. 23–39.

Pejtersen, A. M. (1988): Search Strategies and Database Design for Information Retrieval in Libraries. In Goodstein, L. P., Andersen, H. B., and Olsen, S. E. (Eds.): Tasks, Errors and Mental Models. London: Taylor and Francis. pp. 171–192.

Pejtersen, A. M. (1989): The BOOK HOUSE: Modelling Users' Needs and Search Strategies as a Basis for System Design. Technical report. Risø-M-2794. Roskilde: Risø National Laboratory.

Pejtersen, A. M. (1991a): Icons for Organization and Representation of Domain Knowledge in Interfaces. In Fugman, R. (Ed.): Proceedings 1st International ISKO Conference. Advances in Knowledge Organisation. Tools for Knowledge Organisation and the Human Interface. Vol. 2, Frankfurt/Main: Indeks Verlag. Germany, pp. 175–193

Pejtersen, A. M. (1991b): Interfaces Based on Associative Semantics for Browsing in Information Retrieval. Technical report. Risø M-2794. Roskilde: Risø National Laboratory.

Pejtersen, A. M. (1992a): The Book House. An Icon Based Database System for Fiction Retrieval in Public Libraries. In Cronin. B., (Ed.): The Marketing of Library and Information Services 2. London: ASLIB. pp. 572–591.

Pejtersen, A. M. (1992b): New Model for Multimedia Interfaces to Online Public Access Catalogues. The Electronic Library, International Journal for Minicomputer, Microcomputer and Software Applications in Libraries. 10, (6). pp. 359–366.

Pejtersen, A. M. (1993a): Designing Hypermedia Representations from Work Domain Properties. In Frei, H. P. and Schauble, P. (eds): Hypermedia'93. Conference Proceedings. Zurich: Springer Verlag, pp. 273–289.

Pejtersen, A. M. (1993b): A New Approach to Design of Document Retrieval and Indexing Systems for OPAC Users. In Rouff, D. (Ed.): Online Information Conference Proceedings. London: Learned Information, pp. 273–290.

Pejtersen, A. M. and Austin, J. (1984): Fiction Retrieval: Experimental Design and Evaluation of a Search System based on Users' Value Criteria. Parts 1 and 2. Journal of Documentation, 39, (4) pp. 230–246 and 40 (1) pp. 25–35.

Pejtersen, A. M. and Cramer, 1. (1976): User-Librarian Negotiations in the Children's Library. (in Danish). Copenhagen, The Royal School of Librarianship.

Pejtersen, A. M. and Goodstein, L. P. (1990): Beyond the Desk Top Metaphor: Information Retrieval with an Icon Based Interface. In Tauber, M. J. and Gorny, P. (eds): Proceedings of the 7th Interdisciplinary Workshop on Informatics and Psychology on Visualization in Human-Computer Interaction. Berlin: Springer Verlag. pp. 149–193.

Pejtersen, A. M., Olsen, Sv. E. and Zunde, P. (1987): A Retrieval Aid for Browsing Strategies in Bibliografic Databases Based on Users' Associative Semantics: A Term Association Thesaurus. In Wormell, I. (Ed.): Knowledge Engineering: Expert Systems and Information Retrieval. London: Taylor Graham. pp. 92–112.

Pejtersen, A. M. and Rasmussen, J. (1986): Design and Evaluation of User–System Interfaces or of User–Task Interaction? A Discussion of Interface Design Approaches in Different Domains for Application of Modern Information Technology. In Agrawal, J. C. and Zunde, P. (Eds.): Empirical Foundations of Information and Soft Ware Science. New York: Plenum Press.

Pejtersen, A. M. and Rasmussen, J. (1989): Information Retrieval In Integrated Work Stations. In Fujiwara, Y. (Ed.): ICIK '87, '88, International Conference on Information and Knowledge. Yokohama, Japan: Kanagawa University, pp. 245–265.

Pejtersen, A. M. and Rasmussen, J. (1990): Intelligent Systems for Information Retrieval and Decision Support in Complex Work Domains. In Skov, F. et al. (Eds.): Expert Systems in Agricultural Research. Denmark. Statens Planteavlsforsøg og Statens Husdyrbrugsforsøg. pp. 103–117.

Peplerm, R. D. and Wohl, J. G. (1964): Display Requirements for Prelaunch Checkout of Advanced Space Vehicles. Tech. Report 409-1, Darien, Conn.: Dunlap and Associates.

Pew, R. W. (1974): Human Perceptual-Motor Performance. In Kantowitz, B. H. (Ed.): Human Information Processing: Tutorials in Performance and Cognition. New York: Erlbaum.

Piaget, J. and Inhelder, B. (1958): The Growth of Logical Thinking from Childhood to Adolescence. New York: Basic Books.

Poincare, H. (1904): Science and Method. Book I: The Scientist and Science. Translated by Francis Maitland, New York: Dover Publications.

Polanyi, M. (1958): Personal Knowledge. London: Routledge & Kegan Paul.

Polanyi, M. (1967): The Tacit Dimension. New York: Doubleday.

Ramelli, A. (1588): Le Diverse et Artificiose Machine, Paris. See: M. T. Gnudi and E. S. Ferguson (Trans.); Baltimore: John Hopkins University Press, 1976.

Ranganathan (1967): Prologomena to Library Classification. Bombay: Asia Publishing House.

Rasmussen, J. (1969): Man–Machine Communication in the Light of Accident Record. Presented at International Symposium on Man–Machine Systems, Cambridge, September 8–12. In IEEE Conference Records, 69C58–MMS, Vol. 3.

Rasmussen, J. (1980): What Can Be Learned from Human Error Reports. In Duncan, K., Gruneberg, M., and Wallis, D. (Eds.): Changes in Working Life. New York: Wiley.

Rasmussen, J. (1981): Models of Mental Strategies in Process Plant Diagnosis. In: Rasmussen, J. and Rouse, W. B. (Eds.): Human Detection and Diagnosis of System Failures. New York: Plenum Press.

Rasmussen, J. (1982): Human Errors. A Taxonomy for Describing Human Malfunction in Industrial Installations. Journal of Occupational Accidents, 4, (2–4), pp. 311–333.

Rasmussen, J. (1983): Skill, Rules and Knowledge; Signals, Signs, and Symbols, and Other Distinctions in Human Performance Models. IEEE Transactions on Systems, Man, and Cybernetics. SMC-13, (3).

Rasmussen, J. (1984): Strategies for State Identification and Diagnosis. In Rouse, W. B. (Ed.): Advances in Man–Machine Systems—Vol. 1: Greenwich: J. A. I. Press.

Rasmussen, J. (1985): The Role of Hierarchical Knowledge Representation in Decision Making and System Management. IEEE Transactions on Systems, Man and Cybernetics. SMC-15, (2). 234–243.

Rasmussen, J. (1986): Information Processing and Human–Machine Interaction: An Approach to Cognitive Engineering, New York: North-Holland.

Rasmussen, J. (1990a): The Role of Error in Organizing Behaviour. Ergonomics, 33, (10/11), 11851190.

Rasmussen, J. (1990b): Human Error and the Problem of Causality in Analysis of Accidents. Philosophical Transactions of the Royal Society, London, B 327, 449–462.

Rasmussen, J. (1993): Deciding and Doing: Decision Making in Natural Context. In Klein, G., Orasanu, J., Calderwood, R., and Zsambok, C. E. (Eds.): Decision Making in Action: Models and Methods. Norwood, NJ: Ablex.

Rasmussen, J. (1994a): Taxonomy for Work Analysis. In Salvendy, G. and Karwowski, W. (Eds.): Design of Work and Development of Personnel in Advanced Manufacturing. New York: Wiley-Interscience.

Rasmussen, J. (1994b): Risk Management, Adaptation, and Design for Safety. In: Sahlin, N. E. and Brehmer, B. (Eds.): Future Risks and Risk Management. Dordrecht: Kluwer. 1994.

Rasmussen, J. (1994c): Market Economy, Management Culture and Accident Causation: New Research Issues? Proceedings Second International Conference on Safety Science. Budapest, November, 1993.

Rasmussen, J. and Goodstein, L. P. (1988): Information Technology and Work, In Helander, M. (Ed): Handbook of Human–Computer Interaction. Amsterdam: Elsevier/North Holland, pp. 175–201.

Rasmussen, J. and Jensen, A. (1974): Mental Procedures in Real Life Tasks: A Case Study of Electronic Trouble Shooting. Ergonomics, Vol. 17, No. 3, pp. 293–307.

Rasmussen, J. and Jensen, A. (1973): A Study of Mental Procedures in Electronic Trouble Shooting. Risø-M-1582, Roskilde, Denmark: Risø National Laboratory.

Rasmussen, J. and Rouse, W. B. (Eds.) (1981): Human Detection and Diagnosis of System Failures. New York: Plenum Press.

Rasmussen, J. and Vicente, K. J. (1990): Ecological Interfaces: A Technological Imperative in High Tech Systems? International Journal of Human–Computer Interaction, 2, 93–111.

Rasmussen, J., Kaavé, B., and Rindom, O. (1991): Experimental Distributed Decision Making Scenarios in Manufacturing. Proceedings of Workshop of CEC Basic Research Project MOHAWK. May '91, Stresa, Italy. Roskilde: Risø National Laboratory.

Rasmussen, J., Pedersen, O. M., Mancini, G., Carnino, A., Griffon, M., and Gagnolet, P. (1981b): Classification System for Reporting Events Involving Human Malfunction. Roskilde, Denmark: Risø National Laboratory, Risø–M-2240.

Rasmussen, J., Pejtersen, A. M., and Schmidt, K. (1991): Taxonomy for Cognitive Work Analysis. Roskilde, Denmark: Risø National Laboratory, Risø–M-2871

Rasmussen, J., Pedersen, O. M., Mancini, G., Carnino, A., Griffon, M., and Gagnolet, P. (1981b): Classification System for Reporting Events Involving Human Malfunction. Roskilde, Denmark: Risø National Laboratory, Risø–M-2240.

Reason, J. T. (1990): Human Error. Cambridge, MA: Cambridge University Press.

Reason, J. T. and Mycielska, K. (1982): Absent-Minded? The Psychology of Mental Lapses and Everyday Errors. Englewood Cliffs, NJ: Prentice-Hall.

Rindom, O. (1990): Design of Work Station in an Integrated Environment Based on Field

Studies. In Helander, M.G. and Nagamachi, M. (Eds.): Human Factors in Design for Manufaturability and Process Planning. Proceedings of the International Ergonomics Association, Hawaii, 9–11 August.

Rizzi, D. (1990): Complications in Anesthesia: An Overview. Unpublished working paper. Roskilde, Denmark: Risø National Laboratory.

Robertson, S.E., Maron, M.E., and Cooper, V.S. (1982): Probability of Relevance: A Unification of Two Competing Models for Document Retrieval. Information Technology, Research & Development, 1 (1), 1–21.

Rochlin, G. I., La Porte, T. R., and Roberts, K. H. (1987): The Self-Designing High-Reliability Organization: Aircraft Carrier Flight Operations at Sea, Naval War College Review, Autom.

Rosch, E. (1975): Human Categorization. In: Warren, N. (Ed.): Advances in Cross-Cultural Psychology. New York: Halsted Press.

Rosenbluth, A., Wiener, N., and Biegelow, J. (1943): Behavior, Purpose, and Teleology. Philosophy of Science, 10, 18–24.

Ross, D. T. (1977): Structured Analysis (SA): A Language for Communicating Ideas. IEEE Transactions on Software Reliability, 3, (1), pp. 16–33.

Rouse, W. B. (1981): Experimental Studies and Mathematical Models of Human Problem Solving Performance in Fault Diagnosis Tasks. In Rasmussen, J. and Rouse, W. B. (Eds.): Human Detection and Diagnosis of System Failures. New York: Plenum Press.

Rouse, W. B. (1982): A Mixed Fidelity Approach to Technical Training. Journal of Educational Technology Systems 11 (2), 103–115.

Rouse, W. B. (1984): Developing an Evaluation Plan: Computer-generated Display System Guidelines, Vol. 2: Atlanta: Search Technology. EPRI-NP-3701, Vol. 2.

Rouse, W. B., Frey, P. R., and Rouse, S. H. (1984): Classification and Evaluation of Decision Aids for Nuclear Power Plant Operators. Search Technology Inc. Norcross GA. : Report 8303–1.

Rousseau, J. J. (1742): Les Confessions. Vol. 2, Book 7. For English translation see Crocker, L. (Ed.): The Confessions. New York: Washington Square Press, 1965.

Rubin, E. (1920): Vorteile der Zweckbetrachtung fur die Erkentnis, Zeitschrift für Psychologie, 85, pp. 210–223. (Also in Experimenta Psychologica. Copenhagen: Munksgaard, 1949.)

Russell, B. (1913): On the Notion of Cause. Proceedings of the Aristotelean Society, 13, 1–25.

Sage, A. P. (1981a): A Methodological Framework for Systemic Design and Evaluation of Computer Aids for Planning and Decision Support. Computers and Electrical Engineering, 8 (2), 87–101.

Sage, A. P. (1981b): Behavioral and Organizational Considerations in the Design of Information Systems and Processes for Planning and Decision Support. IEEE Transactions on Systems, Man, and Cybernetics: SMC-11, 9, 640–678.

Sage, A. P. (1987a): Decision Support Systems. In Singh, M. G. (Ed.): Systems and Control Encyclopedia, Vol. 2: Oxford: Pergamon Press, pp. 943–947.

Sage, A. P. (1987b): Information Systems Engineering for Distributed Decisionmaking. IEEE Transactions on Systems, Man, and Cybernetics: SMC-17, 920–936.

Sanderson, P. M. (1989): Verbalizable Knowledge and Skilled Task Performance, Association, Dissociation and Mental Models. Journal of Experimental Psychology, Learning,

Memory and Cognition, 15, 729–747.

Sanderson, P. M. (1990): Knowledge Acquisition and Fault Diagnosis: Experiments With PLAULT. IEEE Transactions on Systems, Man, and Cybernetics, SMC-20, 225–242.

Sanderson, P. M. (1991): Towards the Model Human Scheduler. Human Factors in Manufacturing, 1, pp. 195–220.

Sanderson, P. and Fisher, C. (1994): Exploratory Sequential Analysis: Foundations. To appear in: Journal of Man–Machine Studies.

Sanderson, P. M., Flach, J. M., Buttigieg, M. A., and Casey, E. J. (1989): Object Displays Do Not Always Support Better Integrated Task Performance, Human Factors, 31, nr. 2, 183–198.

Sanderson, P. M., James, J., and Seidler, K. S. (1989): SHAPA: An Interactive Software Environment for Protocol Analysis. Ergonomics, 32, 1271–1302.

Sanderson, P. M. and Moray, N. (1990): The Human Factors of Scheduling Behavior, Engineering Psychology Research Laboratory, University of Illinois at Urbana-Champaign, Report EPRL-90–09.

Sanderson, P. M., Watanabe, LM., James, J. M., and Scott, J. J. P. (1991): Visualization and Analysis of Complex Sequential Data Records with MacSHAPA. Proceedings of the European Annual Conference on Cognitive Science Approaches to Process Control. Cardiff, Wales, UK, 12–14 Sept.

Saracevic, T., Mokros, H., and Su, L. (1990): Nature of Interaction between Users and Intermediaries in Online Searching: A Qualitative Analysis. In Henderson, D. (Ed.): ASIS«90: Information in the Year 2000: From Research to Applications: Proceedings of the American Society for Information Science (ASIS) 53rd Annual Meeting: Vol. 27; November 4–8; Toronto, Canada. Medford, NJ: Learned Information, Inc. for ASIS; 47–54.

Savage, C. M. and Appleton, D. (1988): CIM and Fifth Generation Management. In Fifth Generation Management for Fifth Generation Technology: A Round Table Discussion. SME Blue Book Series. Dearborn, Michigan: Society of Manufacturing Engineers.

Scott, W. R. (1987): Organizations: Rational, Natural, and Open Systems, Englewood Cliffs, NJ: Prentice-Hall.

Scribner, S. (1984): Studying Working Intelligence. In Rogoff, B. and Lave, J. (Eds.): Everyday Cognition: Its Development in Social Context, Cambridge, MA: Harvard University Press, pp. 9–40.

Selznick, Philip (1957): Leadership in Administration, New York: Harper.

Senge, P. M. (1990a): The Leader's New Work: Building Learning Organizations. Sloan Management Review, No. 7, Fall. pp. 7–24.

Senge, P. M. (1990b): The Fifth Discipline: The Art and Practice of The Learning Organization. New York: Doubleday Currency.

Sheridan, T. B. and Ferrell, W. R. (1974): Man–Machine Systems: Information, Control, and Decision Models of Human Performance. Cambridge, MA: MIT Press.

Shneiderman, B. (1983): Direct Manipulation: A Step Beyond Programming Languages. IEEE Computer, 16(8), 57–69.

Simon, H. A. (1969): The Sciences of the Artificial. Cambridge, MA: MIT Press.

Singer, E. A. (1959): Experience and Reflection. Philadelphia: University of Pennsylvania Press.

Smith, B. (1992): The Flowsheet: Animation Used to Analyze and Present Information About Complex Systems. Proceedings of the EDPPMA Virtual Reality Meeting, Arlington, Virginia, June 1–2, 1992.

Smith, S. L. and Aucella, A. F. (1983): Design Guidelines for the User Interface to Computer-based Information Systems. USAF Electronic System Division; Tech. Report No. ESD-TR-83-122 Massachussets: Hanscom Air Force Base.

Smith, S. L. (1988): Standards Versus Guidelines for Designing User Interface Software. In Helander, M. (Ed.): Handbook of Human–Computer interaction. Amsterdam: North-Holland, (pp. 877–889).

SoftTech (1981): Integrated Computer Aided Manufacturing (ICAM). Final Report: IDEFO Functional Modelling Manual. Dayton, OH: SoftTech Ltd.

Sonnenwald, D. (1993): Communication in Design. Ph. D. Thesis. Rutgers, NJ: The State University of New Jersey.

Sosa, E. (1975) (Ed.): Causation and Conditionals, New York: Oxford University Press.

Sperber, D. and Wilson, D. (1986): Relevance. Oxford, UK: Basil Blackwell.

Stokes, A. E. and C. Wickens (1988): Aviation Displays. In Human Factors in Aviation, New York: Academic Press.

Streitz, N. A. (1990): Hypertext. Ein Innovatives Medium zur Kommunikation von Wissen. In Gloor, P. A. and Streitz, N. A. (Eds.): Hypertext und HyperMedia. Von Theoretischen Konzeptem zur Praktischen Anwendung. Informatik-Fachberichte 249. Berlin: Springer.

Suchman, L. A. (1982): Systematics of Office Work: Office Studies for Knowledge-based Systems. Office Automation Conference, Moscone Center, San Francisco, April 5–7.

Suchman, L. A. (1983): Office Procedures as Practical Action: Models of Work and Systems Design. ACM Transactions on Office Information Systems, 1 (4), 320–328.

Sutcliffe, A. C. (1909): Robert Fulton and the Clairmont. New York: The Century Co.

Swain, A. D. and Guttmann, H. E. (1983): Handbook on Human Reliability Analysis with Emphasis on Nuclear Power Plant Applications. NUREG/CR-1278, US-NRC.

Tanabe, F. (1991): Critical Issues in Man–Machine Evaluation. SMIRT 11th Post-Conference Seminar on Probabilistic Safety Assessment Methodology. (To appear in Reliability Engineering & System Safety.)

Taylor, R. S. (1968): Question Negotiation and Information Seeking in Libraries. In college and Research Libraries 29, pp. 178–194.

Thompson, J. D. (1967): Organizations in Actions. New York: McGraw-Hill.

Thompson, R.H. and Croft, W.B. (1989): Support for Browsing in an Intelligent Text Retrieval System. International Journal of Man–Machine Studies, 30, 639–668.

Truxal, J.G. (1955): Automatic Feedback Control System Synthesis. New York: McGraw-Hill.

Tulving, E. (1983): Elements of Episodic Memory. Oxford, UK: Clarendon Press.

van Rijsbergen, C.J. (1979): Information Retrieval. 2. ed. London: Butterworths.

Vicente, K. J. (1991): Supporting Knowledge-Based Behavior through Ecological Interfaces (unpublished dissertation). Urbana, IL: University of Illinois at Urbana-Champaign.

Vicente, K. J. (1992a): Multilevel Interfaces For Power Plant Control Rooms I: An Integrative Review. Nuclear Safety. 33 (3), pp. 391–397.

Vicente, K. J. (1992b): Multilevel Interfaces For Power Plant Control Rooms II: A Preliminary Design Space. Nuclear Safety. 33 (4), pp. 543–548.

Vicente, K. J. and Rasmussen, J. (1990): The Ecology of Human–Machine Systems II: Meadiating Direct Perception in Complex Work Domains. Ecological Psychology, 2(3), 207–249.

Vicente, K. J. and Rasmussen, J. (1988a): On Applying the Skills, Rules and Knowledge Framework to Interface Design: Proceedings of the 32nd Annual Meeting of the Human Factors Society; Anaheim, CA. 24–28 Oct. New York: Human Factors Society.

Vicente, K. J. and Rasmussen, J. (1988b): A Theoretical Framework for Ecological Interface Design, Roskilde, Denmark: Risø National Laboratory, Risø–M-2736.

Vicente, K. J. and Rasmussen, J. (1992): Ecological Interface Design: Theoretical Foundations. IEEE Transactions of SMC, July/August, 22, (4), 589–607.

Visser, J. P. (1991): Development of Safety Management in Shell Exploration and Production. Contribution to '91 Bad Homburg Workshop on Risk Management. In Brehmer, B. and Reason, J. T. (Eds.): Control of Safety. Hove, UK: Lawrence Earlbaum. In press.

Wagenaar, W. (1989): Risk Evaluation and the Causes of Accidents. Invited Contribution to the CEC Workshop on Errors in Operation of Transport Systems; MRC-Applied Psychology Unit, Cambridge, UK May: Proceedings published in Ergonomics, 1990, 33, (10/11).

Wason, P. C. and Johnson-Laird, P. N. (1972): Psychology of Reasoning. Cambridge, MA: Harvard University Press.

Whiteside, J., Bennett, J., and Holtzblatt, K. (1988): Usability Engineering: Our Experience and Evolution. In Helander, M. (Ed.): Handbook of human–computer interaction. Amsterdam: North-Holland, pp. 791–817.

Wilde, G. J. S. (1976): Social Interaction Patterns in Driver Behaviour: An Introductory Review. Human Factors, 18(5), 477–492.

Wilde, G. S. (1988): Risk Homeostasis Theory and Traffic Accidents: Propositions, Deductions, and Discussion in Recent Reactions. Ergonomics. 31, 441–468.

Willet, P. (1988): Recent Trends in Hierarchic Document Clustering: A Critical Review, Information Processing and Management, 24, 577–597.

Woods, D. D. (1984): Visual Momentum: A Concept to Improve the Cognitive Coupling of Person and Computer. International Journal of Man–Machine Studies, 21, 229–244.

Woods, D. D. and Eastman, M. C. (1989): Integrating Principles for Human–Computer Interaction Into the Design Process: Heterarchically Organized HCI Principles. In Proceedings of the 1989 International Conference on Systems, Man, and Cybernetics, New York: IEEE (Institute of Electrical and Electronic Engineers) 29–34.

Woodward, J. (1965): Industrial Organization. Theory and Practice, London: Oxford University Press.

Wulff, H. R., Pedersen, S. A., and Rosenberg, R. (1986): Philosophy of Medicine: An Introduction. Oxford, UK: Blackwell Scientific Publications.

Index